CW01151793

Iterative Learning Control with Passive Incomplete Information

Dong Shen

Iterative Learning Control with Passive Incomplete Information

Algorithms Design and Convergence Analysis

Springer

Dong Shen
College of Information Science
  and Technology
Beijing University of Chemical Technology
Beijing
China

ISBN 978-981-10-8266-5   ISBN 978-981-10-8267-2 (eBook)
https://doi.org/10.1007/978-981-10-8267-2

Library of Congress Control Number: 2018931475

© Springer Nature Singapore Pte Ltd. 2018
This work is subject to copyright. All rights are reserved by the Publisher, whether the whole or part of the material is concerned, specifically the rights of translation, reprinting, reuse of illustrations, recitation, broadcasting, reproduction on microfilms or in any other physical way, and transmission or information storage and retrieval, electronic adaptation, computer software, or by similar or dissimilar methodology now known or hereafter developed.
The use of general descriptive names, registered names, trademarks, service marks, etc. in this publication does not imply, even in the absence of a specific statement, that such names are exempt from the relevant protective laws and regulations and therefore free for general use.
The publisher, the authors and the editors are safe to assume that the advice and information in this book are believed to be true and accurate at the date of publication. Neither the publisher nor the authors or the editors give a warranty, express or implied, with respect to the material contained herein or for any errors or omissions that may have been made. The publisher remains neutral with regard to jurisdictional claims in published maps and institutional affiliations.

Printed on acid-free paper

This Springer imprint is published by the registered company Springer Nature Singapore Pte Ltd. part of Springer Nature
The registered company address is: 152 Beach Road, #21-01/04 Gateway East, Singapore 189721, Singapore

To May (Lamei Chen) and Maggie (Moqi Shen).

Preface

Iterative learning control (ILC) has been developed over three decades since proposed, as it is a simple and effective control methodology for repetitive systems. For such type of systems, the tracking information of previous iterations is fully utilized to generate the control command for the next iteration, so that the tracking performance is expected to be gradually improved. The system is required to repeat some given tracking task in a finite time interval, and then the input for the next iteration is generated as a function of inputs and tracking errors of previous iterations. Consequently, ILC is an iteration-based control strategy, differing from other control methodologies such as feedback control, adaptive control, and robust control, which mainly concentrate on the time-based evolution. In other words, the precise tracking performance is gradually achieved along the iteration axis rather than time axis in ILC. Indeed, ILC mimics the learning ability of human being, which is a trying and correcting process for a given task. Thus, ILC behaves an outstanding tracking performance for repetitive systems comparing with traditional control methodologies.

With the fast development of communication and network technology, more and more systems choose the networked implementation. That is, the plant and the controller are located at different sites and communicate with each other through wired/wireless networks. Such configuration is advantageous in flexibility, convenience, and robustness in both implementation and control performance, which therefore has gained wide applications in intelligent transportation, industrial automation, unmanned aerial vehicles, etc. However, in the networked configuration, due to network congestion, broken linkages, and transmission errors, the information exchange may suffer various types of randomness such as data dropouts and communication delays. These random phenomena can critically influence the control performance. Consequently, it is of great significance to make an in-depth exploration of the learning control under a networked structure, especially in the aspects of algorithms design and convergence analysis.

The primary objective of this monograph is to present a systematic framework of ILC algorithms design and analysis for stochastic systems with passive incomplete information. By passive incomplete information, we mean the incomplete operation information and data caused by the system and transmission limitations during data collecting, storing, transmitting, and processing stages. For example, when applying ILC to practical systems, the operation may end early in consideration of safety when the system output largely deviates from the desired operation zone, which yields an incomplete iteration. For another example, when transmitting the data through wireless networks, the communication channel may suffer data dropouts, communication delays, fading, and data disordering, which surely degrades the data quality and induces incomplete information. In addition, limited transmission bandwidth and memory capacity will apparently exclude part data as we cannot accommodate all the information, whence only incomplete information is available for learning update. In consideration of all the mentioned passive incomplete information problems, we have established a unified framework for the design and analysis of ILC schemes based on stochastic approximation theory in this monograph. Indeed, the stochastic approximation is a quite effective tool for solving the stochastic control and optimization problems, which inspires us to consider the application of stochastic approximation in dealing with stochastic ILC problems under various incomplete information environments. We anticipate that the techniques provided in this monograph can help to solve more networked ILC problems.

This monograph summarizes results on this topic mostly distributed over journal papers and partly contained in unpublished materials. We have arranged the whole structure of the monograph in a systematical way: we start with the basic data dropout problem, extend it to general dropout environments, and then proceed to address general incomplete information problems. Strict proofs of main theorems and various intuitive simulations are also provided. This monograph contains an introductory chapter and three specific parts.

In Chap. 1, we present the introduction of ILC design and analysis backgrounds, where the fundamental formulation of ILC schemes in both discrete-time and continuous-time systems is presented. The in-depth literature review on ILC under data dropouts and other incomplete information environments including random iteration-varying lengths and communication asynchronization is also provided.

Part I consists of six chapters concentrating on the elaboration of ILC under one-side data dropouts. That is, only the data dropout at the measurement side is considered and the channel at the actuator side is assumed to work well. In this part, we aim to present an in-depth exploitation in three models of random data dropouts, namely, random sequence model, Bernoulli variable model, and Markov chain model. In Chaps. 2–4, the random sequence model is proposed and investigated for linear stochastic systems, nonlinear stochastic systems, and nonlinear systems with unknown control direction, respectively. The almost sure convergence of the intermittent update scheme is obtained. In Chaps. 5–6, the Bernoulli variable model is revisited and two ILC schemes are fully discussed. One is the intermittent update scheme, which only updates its control signal when data is successfully transmitted, and the other is the successive update scheme, which continuously updates its control signal with the latest available data in each iteration no matter whether the output information of the last iteration is successfully transmitted or lost.

The almost sure convergence of the two schemes is given for linear and nonlinear stochastic systems, respectively. In Chap. 7, the Markov chain model is considered for providing a general description of the data dropouts in consideration of iteration-dependence and time-dependence. Both almost sure and mean square convergence properties of the proposed schemes are strictly proved. We note that the mean square convergence is also presented for random sequence model and Bernoulli variable model regarding linear systems.

Part II consists of three chapters extending the one-side data dropout problem to two-side data dropout problem. Particularly, in this part, the data dropouts are assumed to occur at both the measurement and actuator sides simultaneously. In such case, the data dropout at the measurement side can adopt similar techniques provided in Part I; however, the data dropout at the actuator side introduces a new random asynchronism between the computed control, which is generated by the learning controller, and the actual control, which is fed to the plant, due to random data dropout in transmitting the computed control signal to the plant. In fact, this asynchronism is the main issue in this part and it is addressed for linear deterministic systems in Chap. 8, linear stochastic systems in Chap. 9, and affine nonlinear systems in Chap. 10. Various techniques are presented according to the specific system formulation. In Chaps. 8–9, the path behavior of the input sequence is formulated as a Markov chain in line with the asynchronization and synchronization states. Then, the almost sure and mean square convergence of the proposed update algorithms is obtained based on this Markov chain model. In Chap. 10, an augmented input error is introduced so that the asynchronization between the computed control and the actual control is reformulated as internal dynamics. Then, the newly defined regression of the augmented input error is analyzed to establish the convergence using the conventional contraction mapping method.

Part III consists of four chapters considering various general incomplete information problems. The multiple communication constraints including data dropouts, communication delays, and transmission disordering are addressed in Chap. 11, where the memory capacity is meanwhile limited to be one iteration storage. A renewal mechanism is added to regulate the packet arrivals, and a recognition mechanism is introduced for packet selections. Both intermittent and successive update schemes are proposed with their almost sure convergence. The randomly iteration-varying length problem is considered in Chaps. 12–13 for linear and nonlinear systems, respectively. By strengthening the design conditions of learning gain matrix, the prior knowledge on the probability distribution of random lengths is removed from the design of ILC update laws. Novel techniques are provided in these two chapters for the convergence analysis. The large-scale system consisting of several inner-connected subsystems is investigated in Chap. 14, where the operation asynchronism and communication delays among different subsystems are the critical issues to be resolved. A novel decentralized ILC update scheme is proposed to generate the input sequences such that each subsystem is driven to track the desired reference.

The monograph is self-contained in the sense that there are only a few points using knowledge for which we refer to other sources, and these points can be ignored when reading the main body of the monograph. The monograph is written for students, engineers, and researchers working in the field of ILC.

I would like to take the opportunity to thank my collaborators and students as some of the materials in this monograph arise from the joint work with them. I am indebted to my Ph.D. advisor, Prof. Han-Fu Chen from Academy of Mathematics and System Science, Chinese Academy of Sciences, for leading me into the research of ILC and guiding me to the right direction. I sincerely thank Prof. Jian-Xin Xu from National University of Singapore for his invitation for visiting and for his insightful discussions on many research problems. My appreciation also goes to Prof. Chiang-Ju Chien from Huafan University, Prof. Youqing Wang from Shandong University of Science and Technology, and Prof. Samer S. Saab from Lebanese American University, who have provided me with fruitful collaborations and discussions. I also thank my students, Jian Han, Wei Zhang, Yun Xu, Fanshou Zhang, Lanjing Wang, Chao Zhang, Chun Zeng, Chen Liu, and Yanqiong Jin, for our collaboration in the research.

I am pleased to thank the support of the National Natural Science Foundation of China under grants 61673045 and 61304085, and Beijing Natural Science Foundation under grant 4152040. I sincerely appreciate Elsevier, IEEE, John Wiley & Sons, and Taylor & Francis for granting the permission to reuse materials in the papers copyrighted by these publishers.

Beijing, P.R. China
January 2018

Dong Shen

Contents

1 **Introduction** 1
  1.1 Iterative Learning Control—Why and How 1
  1.2 Basic Formulation of ILC 3
    1.2.1 Discrete-Time Case 3
    1.2.2 Continuous-Time Case 6
  1.3 ILC with Random Data Dropouts 7
    1.3.1 Data Dropout Models 10
    1.3.2 Data Dropout Positions 10
    1.3.3 Convergence Meanings 12
  1.4 ILC with Other Incomplete Information 14
    1.4.1 Communication Delay and Asynchronism 14
    1.4.2 Iteration-Varying Lengths 15
  1.5 Structure of This Monograph 16
  1.6 Summary 18
  References 18

Part I One-Side Data Dropout

2 **Random Sequence Model for Linear Systems** 23
  2.1 Problem Formulation 23
  2.2 Intermittent Update Scheme and Its Almost Sure
     Convergence 27
  2.3 Extension to Arbitrary Relative Degree Case
     with Mean Square Convergence 31
    2.3.1 Noise-Free System Case 34
    2.3.2 Stochastic System Case 40
  2.4 Illustrative Simulations 46
  2.5 Summary 49
  References 50

3	**Random Sequence Model for Nonlinear Systems**	51
	3.1 Problem Formulation.	51
	3.2 Intermittent Update Scheme and Its Convergence	54
	3.3 Successive Update Scheme and Its Convergence.	56
	3.4 Illustrative Simulations	61
	3.5 Summary	64
	References	64
4	**Random Sequence Model for Nonlinear Systems with Unknown Control Direction**	65
	4.1 Problem Formulation.	65
	4.2 Intermittent Update Scheme and Its Almost Sure Convergence.	68
	4.3 Proofs of Lemmas.	71
	4.4 Illustrative Simulations	78
	4.5 Summary	82
	References	82
5	**Bernoulli Variable Model for Linear Systems**	83
	5.1 Problem Formulation.	83
	5.2 Intermittent Update Scheme and Its Almost Sure Convergence.	86
	5.3 Successive Update Scheme and Its Almost Sure Convergence.	91
	5.4 Mean Square Convergence of Intermittent Update Scheme	97
	5.4.1 Noise-Free System Case	99
	5.4.2 Stochastic System Case	102
	5.5 Illustrative Simulations	104
	5.5.1 System Description.	104
	5.5.2 Tracking Performance of both Schemes.	106
	5.5.3 Comparison of Different Data Dropout Rates	108
	5.5.4 Comparison of Different Learning Gains	110
	5.5.5 Comparison with Conventional P-Type Algorithm	110
	5.6 Summary	113
	References	114
6	**Bernoulli Variable Model for Nonlinear Systems**	115
	6.1 Problem Formulation.	115
	6.2 Intermittent Update Scheme and Its Almost Sure Convergence.	118
	6.3 Successive Update Scheme and Its Almost Sure Convergence.	122

		6.4	Illustrative Simulations	125
		6.5	Summary	130
		References		131
7	**Markov Chain Model for Linear Systems**			133
		7.1	Problem Formulation	133
		7.2	ILC Algorithms	137
		7.3	ILC for Classical Markov Chain Model Case	139
		7.4	ILC for General Markov Data Dropout Model Case	147
		7.5	Illustrative Simulations	153
		7.6	Summary	159
		References		159

Part II Two-Side Data Dropout

8	**Two-Side Data Dropout for Linear Deterministic Systems**			163
		8.1	Problem Formulation	163
		8.2	ILC Algorithms	165
		8.3	Markov Chain Model of Input Evolution	166
		8.4	Convergence Analysis	169
		8.5	Illustrative Simulations	174
		8.6	Summary	178
		References		178
9	**Two-Side Data Dropout for Linear Stochastic Systems**			179
		9.1	Problem Formulation	179
		9.2	Markov Chain of Input Evolution	182
		9.3	Convergence Analysis	185
		9.4	Discussions on Convergence Speed	190
		9.5	Illustrative Simulations	191
		9.6	Summary	195
		References		195
10	**Two-Side Data Dropout for Nonlinear Systems**			197
		10.1	Problem Formulation	197
		10.2	Convergence Analysis of ILC Algorithms	200
		10.3	Extensions to Non-affine Nonlinear Systems	207
		10.4	Illustrative Simulations	210
		10.5	Summary	214
		References		214

Part III General Incomplete Information Conditions

11 Multiple Communication Conditions and Finite Memory 217
    11.1   Problem Formulation. 217
    11.2   Communication Constraints. 219
    11.3   Control Objective and Preliminary Lemmas 221
    11.4   Intermittent Update Scheme and Its Almost Sure
           Convergence. 222
    11.5   Successive Update Scheme and Its Almost Sure
           Convergence. 226
    11.6   Illustrative Simulations 228
           11.6.1   Intermittent Update Scheme Case 229
           11.6.2   Successive Update Scheme Case 230
           11.6.3   Intermittent Update Scheme Versus Successive
                    Update Scheme 232
    11.7   Proofs of Theorems. 234
    11.8   Summary 240
    References 240

12 Random Iteration-Varying Lengths for Linear Systems 241
    12.1   Problem Formulation. 241
    12.2   ILC Design 243
    12.3   Strong Convergence Properties 247
    12.4   Illustrative Simulations 250
    12.5   Summary 253
    References 253

13 Random Iteration-Varying Lengths for Nonlinear Systems. 255
    13.1   Problem Formulation. 255
    13.2   ILC Design 257
    13.3   Convergence Analysis. 259
    13.4   Illustrative Simulations 265
    13.5   Summary 268
    References 269

14 Iterative Learning Control for Large-Scale Systems 271
    14.1   Problem Formulation. 271
    14.2   Optimal Control 273
    14.3   Optimal ILC Algorithms and Convergence Analysis 275
    14.4   Illustrative Example 281
    14.5   Summary 284
    References 284

Appendix. 287

Index 293

Chapter 1
Introduction

1.1 Iterative Learning Control—Why and How

Learning is a basic skill of human being so that we can survive against the severe environments in the ancient times. Learning also helps us behave better and better in our daily lives and work nowadays. Through the learning process, we can improve our behavior by utilizing the experience from previous executions. Take the basketball shooting from a fixed point as an example. For the first shot, the shooter may fail to hit the basket and even the basketball may be far away from hitting. Nevertheless, the shooter would learn how the final behavior is and how he/she should do in the next shot. Thus, as the number of attempts increases, the shooter is able to increase the hit ratio since he/she may adjust the angle and speed to reduce the shooting deviation shot by shot. Such learning scenes are very common in our daily lives, for examples, we learn to speak, to write, and to drive, all by repeating. That is why we believe learning is a basic skill of human being.

Then, one is interested in whether such basic skill, i.e., learning, could be introduced to the design of automatic control so that the machine can own certain ability of learning. The answer is yes. As a matter of fact, this basic cognition motivates the introduction and developments of a novel learning control method, iterative learning control (ILC). Before proceeding to the formal formulation of ILC, let us first recall some observations on the learning process of human being. In order to learn from experience, repetition is a fundamental requirement of the process. That is, the learning objective should be kept the same in different iterations when we are learning; otherwise, the learning experience is difficult to be applied to a new task. Besides, a repeatable learning process provides us the space for asymptotical improvement so that we could continuously perfect our performance. In short, repetition is closely connected with learning.

When considering the control method where the idea of learning is involved, such repetition requirement is also reserved. As to ILC, it is designed for the systems that are able to complete certain task over a fixed time interval and perform them repeatedly. In such systems, the input and output information of past itera-

© Springer Nature Singapore Pte Ltd. 2018
D. Shen, *Iterative Learning Control with Passive Incomplete Information*,
https://doi.org/10.1007/978-981-10-8267-2_1

tions/cycles/batches,[1] as well as the tracking objective, are used to formulate the input signal for the next iteration/cycle/batch, so that the tracking performance can be improved as the iteration number increases to infinity.

Thus, ILC has the following features: (1) the system can complete a task in a finite time interval, (2) the system can be reset to the same initial value, and (3) the tracking objective is iteration-invariant. We give a brief explanation for such features. First of all, the feature of finite operation length is a specified character for ILC, which is satisfied in many practical applications such as the production line automation and pick and place robot. If such feature is removed, i.e., the system would keep running along the time axis, certain learning mechanisms may be applicable if the system dynamics is periodic. Moreover, from the generic requirement of repetition, the resetting condition, i.e., the second feature, is a natural but specific formulation of ILC. Meanwhile, it is also a major criticizing point on ILC for practical applications. In the past three decades, such condition has attracted much attention and much literature has been published to relax it. Last but not least, the invariance of tracking objective is also a specific formulation of repetition. Noticing that the human being could apply the knowledge learned from other target to a new target but with similarities, we also expect such feature to be relaxed to the iteration-varying case. This issue has been addressed in many existing papers.

The main concept of ILC is shown in Fig. 1.1, where y_d denotes the reference trajectory. At the kth iteration, the input u_k is fed to the plant so that the corresponding output is y_k and the tracking error is $e_k = y_d - y_k$. This is a causal process. Since the tracking error e_k is nonzero, implying that the input u_k is not good enough, we will make an improvement to the input in the next iteration (i.e., the $(k+1)$th iteration). The specific construction for the improvement actually is a function of u_k and e_k, although it is usually specified as a linear combination. In other words, the input u_{k+1} for the $(k+1)$th iteration is generated before the running of the $(k+1)$th iteration. Then, the newly generated input is fed to the plant for the next iteration. Meanwhile, the input u_{k+1} is also stored in the memory for updating the input for the $(k+2)$th iteration. As a result, a closed-loop feedback is formed along the iteration axis.

By comparing ILC with our daily lives, we find that the previous information on inputs and outputs of the plant corresponds to the experiences accumulated in our daily lives. Persons usually decide a strategy for a given task based on previous experiences, while the strategy here corresponds to the input signal of ILC. Note that the previous experiences would help us to improve our behavior; thus, it is reasonable to believe that information on the previous operation may help improve the control performance to some extent.

The major advantage of ILC is that the design of control laws mainly requires the tracking references and input/output signals. In other words, little information about the plant is required and the system matrices may even be completely unknown. However, the algorithm is simple and effective.

---

[1] When we refer one operation process, the terminologies iteration, cycle, and batch are equivalent to each other.

Fig. 1.1 Framework of ILC

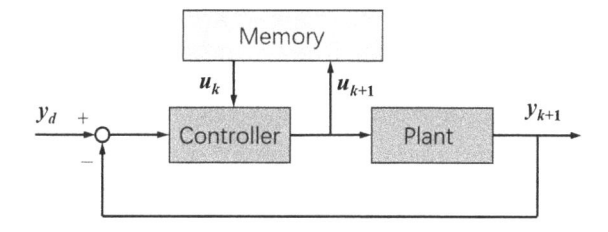

It is important to note that ILC adjusts the control along the iteration index rather than the time index, which is the main difference with other control methods such as proportional–integral–derivative (PID) control. PID control is a widely used feedback control. However, for repetitive systems, PID generates the same tracking error for all iterations since no previous information is used, whereas ILC reduces the tracking error iteration by iteration. Additionally, ILC differs from adaptive control, which also learns from previous operation information. Adaptive control aims to adjust the parameter of a structure-fixed controller, while ILC aims to construct the input signal directly.

The concept of ILC may be traced back to a paper published in 1978 by Uchiyama [1]. However, this paper failed to attract widespread attention as it was written in Japanese. Three papers that were published in 1984 [2–4] resulted in the further research on ILC. Subsequently, large amounts of literature have been published on various related issues, such as research monographs [5–9], survey papers [10–12], and special issues of academic journals [13–16]. ILC has recently become an important branch of intelligent control, and its use is widespread in many practical applications such as robotics [17–20], hard disk drives [21, 22], and industrial processes [23, 24].

1.2 Basic Formulation of ILC

We now present a basic formulation of ILC, followed by some traditional convergence results. Note that computer has been widely applied in the control of practical systems, where the sampled control systems or equivalently discrete-time systems are investigated. Moreover, this monograph concentrates on the discrete-time system. Thus, we first present the discrete-time case and then give the continuous-time counterpart briefly.

1.2.1 Discrete-Time Case

Consider the following discrete-time linear time-invariant system:

$$\begin{aligned} x_k(t+1) &= Ax_k(t) + Bu_k(t), \\ y_k(t) &= Cx_k(t), \end{aligned} \qquad (1.1)$$

where $x \in \mathbf{R}^n$, $u \in \mathbf{R}^p$, and $y \in \mathbf{R}^q$ denote the system state, input, and output, respectively. Matrices A, B, and C are system matrices with appropriate dimensions. t denotes an arbitrary time instant in an operation iteration, $t = 0, 1, \ldots, N$, where N is the length of the operation iteration. For simplicity, $t \in [0, N]$ is used in the following. $k = 0, 1, 2, \ldots$ denote different iterations.

Since it is required that a given tracking task should be repeated, the initial state needs to be reset at each iteration. The following is a basic reset condition, called identical initialization condition (i.i.c.), which has been used in many publications:

$$x_k(0) = x_0, \quad \forall k. \tag{1.2}$$

The reference trajectory is denoted by $y_d(t)$, $t \in [0, N]$. With regard to the reset condition, it is usually required that $y_d(0) = y_0 \triangleq Cx_0$. The control purpose of ILC is to design a proper update law for the input $u_k(t)$, so that the corresponding output $y_k(t)$ can track $y_d(t)$ as closely as possible. To this end, for any t in $[0, N]$, we define the tracking error as

$$e_k(t) = y_d(t) - y_k(t). \tag{1.3}$$

Then, the update law is a function of $u_k(t)$ and $e_k(t)$ to generate $u_{k+1}(t)$, whose general form is as follows:

$$u_{k+1}(t) = h(u_k(\cdot), \cdots, u_0(\cdot), e_k(\cdot), \cdots, e_0(\cdot)). \tag{1.4}$$

When the above relationship depends only on the last iteration, it is called first-order ILC update law; otherwise, it is called high-order ILC update law. Generally, to achieve the algorithm simplicity, most update laws are first-order laws, i.e.,

$$u_{k+1}(t) = h(u_k(\cdot), e_k(\cdot)). \tag{1.5}$$

Additionally, the update law is usually linear. The simplest update law is as follows:

$$u_{k+1}(t) = u_k(t) + Ke_k(t+1), \tag{1.6}$$

where K is the learning gain matrix, which is also the designed parameter. In (1.6), $u_k(t)$ is the input of the current iteration, while $Ke_k(t+1)$ is the innovation term. The update law (1.6) is called P-type ILC update law. If the innovation term is replaced by $K(e_k(t+1) - e_k(t))$, the update law is called D-type.

For system (1.1) and update law (1.6), a basic convergence condition is that K satisfies

$$\|I - CBK\| < 1. \tag{1.7}$$

Then, one has $\|e_k(t)\| \xrightarrow[k \to \infty]{} 0$, where $\|\cdot\|$ denotes the matrix or vector norm.

From this condition, one can deduce that the design of K needs no information with regard to the system matrix A, but requires information of the coupling matrix

1.2 Basic Formulation of ILC

CB only. This fact illustrates the advantage of ILC from the perspective that ILC has little dependence on the system information. Thus, ILC can handle tracking problems that have more uncertainties.

Remark 1.1 From the formulation of ILC, one can see that the model takes the classic features of a 2D system. That is, the system dynamics (1.1) and the update law (1.6) evolve along the time and iteration axes, respectively. Many researchers have made contributions from this point of view, and developed a 2D system-based approach, which is one of the principal techniques for ILC design and analysis.

Note that the operation length is limited by N and is then repeated multiple times. Thus, one could use the so-called lifting technique for discrete-time systems, which implies lifting all of the inputs and outputs as supervectors,

$$U_k = [u_k^T(0), u_k^T(1), \cdots, u_k^T(N-1)]^T, \tag{1.8}$$

$$Y_k = [y_k^T(1), y_k^T(2), \cdots, y_k^T(N)]^T. \tag{1.9}$$

Denote

$$G = \begin{bmatrix} CB & 0 & 0 & \cdots & 0 \\ CAB & CB & 0 & \cdots & 0 \\ \vdots & \vdots & \vdots & \ddots & \vdots \\ CA^{N-1}B & CA^{N-2}B & \cdots & \cdots & CB \end{bmatrix}, \tag{1.10}$$

then, one has

$$Y_k = GU_k + \mathbf{d}, \tag{1.11}$$

where

$$\mathbf{d} = [(CAx_0)^T, (CA^2x_0)^T, \ldots, (CA^N x_0)^T]^T. \tag{1.12}$$

Similar to (1.8) and (1.9), define

$$Y_d = [y_d^T(1), y_d^T(2), \cdots, y_d^T(N)]^T,$$
$$E_k = [e_k^T(1), e_k^T(2), \cdots, e_k^T(N)]^T,$$

then it leads to

$$U_{k+1} = U_k + \mathbf{K}E_k, \tag{1.13}$$

where $\mathbf{K} = \text{diag}\{K, K, \cdots, K\}$. By simple calculation, one has

$$\begin{aligned} E_{k+1} &= Y_d - Y_{k+1} = Y_d - GU_{k+1} - \mathbf{d} \\ &= Y_d - GU_k - G\mathbf{K}E_k - \mathbf{d} \\ &= E_k - G\mathbf{K}E_k \\ &= (I - G\mathbf{K})E_k. \end{aligned}$$

Therefore, we obtain a condition (1.7) that is sufficient to guarantee the convergence of ILC. Actually, the lifting technique not only helps us to obtain the convergence condition but also provides us with an intrinsic understanding of ILC. In the lifted model (1.11), the evolutionary process of an operation iteration has been integrated into G, whereas the relationship between adjacent iterations is highlighted. That is, the lifted model (1.11) depends on the k-axis only.

Remark 1.2 Note that the focus of ILC is how to improve the tracking performance gradually along the iteration axis, as one can see from the design of the update law (1.13) and lifted model (1.11). Therefore, it would not cause additional difficulties when the system is extended from the linear time-invariant case to the linear time-varying case. This is because, for any fixed time, the updating process along the iteration axis is a time-invariant case.

It is usually assumed that the reference trajectory $y_d(t)$ is realizable. That is, there exists an appropriate initial state x_0 and input $u_d(t)$ such that the expression (1.1) still holds with k replaced by d. In other words, $Y_d = GU_d + \mathbf{d}$, where U_d is defined in a similar manner of (1.8). Then, the discussion that the system output converges to the reference trajectory, i.e., $\lim_{k \to \infty} Y_k = Y_d$, becomes one that the system input converges to the objective input, i.e., $\lim_{k \to \infty} U_k = U_d$. For the system with stochastic noises, this transformation is more convenient for convergence analysis.

Remark 1.3 One may be interested in the case that the reference trajectory is not realizable. In other words, there is no control input producing the reference trajectory; thus, entirely accurate tracking is impossible. Then, the design objective of the ILC algorithm is no longer to guarantee asymptotically accurate tracking, but to converge to the nearest trajectory of the given reference. Consequently, the tracking problem has become an optimization problem. On the other hand, from the point of view of practical applications, the reference trajectory is usually realizable; thus, the assumption is not rigorous.

1.2.2 Continuous-Time Case

Let us consider the following linear continuous-time system:

$$\begin{aligned} \dot{x}_k(t) &= Ax_k(t) + Bu_k(t), \\ y_k(t) &= Cx_k(t), \end{aligned} \quad (1.14)$$

where the notations have similar meaning to the discrete-time formulation.

The control task is to servo the output y_k to track the desired reference y_d on a fixed time interval $t \in [0, T]$ as the iteration number k increases. If the system has relative degree one, an ILC scheme of Arimoto-type can be given as

$$u_{k+1} = u_k + \Gamma \dot{e}_k, \quad (1.15)$$

1.2 Basic Formulation of ILC

where $e_k(t) = y_d(t) - y_k(t)$ and Γ is the diagonal learning gain matrix. Similarly, if the learning gain matrix satisfies

$$\|I - CB\Gamma\| < 1, \tag{1.16}$$

then the control purpose is ensured, i.e., $\lim_{k\to\infty} y_k(t) \to y_d(t)$. Note that the basic formula for selecting the learning gain matrix given in (1.16) requires no information about the system matrix A, which implies that ILC can be effective for model-uncertain systems.

Moreover, a "PID-like" update law can be formulated as

$$u_{k+1} = u_k + \Phi e_k + \Gamma \dot{e}_k + \Psi \int e_k dt, \tag{1.17}$$

where Φ, Γ, and Ψ are learning gain matrices. The high-order PID-like update law can be formulated as

$$u_{k+1} = \sum_{i=1}^{n}(I - \Lambda)P_i u_{k+1-i} + \Lambda u_0 + \sum_{i=1}^{n}\left(\Phi_i e_{k+1-i} + \Gamma_i \dot{e}_{k+1-i} + \Psi_i \int e_{k+1-i} dt\right), \tag{1.18}$$

where $\sum_{i=1}^{n} P_i = I$.

1.3 ILC with Random Data Dropouts

We first provide three common models of data dropouts and detail the differences among the models. Throughout this monograph, the data dropout occurring or not can be regarded as a switch that opens and closes the network in a random manner. It is denoted by a random variable $\gamma_k(t)$. Therefore, there are two possible states of the variable $\gamma_k(t)$. Specifically, we let $\gamma_k(t)$ be equal to 1 if the corresponding data packet is successfully transmitted, and let $\gamma_k(t)$ be equal to 0 otherwise.

Generally, we have the following three most common models of data dropouts.

- Random sequence model (RSM): For each t, the measurement packet loss is random without obeying any certain probability distribution, but there is a positive integer $K \geq 1$ such that at least in one iteration the measurement is successfully sent back during the successive K iterations.
- Bernoulli variable model (BVM): The random variable $\gamma_k(t)$ is independent for different time instant t and iteration number k. Moreover, $\gamma_k(t)$ obeys a Bernoulli distribution with

$$\mathbb{P}(\gamma_k(t) = 1) = \bar{\gamma}, \quad \mathbb{P}(\gamma_k(t) = 0) = 1 - \bar{\gamma}, \tag{1.19}$$

where $\bar{\gamma} = \mathbb{E}\gamma_k(t)$ with $0 < \bar{\gamma} < 1$.

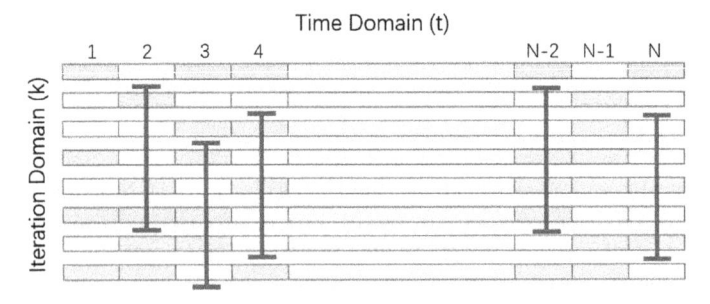

Fig. 1.2 Illustration of the RSM

- Markov chain model (MCM): The random variable $\gamma_k(t)$ is independent for different time instant t. Moreover, for each t, the evolution of $\gamma_k(t)$ along the iteration axis follows a two-state Markov chain, of which the probability transition matrix is

$$P = \begin{bmatrix} P_{11} & P_{10} \\ P_{01} & P_{00} \end{bmatrix} = \begin{bmatrix} \mu & 1-\mu \\ 1-\nu & \nu \end{bmatrix} \quad (1.20)$$

with $0 < \mu, \nu < 1$, where $P_{11} = \mathbb{P}(\gamma_{k+1}(t) = 1 \mid \gamma_k(t) = 1)$, $P_{10} = \mathbb{P}(\gamma_{k+1}(t) = 0 \mid \gamma_k(t) = 1)$, $P_{01} = \mathbb{P}(\gamma_{k+1}(t) = 1 \mid \gamma_k(t) = 0)$, $P_{00} = \mathbb{P}(\gamma_{k+1}(t) = 0 \mid \gamma_k(t) = 0)$.

Remark 1.4 The RSM is illustrated in Fig. 1.2 where a horizontal bar denotes an iteration process. In any bar, the white rectangle and light gold rectangle denote the lost packet and successfully transmitted packet, respectively. The gray part of each horizontal bar denotes the omission part. The RSM implies that, for arbitrary time instant t, the corresponding output information can be received at least once for any successive K iterations. As shown in Fig. 1.2, we take $K = 5$ for example. It is seen that there is at least one colored rectangle for any successive K horizontal bars and for any time instant. Moreover, this model can be formulated using the random variable $\gamma_k(t)$ as follows: for each t, $\sum_{i=0}^{K-1} \gamma_{k+i}(t) \geq 1$ for all $k \geq 1$. It is worth pointing out that we only require the existence of the number K rather than its specific value; that is, the number K is not necessary to be known prior and it is not involved in the design of ILC update law later. In fact, this model means that the output information should not be lost too much to ensure the learning ability in a somewhat deterministic point of view.

Remark 1.5 The number K of the RSM indicates that the maximum length of successive data dropouts is $K - 1$. Thus, the case $K = 1$ means no data dropout occurring, while the case $K = 2$ means no successive data dropout occurring for any two subsequent iterations. Moreover, the value of the successive iteration number K is a reflection of the rate of data dropouts. However, it is not equivalent to the data dropout rate (DDR), which can be formulated as $\lim_{n \to \infty} 1/n \times \left[\sum_{k=1}^{n}(1 - \gamma_k(t))\right]$.

1.3 ILC with Random Data Dropouts

In fact, DDR denotes the average level of data dropouts along the iteration axis, while K implies the worst case of successive data dropouts. In other words, a larger K usually corresponds to a higher DDR, while a smaller K usually corresponds to a lower DDR. But the connection between K and DDR is not necessarily to be positively related.

Remark 1.6 The mathematical expectation $\overline{\gamma}$ of the BVM is closely related to the DDR in the light of the law of large numbers; that is, DDR is equal to $1 - \overline{\gamma}$. Specifically, the data dropout is independent along the iteration axis, thus, $\lim_{n\to\infty} 1/n \times \left[\sum_{k=1}^{n} \left(1 - \gamma_k(t)\right)\right] = 1 - \mathbb{E}\gamma_k(t) = 1 - \overline{\gamma}$. If $\overline{\gamma} = 0$, implying that the network is completely broken down, then no information can be received from the plant, and thus no algorithm can be applied to improve the tracking performance. If $\overline{\gamma} = 1$, implying that no data dropout occurs, then the framework converts into the classical ILC problem. In this monograph, we simply assume $0 < \overline{\gamma} < 1$. Moreover, the statistics property of $\gamma_k(t)$ is assumed to be identical for different time instants for concise expression. The extension to the time-dependent case, i.e., the case that $\mathbb{E}\gamma_k(t) = \overline{\gamma}_t$, is straightforward without additional efforts.

Remark 1.7 The MCM is general for modeling the data dropouts. The transition probabilities μ and ν denote average levels of retaining the same state for successful transmission and loss, respectively. If $\mu + \nu = 1$, then MCM converts into BVM. That is, BVM is a special case of MCM. It is worth pointing out that all the three models are widely investigated in the field of networked control systems, such as [25] for RSM, [26] for BVM, and [27] for MCM.

Remark 1.8 In this remark, we comment the differences among the three models. RSM differs from both BVM and MCM as it requires no probability distribution or statistics property of the random variable $\gamma_k(t)$. However, RSM pays the price that the successive data dropouts length is bounded comparing with BVM and MCM. Specifically, both BVM and MCM admit arbitrary successive data dropouts associated with a suitable occurring probability. Consequently, RSM cannot cover BVM/MCM and vice versa. It should be pointed out that RSM implies that the data dropout is not totally stochastic. Moreover, the difference between BVM and MCM lies in the point that the data dropout occurs independently along the iteration axis for BVM, while dependently for MCM. The independence of data dropout admits some specific computations such as mean and variance, and then derives the convergence analysis. Such technique is not applicable for MCM.

Next, we give a brief literature review on ILC for systems with random data dropouts and classify the contributions of the existing papers from three aspects, namely, random data dropout models, data dropouts positions, and the convergence meaning. From these three dimensions, we can get a comprehensive picture view of the state of the art.

1.3.1 Data Dropout Models

The most popular model for the data dropout should go to BVM. In this model, the random variable takes the value 1 with success probability $\bar{\gamma}$ and the value 0 with failure probability $1 - \bar{\gamma}$. Besides, the data dropouts for different packets occur independently. In other words, this model has a clear probability distribution and good independence. Therefore, it is widely used in many existing papers addressing data dropout topic. Most ILC papers adopt this model [28–41] with/without extra requirements on data dropouts.

There are a few ILC papers dropping this model. In [42], the authors contributed to give an elaborate investigation of the effect of data dropouts. Thus, the authors mainly considered the case that only a single packet was lost during the transmission and provided a specific derivation for the effect on the input error and tracking performance. As to the multiple packet loss case, a general discussion was given instead of strict analysis and description. Specifically, the authors claimed that the data dropout level should be far smaller than 100% to ensure a satisfied tracking performance.

The papers [43, 44] provided RSM for data dropouts. Specifically, the sequence of the data dropout variables along the iteration axis was not assumed to be with any specific probability distribution. In other words, the statistical property of the data dropouts can vary along the iteration axis. Thus, the steady distribution in the Bernoulli model is removed. However, in order to ensure an asymptotical convergence of the input sequence, an additional requirement was imposed to the data dropout model in [43, 44] that there should exist a sufficiently large number K such that during any successive K iterations, at least one data dropout variable takes value 1. In other words, the data should be successfully transmitted from time to time.

There is another model for data dropouts, MCM, which has been used in some papers addressing other control strategies. In this model, the data dropouts have some dependence on the previous event. That is, the loss or not of the current packet would affect the probability of successful transmission for the next packet. In the ILC under data dropouts, this model has been discussed in very few papers.

1.3.2 Data Dropout Positions

In the networked ILC, the plant and the learning controller are separated in different sites and communicate with each other through wired/wireless networks. Thus, there are two channels connecting the plant and the learning controller. One is at the measurement side to transmit the measured output information back to the learning controller. The other one is at the actuator side to transmit the generated input signal to the plant so that the operation process can continuously run.

When considering the data dropouts problem for ILC, the position at which data dropout occurs is usually assumed to be the measurement side. In other words, only

1.3 ILC with Random Data Dropouts

the network at the measurement side is assumed to be lossy and the network at the actuator side is assumed to work well in most papers such as [28–31, 33, 34, 37–40, 43, 44]. In these papers, the generated input signal can be always sent to the plant without any loss. Although some papers claimed that their results can be extended to the general case where both the networks at the measurement and actuator sides suffered random data dropouts, it is actually not a trivial extension.

Specifically, when the network at the measurement side suffers random data dropouts, the output signal of the plant may or may not be successfully transmitted. One simple mechanism for treating the measured data is as follows: if the measured output is successfully transmitted, then the learning controller would employ such information for updating; if the measured output is lost during the transmission, then the learning controller would stop updating until the corresponding output information is successfully transmitted. One may find that the lost data is simply replaced by 0 in this mechanism. For the case that data dropout occurs only at the measurement side, such simple mechanism is sufficient to ensure the learning process as long as the network is not completely broken down. However, when considering the data dropout at the actuator side, it is clear that the lost input signal cannot be simply replaced by 0 as it would greatly damage the tracking performance. That is, if the network at the actuator side suffers data dropouts, the lost input signal must be compensated with a suitable packet to maintain the operation process of the plant. This observation motivates the investigation on compensation mechanisms for the lost data [32, 35, 36, 42].

In [42], the authors gave an earlier attempt on compensating the lost data. When one packet of the input signal is lost at the actuator side, the one-time-instant ahead input signal is applied to compensate the lost one. That is, if the input at time instant t is lost, it would be compensated with the input at time instant $t-1$. When one packet of the output signal is lost at the measurement side, a similar compensation mechanism is applied. It is worth noting that the data dropouts at the measurement side and actuator side are separately discussed in [42]. Moreover, this mechanism was then adopted by [32] for a Bernoulli model of random data dropouts occurring at both the measurement and actuator sides simultaneously. We should emphasize that, as a natural consequence, the data at adjacent time instants of the same iteration cannot be dropped simultaneously due to the inherent compensation requirement. Another compensation mechanism is to apply the corresponding data from the last iteration as shown in [35, 36]. That is, if the data packet at the kth iteration is lost during the transmission, it is compensated with the packet at the $(k-1)$th iteration with the same time instant label. In such assumption, the successive data dropouts along the time axis are allowed; however, it restricts that there was no simultaneous data dropout at the same time instant across any two adjacent iterations. In other words, no successive data dropouts along the iteration axis are allowed.

In short, the contributions in [32, 35, 36, 42] show that the newly introduced compensation mechanisms impose additional limitations on the data dropout models. In fact, the inherent difficulty of convergence analysis lies in the asynchronism between the computed input of the learning controller and the actual input fed to the plant. A recent paper [41] solved this problem according to the Bernoulli model allowing successive data dropouts along both time and iteration axes and provided a simple compensation mechanism with the iteration-latest available packet.

1.3.3 Convergence Meanings

In this subsection, we review the analysis techniques and the related convergence results, especially the convergence meanings in consideration of the randomness of data dropouts besides optional stochastic noises.

Ahn et al. provided earlier attempts to the ILC for linear systems in the presence of data dropouts [28–30]. The Kalman filtering based technique, which was first proposed by Saab in [45], was applied and thus the mean square convergence of the input sequence was obtained. The main difference among the papers [28–30] lies in the position where data dropouts occur. Specifically, in the first paper [28], the output vector was assumed to be lossy, and in [29] this assumption was relaxed to the case only partial dimensions of the output may suffer data dropouts. Last, in [30], the data dropouts at both the measurement and actuator sides were taken into account. In short, the Kalman filtering based technique was deeply investigated in these papers.

Bu et al. gave some different angles to solve this problem in [31–34]. In [31], the exponential stable result of asynchronous dynamical systems [46] was referred to establish the convergence condition of ILC under data dropouts. As a result, the randomness of data dropouts was not involved in the analysis steps. In [32], such randomness was eliminated from the recursion by taking mathematical expectation, thus the algorithm was converted into deterministic type and then the design and analysis of the convergence followed the conventional way. Therefore, the convergence was clear in the mathematical expectation sense. In [33], a new H-∞ framework was defined along the iteration axis and then the related control problem was solved in the newly defined framework. That is, the kernel objective was to satisfy an H-∞ performance index in the mean square sense. An LMI design condition for the learning gain matrix was also provided. In [34], the widely used 2D system approach was revisited to deal with data dropouts. A mean square asymptotically stable result was obtained and the design condition for the learning gain matrix was solved through LMI techniques. In short, the evolution dynamics along the iteration axis was carefully studied and related techniques are applied for the design and analysis of ILC.

There are several scattered results on this topic including [35–37, 42]. The paper [42] proposed a detailed analysis of the effect of packet loss for the sampled ILC. Specifically, a single packet loss at the measurement side and the actuator side was evaluated separately to study the inherent influence of data dropout on the tracking performance. In other words, a deterministic analysis was given according to the input error. The results in [42] revealed that neither contraction nor expansion occurred for the input error if the corresponding packet was lost during the transmission. Such technique was further exploited in [35] to study the general data dropout case.

1.3 ILC with Random Data Dropouts

In [36], a mathematical expectation was taken to the recursive inequality of input error to eliminate the randomness of data dropouts similar to [32] and then the conventional contraction mapping method was used to derive the convergence results. Moreover, to construct an explicit contraction mapping, the conditions in [36] were much conservative and it may be further relaxed. Similar techniques were also used in [37] incorporated with the conventional λ-norm technique to derive a convergence in mathematical expectation sense.

We mainly contribute the almost sure convergence results of ILC under data dropouts environments. In [38], a simple case that the whole iteration was packed and transmitted as a single packet was investigated by a switched system approach. Specifically, the evolution along the iteration axis was formulated as a switched system and the statistical properties were recursively computed. Then, the convergence in the sense of expectation, mean square, and almost sure was established in turn. In [43], based on stochastic approximation theory, the almost sure convergence of the input sequence was proved for the case that the data dropouts were modeled by a random sequence. This result was then extended to the unknown control direction case in [44]. For the traditional Bernoulli model of data dropouts, the essential difficulty in obtaining the almost sure convergence lies in the random successive data dropouts along the iteration axis. This problem was solved in [39] and [40] for linear and nonlinear stochastic systems, respectively. We then proceeded to investigate the general data dropouts at both measurement and actuator sides without any additional requirements except the Bernoulli assumption in [41]. When data dropouts occur at the actuator sides, there is a newly introduced asynchronism between the computed control generated by the learning controller and the actual control fed to the plant. Such asynchronism was characterized by a Markov chain in [41] and then the mean square and almost sure convergence were established.

The recent progress on ILC in the presence of data dropouts is classified in Table 1.1 according to data dropout model, data dropout position, and convergence meaning. From this table, we observe the following points:

- In most papers, the data dropout is modeled by the Bernoulli random variable, whereas the results according to random sequence model are rather limited. Moreover, for the Markov chain model, few results have been reported.
- All the papers consider the data dropout occurring at the measurement side and only a few papers address the case at the actuator side. As we have explained above, the latter case would involve an essential influence on the controller design and convergence analysis.
- The convergence meanings are scattered in different papers. Both mean square and almost sure convergence imply the convergence of mathematical expectation sense. However, they cannot imply each other according to the probability theory. Thus, it is of interest to propose an in-depth framework for the design and analysis of ILC in both senses simultaneously.

Table 1.1 Classification of the papers on ILC under data dropouts

Refs.	Model			Position		Convergence			
	RSM	BVM	MCM	Measurement	Actuator	M.E.	M.S.	A.S.	D.A.
[28]		•		•			•		
[29]		•		•			•		
[30]		•		•	•		•		
[31]		•		•					•
[32]		•		•	•	•			
[33]		•		•			•		
[34]		•		•			•		
[35]		•		•	•				•
[36]		•		•	•	•			
[37]		•		•		•			
[38]		•		•		•	•	•	
[39]		•		•				•	
[40]		•		•				•	
[41]		•		•	•		•	•	
[42]		•		•	•				•
[43]	•			•			•		
[44]	•			•			•		

RSM: random sequence model, BVM: Bernoulli variable model, MCM: Markov chain model, M.E.: mathematical expectation, M.S.: mean square, A.S.: almost sure, D.A.: deterministic analysis

1.4 ILC with Other Incomplete Information

1.4.1 Communication Delay and Asynchronism

In ILC, very few papers have been published on the issue of communication delay or communication asynchronism. A recent paper [36] considered this problem for discrete-time systems, where a P-type networked ILC scheme was proposed. In the scheme, the delayed data was compensated by the data from the previous iteration, and as a result, successive delays along the iteration axis were not allowed. Based on such assumption of the communication delay, a deterministic convergence analysis was given in the paper following the conventional contraction mapping principle. Indeed, the problem formulated in [36] can be transformed as an ILC problem with data dropout, which has been briefed in the last section.

A general model of communication delay or communication asynchronism was addressed in [47] for a large-scale system consisting of several subsystems, where the communication among different subsystems suffered random and possibly asynchronous communication delays. In particular, for large-scale systems, it is hard for each subsystem to learn all information of the whole system when generating control signals; hence, decentralized control is more suitable. Due to potentially different working efficiency values among subsystems, the control action may not be updated for all subsystems at each iteration step. Therefore, the update of all subsystems would introduce random asynchronism. We note that such problem can be solved following a similar method for the data dropout problem described by a random sequence model.

Time delay was widely investigated as it may reduce the performance of systems, which has been reported in many traditional control methodologies. In ILC, some results can be found in [48, 49]. However, the essence of ILC is to adjust the input signal using the input and output information of previous iterations; thus, the repetitive information along the iteration axis would not significantly affect ILC. As is well known, one of the main advantages of ILC is its reduced dependence on system information; thus, unknown but fixed time delays would have no impact on control performance since it could be regarded as a part of the information of the system. This intuition was verified in [50], where a class of affine nonlinear systems with time delays were studied. Generally, we believe that the random time delay would make a significant effect on system performance. However, there is yet no complete or explicit reply to this question. Thus, further explorations should be made regarding the basic influence of random time delays on ILC performance.

1.4.2 Iteration-Varying Lengths

In many practical applications, the inherent iteration-invariance is often violated due to unknown uncertainties or unpredictable factors. In this subsection, we concentrate on necessity of identical trial lengths, which have been found to be invalid in certain biomedical application systems. For example, while applying ILC in a functional electrical stimulation for upper limb movement and gait assistance, it has been seen that the operation processes end early for at least the first few passes due to safety considerations that the output may significantly deviate from the desired trajectory [51]. The associated variable-trial-length problem is detailed in [52] and [53], which clearly demonstrates the violation of the identical-trial-length assumption typically used in ILC.

There were some early research attempts to provide a suitable design and analysis framework for iteration-varying-length ILC that formed the groundwork for subsequent investigations [51–53]. For example, based on the experimental verifications and primary analysis of the convergence property given in [51–53], a systematic proof of the monotonic convergence in different norm senses was elaborated in [54] for linear systems. In particular, the necessary and sufficient conditions for monotonic convergence were discussed, as well as other issues including the controller design guidelines and influence of disturbances. Most importantly, no specific formulation of varying length was imposed in this framework. The first random model of varying-length iterations was proposed in [55] for discrete-time systems, and it was then extended to continuous-time systems in [56]. In these models, a stochastic variable was used to represent the occurrence of the output at each time instant and iteration, and it was then multiplied with the tracking error, which denoted the actual information of the updating process. To compensate for the information loss caused by randomly varying trial lengths, an iteration-average operator of all historical data was introduced to the ILC algorithm in [55], whereas in [56], this average operator was replaced by an iteration-moving-average operator to reduce the influence of very old data. Moreover, a lifted framework of ILC for a discrete-time linear system was provided in [57] to avoid the conservatism of the conventional λ-norm-based analysis in [55, 56]. We note that all of the contributions in [55–57] have two disadvantages: (1) they only obtained asymptotical convergence with respect to the expected value, which is a very weak convergence criteria for the control of stochastic models, and (2) the distribution of the introduced stochastic variable should be known to the controller. Therefore, one would wonder whether stronger convergence could be obtained. The answer to this question was given in [58] for a simpler ILC update law. Specifically, the discrete-time linear system was revisited, and the traditional P-type ILC law was employed. The authors transformed the error evolution along the iteration axis by modeling it as a switching system and then established the statistical properties of input errors (i.e., the mathematical expectations and covariances) in a recursive form. The convergence in the mathematical expectation, mean square, and almost sure senses was derived simultaneously. The results were then extended to a class of affine nonlinear systems in [59] using different analysis techniques. A recent work [60] further proposed two novel and improved ILC schemes based on the iteration-moving-average operator, in which a searching mechanism was additionally introduced to collect useful information while avoiding redundant tracking information from the past. We should note that, as opposed to [55–57], no specific distribution of the stochastic variable was required in these works.

In addition, some extensions have also been reported in the existing literature. In particular, nonlinear stochastic systems were taken into account in [61], and the bounded disturbances were included. In that study, the average-operator-based scheme was improved by collecting all available information. Nevertheless, we note that a Gaussian distribution of the variable pass length was required, which limits the possible application range. In [62], the authors extended the method to discrete-time linear systems with a vector relative degree. In this case, one needs to carefully select the output data for the learning algorithms to function. Additionally, this issue was further extended to stochastic impulse differential equations in [63] and fractional order systems in [64]. We would like to note that the convergence analyses derived in these papers are primarily based on the mature contraction mapping method, and thus are similar to [55].

1.5 Structure of This Monograph

In this monograph, we concentrate on ILC with passive incomplete information. Hereafter, by passive incomplete information, we mean the incomplete information and data caused by practical system limitation during data collecting, storing, trans-

1.5 Structure of This Monograph

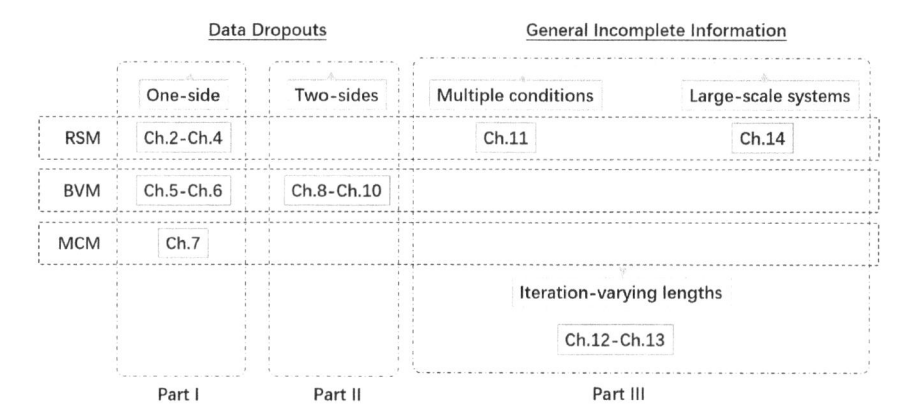

	Data Dropouts		General Incomplete Information	
	One-side	Two-sides	Multiple conditions	Large-scale systems
RSM	Ch.2-Ch.4		Ch.11	Ch.14
BVM	Ch.5-Ch.6	Ch.8-Ch.10		
MCM	Ch.7			
			Iteration-varying lengths	
			Ch.12-Ch.13	
	Part I	Part II	Part III	

Fig. 1.3 Structure of this monograph

mitting, and processing stages, such as data dropouts, delays, disordering, and limited transmission bandwidth. Our primary objective is to design and analyze the corresponding ILC algorithms. In other words, we focus on two aspects, one of which is how to design the data compensation mechanism when certain data is lost and then design the ILC scheme, and the other one is to establish a unified framework for convergence analysis in both almost sure and mean square senses. The investigations in this monograph would help to understand the restrictive relationship between incomplete information and tracking performance quantitatively.

The rest of this monograph consists of three parts, as shown in Fig. 1.3. Part I aims to provide in-depth discussions on ILC under various data dropout models, where the data dropout is assumed to occur at the measurement side only. In particular, three models, i.e., RSM, BVM, and MCM, are all addressed in detail. This part contains six chapters with three chapters on RSM case, two chapters on BVM case, and one chapter on MCM case. Part II devotes to the general data dropout environments that the data dropout occurs at both measurement and actuator sides. This part contains three chapters elaborating linear deterministic systems, linear stochastic systems, and nonlinear systems in sequence. Part III provides more results on ILC with other types of passive incomplete information such as communication delay and iteration-varying lengths. Four chapters are included in this part concerning on multiple communication conditions, iteration-varying lengths, and asynchronism among subsystems of a large-scale system.

The chapters in this monograph can also be viewed from another angle, i.e., the random model angle (see Fig. 1.3). In particular, the convergence analysis was conducted based on RSM in Chaps. 2–4, 11, and 14, and based on BVM in Chaps. 5–10. We note that the model for iteration-varying lengths in Chaps. 12 and 13 can be regarded as a generalized Bernoulli model, and thus the techniques in Chaps. 12 and 13 can be applied to solve the BVM-based problem.

1.6 Summary

In this chapter, we present the fundamental principles of ILC first to clarify why and how we design ILC for repetitive systems. Then, we proceed to formulate the basic framework of ILC in both discrete-time and continuous-time forms. The literature on ILC with various types of passive incomplete information is reviewed and analyzed. Lastly, we provide a visual structure of the whole monograph.

References

1. Uchiyama, M.: Formulation of high-speed motion pattern of a mechanical arm by trial. Trans. SICE(Soc. Instrum. Contr. Eng.) **14**(6), 706–712 (1978)
2. Arimoto, S., Kawamura, S., Miyazaki, F.: Bettering operation of robots by learning. J. Robotic Syst. **1**(2), 123–140 (1984)
3. Casalino, G., Bartolini, G.: A learning procedure for the control of movements of robotic manipulators. In: Proceedings of the IASTED Symposium Robotics and Automation, pp. 108–111 (1984)
4. Craig, J.: Adaptive control of manipulators through repeated trials. In: Proceedings of the American Control Conference, pp. 1566–1573 (1984)
5. Moore, K.L.: Iterative Learning Control Control for Deterministic Systems. Springer, Berlin (1993)
6. Bien, Z., Xu, J.-X.: Iterative Learning Control - Analysis, Design, Integration and Applications. Kluwer Academic Publishers, Dordrecht (1998)
7. Chen, Y.Q., Wen, C.: Iterative Learning Control: Convergence, Robustness and Applications. Springer, London (1999)
8. Xu, J.-X., Tan, Y.: Linear and Nonlinear Iterative Learning Control. Springer, New York (2003)
9. Ahn, H.-S., Moore, K.L., Chen, Y.Q.: Iterative Learning Control: Robustness and Monotonic Convergence for Interval Systems. Springer, Berlin (2007)
10. Bristow, D.A., Tharayil, M., Alleyne, A.G.: A survey of iterative learning control: a learning-based method for high-performance tracking control. IEEE Control Syst. Mag. **26**(3), 96–114 (2006)
11. Ahn, H.-S., Chen, Y.Q., Moore, K.L.: Iterative learning control: survey and categorization from 1998 to 2004. IEEE Trans. Syst. Man Cybern. Part C **37**(6), 1099–1121 (2007)
12. Wang, Y., Gao, F., Doyle III, F.J.: Survey on iterative learning control, repetitive control and run-to-run control. J. Process Control **19**(10), 1589–1600 (2009)
13. Moore, K.L., Xu, J.-X.(Guest eds.): Special issue on iterative learning control. Int. J. Control **73**(10), 819–999 (2000)
14. Special issue on iterative learning control. Asian J. Control **4**(1), 1–118 (2002)
15. Ahn, H.-S., Moore, K.L.(Guest eds.): Special issue on iterative learning control. Asian J. Control **13**(1), 1–212 (2011)
16. Freeman, C.T., Tan, Y.(Guest eds): Special issue on iterative learning control and repetitive control. Int. J. Control **84**(7), 1193–1294 (2011)
17. Tayebi, A., Abdul, S., Zaremba, M.B., Ye, Y.: Robust iterative learning control design: application to a robot manipulator. IEEE/ASME Trans. Mechatron **13**(5), 608–613 (2008)
18. Freeman, C., Lewin, P., Rogers, E., Ratcliffe, J.: Iterative learning control applied to a gantry robot and conveyor system. Trans. Inst. Meas. Control **32**(3), 251–264 (2010)
19. Inaba, K.: Iterative learning control for industrial robots with end effector sensing. Ph.D. dissertation, University of California, Berkeley (2008)

20. Hoelzle, D.J., Alleyne, A.G., Johnson, A.J.W.: Iterative Learning Control for Robotic Deposition Using Machine Vision. In: Proceedings of the American Control Conference, pp. 4541–4547 (2008)
21. Chen, Y.Q., Moore, K.L., Yu, J., Zhang, T.: Iterative learning control and repetitive control in hard disk drive industry–a tutorial. Int. J. Adap. Control Signal Process. **22**(4), 325–343 (2008)
22. Wu, S.-C., Tomizuka, M.: An iterative learning control design for self-servoWriting in hard disk drives. Mechatronics **20**(1), 53–58 (2010)
23. Liu, T., Gao, F.: IMC-based iterative learning control for batch processes with time delay variation. J. Process Control **20**(2), 173–180 (2010)
24. Liu, T., Gao, F.: Robust two-dimensional iterative learning control for batch processes with state delay and time-varying uncertainties. Chem. Eng. Sci. **65**(23), 6134–6144 (2010)
25. Lin, H., Antsaklis, P.J.: Stability and persistent disturbance attenuation properties for networked control systems: switched system approach. Int. J. Control **78**(18), 1447–1458 (2005)
26. Sinopoli, B., Schenato, L., Franceschetti, M., Poolla, K., Jordan, M.I., Sastry, S.S.: Kalman filtering with intermittent observations. IEEE Trans. Autom. Control **49**(9), 1453–1464 (2004)
27. Shi, Y., Yu, B.: Output feedback stabilization of networked control systems with random delays modeled by Markov chains. IEEE Trans. Autom. Control **54**(7), 1668–1674 (2009)
28. Ahn, H.S., Chen, Y.Q., Moore, K.L.: Intermittent iterative learning control. In: Proceedings of the 2006 IEEE International Symposium on Intelligent Control, pp. 832–837 (2006)
29. Ahn, H.S., Moore, K.L., Chen, Y.Q.: Discrete-time intermittent iterative learning controller with independent data dropouts. In: Proceedings of the 2008 IFAC World Congress, pp. 12442–12447 (2008)
30. Ahn, H.S., Moore, K.L., Chen, Y.Q.: Stability of discrete-time iterative learning control with random data dropouts and delayed controlled signals in networked control systems. In: Proceedings the IEEE International Conference Control Automation, Robotics, and Vision, pp. 757–762 (2008)
31. Bu, X., Hou, Z.-S., Yu, F.: Stability of first and high order iterative learning control with data dropouts. Int. J. Control Autom. Syst. **9**(5), 843–849 (2011)
32. Bu, X., Yu, F., Hou, Z.-S., Wang, F.: Iterative learning control for a class of nonlinear systems with random packet losses. Nonlinear Anal. Real World Appl. **14**(1), 567–580 (2013)
33. Bu, X., Hou, Z.-S., Yu, F., Wang, F.: H-∞ iterative learning controller design for a class of discrete-time systems with data dropouts. Int. J. Syst. Sci. **45**(9), 1902–1912 (2014)
34. Bu, X., Hou, Z.-S., Jin, S., Chi, R.: An iterative learning control design approach for networked control systems with data dropouts. Int. J. Robust Nonlinear Control **26**, 91–109 (2016)
35. Huang, L.-X., Fang, Y.: Convergence analysis of wireless remote iterative learning control systems with dropout compensation. Math. Probl. Eng. **2013**, 609284 (2013)
36. Liu, J., Ruan, X.: Networked iterative learning control approach for nonlinear systems with random communication delay. Int. J. Syst. Sci. **47**(16), 3960–3969 (2016)
37. Liu, C., Xu, J.-X., Wu, J.: Iterative learning control for remote control systems with communication delay and data dropout. Math. Probl. Eng. **2012**, 705474 (2012)
38. Shen, D., Wang, Y.: ILC for networked discrete systems with random data dropouts: a switched system approach. In: Proceedings of the 33rd Chinese Control Conference, pp. 8670–8677 (2014)
39. Shen, D., Zhang, C., Xu, Y.: Two compensation schemes of iterative learning control for networked control systems with random data dropouts. Inf. Sci. **381**, 352–370 (2017)
40. Shen, D., Zhang, C., Xu, Y.: Intermittent and successive ILC for stochastic nonlinear systems with random data dropouts. Asian J. Control (2018). https://doi.org/10.1002/asjc.1480
41. Shen, D., Xu, J.-X.: A novel Markov chain based ILC analysis for linear stochastic systems under general data dropouts environments. IEEE Trans. Autom. Control **62**(11), 5850–5857 (2017)
42. Pan, Y.-J., Marquez, H.J., Chen, T., Sheng, L.: Effects of network communications on a class of learning controlled non-linear systems. Int. J. Syst. Sci. **40**(7), 757–767 (2009)
43. Shen, D., Wang, Y.: Iterative learning control for networked stochastic systems with random packet losses. Int. J. Control **88**(5), 959–968 (2015)

44. Shen, D., Wang, Y.: ILC for networked nonlinear systems with unknown control direction through random Lossy channel. Syst. Control Lett. **77**, 30–39 (2015)
45. Saab, S.S.: A discrete-time stochastic learning control algorithm. IEEE Trans. Autom. Control **46**(6), 877–887 (2001)
46. Hassibi, A., Boyd, S.P., How, J.P.: Control of asynchronous dynamical systems with rate constraints on events. In: Proceedings the 38th IEEE Conference on Decision and Control, pp. 1345–1351 (1999)
47. Shen, D., Chen, H.-F.: Iterative learning control for large scale nonlinear systems with observation noise. Automatica **48**, 577–582 (2012)
48. Li, X.D., Chow, T.W.S., Ho, J.K.L.: 2D system theory based iterative learning control for linear continuous systems with time delays. IEEE Trans. Circuits Syst. **52**(7), 1421–1430 (2005)
49. Meng, D., Jia, Y., Du, J., Yu, F.: Robust iterative learning control design for uncertain time-delay systems based on a performance index. IET Control Theory Appl. **4**(5), 759–772 (2010)
50. Shen, D., Mu, Y., Xiong, G.: Iterative learning control for non-linear systems with deadzone input and time delay in presence of measurement noise. IET Control Theory Appl. **5**(12), 1418–1425 (2011)
51. Seel, T., Schauer, T., Raisch, J.: Iterative learning control for variable pass length systems. In: Proceedings of the 18th IFAC world congress, pp. 4880–4885 (2011)
52. Seel, T., Werner, C., Schauer, T.: The adaptive drop foot stimulator - multivariable learning control of foot pitch and roll motion in paretic gait. Med. Eng. Phys. **38**(11), 1205–1213 (2016)
53. Seel, T., Werner, C., Raisch, J., Schauer, T.: Iterative learning control of a drop foot neuroprosthesis - generating physiological foot motion in paretic gait by automatic feedback control. Control Eng. Pract. **48**, 87–97 (2016)
54. Seel, T., Schauer, T., Raisch, J.: Monotonic convergence of iterative learning control systems with variable pass length. Int. J. Control **90**(3), 393–406 (2017)
55. Li, X., Xu, J.-X., Huang, D.: An iterative learning control approach for linear systems with randomly varying trial lengths. IEEE Trans. Autom. Control **59**(7), 1954–1960 (2014)
56. Li, X., Xu, J.-X., Huang, D.: Iterative learning control for nonlinear dynamic systems with randomly varying trial lengths. Int. J. Adap. Control Signal Process. **29**(11), 1341–1353 (2015)
57. Li, X., Xu, J.-X.: Lifted system framework for learning control with different trial lengths. Int. J. Autom. Comput. **12**(3), 273–280 (2015)
58. Shen, D., Zhang, W., Wang, Y., Chien, C.-J.: On almost sure and mean square convergence of p-type ILC under randomly varying iteration lengths. Automatica **63**, 359–365 (2016)
59. Shen, D., Zhang, W., Xu, J.-X.: Iterative learning control for discrete nonlinear systems with randomly iteration varying lengths. Syst. Control Lett. **96**, 81–87 (2016)
60. Li, X., Shen, D.: Two novel iterative learning control schemes for systems with randomly varying trial lengths. Syst. Control Lett. **107**, 9–16 (2017)
61. Shi, J., He, X., Zhou, D.: Iterative learning control for nonlinear stochastic systems with variable pass length. J. the Frankl. Inst. **353**, 4016–4038 (2016)
62. Wei, Y.-S., Li, X.-D.: Varying trail lengths-based iterative learning control for linear discrete-time systems with vector relative degree. Int. J. Syst. Sci. **48**(10), 2146–2156 (2017)
63. Liu, S., Debbouche, A., Wang, J.: On the iterative learning control for stochastic impulsive differential equations with randomly varying trial lengths. J. Comput. Appl. Math. **312**, 47–57 (2017)
64. Liu, S., Wang, J.: Fractional order iterative learning control with randomly varying trial lengths. J. the Frankl. Inst. **354**, 967–992 (2017)

Part I
One-Side Data Dropout

In this part, we concentrate on the case that data dropout occurs at the output side only. In other words, the communication channel at the actuator side is assumed to work well. In such case, the major problem to which we should pay sufficient attention is to propose suitable analysis techniques for various data dropout models. In particular, we will elaborate the random sequence model, Bernoulli variable model, and Markov chain model in sequence for linear and nonlinear systems in this part.

Chapter 2
Random Sequence Model for Linear Systems

2.1 Problem Formulation

Consider the following single-input single-output (SISO) time-varying linear system

$$x_k(t+1) = A_t x_k(t) + \mathbf{b}_t u_k(t) + w_k(t+1),$$
$$y_k(t) = \mathbf{c}_t x_k(t) + v_k(t), \quad (2.1)$$

where $t \in \{0, 1, \ldots, N\}$ denotes the time instant in an iteration of the process, and $k = 1, 2, \ldots$ labels different iterations. $u_k(t) \in \mathbf{R}$, $x_k(t) \in \mathbf{R}^n$, and $y_k(t) \in \mathbf{R}$ denote the input, state, and output, respectively. A_t, $\mathbf{b}_t$, and $\mathbf{c}_t$ denote unknown system information. $w_k(t)$ and $v_k(t)$ are stochastic system noise and measurement noise, respectively.

The setup of the control system is illustrated in Fig. 2.1. For convenience, only data dropout at the measurement side is considered in this chapter, which can be extended to the case that both measurement and control signals are lossy using the techniques in Part II. As shown in Fig. 2.1, the measurement signals are transmitted back through a lossy channel, which could be regarded as a switch that opens and closes in a random manner. In the existing publications, the random data dropout is modeled as a binary Bernoulli random variable. However, in this chapter, a random sequence model of the random packet loss is proposed. To this end, denote $\mathscr{M}_k$ as the random set of time locations at which measurements are lost in the kth iteration. In other words, $t_0 \in \mathscr{M}_k$ if $y_k(t_0)$ is lost.

Let $\mathscr{F}_k \triangleq \sigma\{y_j(t), x_j(t), w_j(t), v_j(t), 0 \leq j \leq k, t \in \{0, \ldots, N\}\}$ be the σ-algebra generated by $y_j(t), x_j(t), w_j(t), v_j(t), 0 \leq t \leq N, 0 \leq j \leq k$. Then the set of admissible controls is defined as

$$U = \{u_{k+1}(t) \in \mathscr{F}_k, \sup_k \|u_k(t)\| < \infty, \text{ a.s.}$$
$$t = 0, \ldots, N-1, \quad k = 0, 1, 2, \ldots\}.$$

Fig. 2.1 Block diagram of networked control system with measurement packet loss

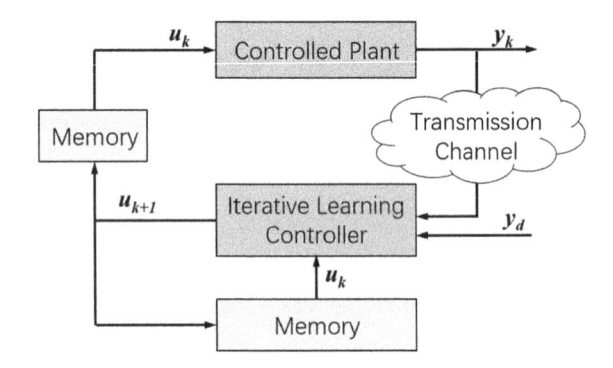

The conventional control objective of ILC for a noise-free system is to construct an update law such that the generated input sequence can guarantee the asymptotically precise tracking to the desired reference; that is, the output $y_k(t)$ can track the given trajectory $y_d(t)$ asymptotically for the specified time instants. However, when dealing with stochastic systems, it is impossible to achieve this control objective because of the existence of the unpredictable stochastic noises $w_k(t)$ and $v_k(t)$. That is, we cannot expect that $y_k(t) \to y_d(t)$, $\forall t$, for stochastic systems, as the iterations increase to infinity. Note that the stochastic noises cannot be eliminated by any algorithm in advance, and thus the best achievable control objective is to ensure that the desired reference can be precisely tracked by the output with removing these stochastic noises. In this chapter, the control objective is to find a control sequence $\{u_k(t), k = 0, 1, 2, \ldots\} \subset U$ under the random data dropout environment to minimize the following averaged tracking errors: $\forall t \in \{0, 1, \ldots, N\}$

$$V(t) = \limsup_{n \to \infty} \frac{1}{n} \sum_{k=1}^{n} \|y_k(t) - y_d(t)\|^2, \tag{2.2}$$

where $y_d(t)$, $t \in \{0, 1, \ldots, N\}$ is the tracking target. The following assumptions are used.

Assumption 2.1 The real number $\mathbf{c}_{t+1}\mathbf{b}_t$ coupling the input and output is an unknown nonzero constant, but its sign, characterizing the control direction, is assumed known.

Without loss of generality, it is assumed that $\mathbf{c}_{t+1}\mathbf{b}_t > 0$ in the rest of this chapter. It is noted that $\mathbf{c}_{t+1}\mathbf{b}_t \neq 0$ is equivalent to that the input/output relative degree is one.

Under Assumption 2.1 and the system (2.1), it is noted there exist suitable initial state value $x_d(0)$ and input $u_d(t)$ such that

$$\begin{aligned} x_d(t+1) &= A_t x_d(t) + \mathbf{b}_t u_d(t), \\ y_d(t) &= \mathbf{c}_t x_d(t). \end{aligned} \tag{2.3}$$

By Assumption 2.1 and (2.3), one has

2.1 Problem Formulation

$$u_d(t) = (\mathbf{c}_{t+1}\mathbf{b}_t)^{-1}(y_d(t+1) - \mathbf{c}_{t+1}A_t x_d(t)).$$

Assumption 2.2 For each time instant t, the data dropout at the measurement side is random without obeying any certain probability distribution, but there exists a number K such that during any successive K iterations, at least in one iteration the measurement is successfully sent back.

The number K is not necessary to be known prior. In other words, only the existence of such number is required. Thus, this condition means that the measurements should not be dropped too much to guarantee the convergence in almost sure sense. It is worth pointing out that this model of data dropout is different from the traditional binary Bernoulli one, which could not be covered by Assumption 2.2 and vice versa. Besides, the new model is another suitable description of practical data dropouts (for illustration, see Fig. 1.2).

Assumption 2.3 For each time instant t, the independent and identically distributed (i.i.d.) sequence $\{w_k(t), k = 0, 1, \ldots\}$ is independent of the i.i.d. sequence $\{v_k(t), k = 0, 1, \ldots\}$ with $\mathbb{E}w_k(t) = 0$, $\mathbb{E}v_k(t) = 0$, $\sup_k \mathbb{E}w_k^2(t) < \infty$, $\sup_k \mathbb{E}v_k^2(t) < \infty$, $\lim_{n\to\infty} \frac{1}{n} \sum_{k=1}^n w_k^2(t) = R_w^t$, and $\lim_{n\to\infty} \frac{1}{n} \sum_{k=1}^n v_k^2(t) = R_v^t$, a.s., where R_w^t and R_v^t are unknown.

Note that the noise assumption is made with respect to the iteration axis rather than the time axis, and thus this requirement is not rigorous as the process would be performed repeatedly.

Assumption 2.4 The sequence of initial values $\{x_k(0)\}$ is i.i.d. with $\mathbb{E}x_k(0) = x_d(0)$, $\sup_k \mathbb{E}x_k^2(0) < \infty$, and $\lim_{n\to\infty} \frac{1}{n} \sum_{k=1}^n x_k^2(0) = R_0$. Further, the sequences $\{x_k(0), k = 0, 1, \ldots\}$, $\{w_k(t), k = 0, 1, \ldots\}$, and $\{v_k(t), k = 0, 1, \ldots\}$ are mutually independent.

To facilitate the expression, denote $w_k(0) = x_k(0) - x_d(0)$. Then, it is easy to define $\lim_{n\to\infty} \frac{1}{n} \sum_{k=1}^n w_k^2(0) = R_w^0$ to satisfy the formulation of Assumption 2.3.

In order to give a control update algorithm, the optimal control which minimizes (2.2) should be first shown as in the following lemma.

Lemma 2.1 *For the stochastic system (2.1) and tracking objective $y_d(t+1)$, assume Assumptions 2.1, 2.3, and 2.4 hold, then for any arbitrary time instant $t+1$, the index (2.2) will be minimized if the control sequence $\{u_k(t)\}$ is admissible and satisfies $u_k(i) \to u_d(i)$ as $k \to \infty$, $i = 0, 1, \ldots, t$. In this case, $\{u_k(t)\}$ is called the optimal control sequence.*

Proof For simplicity of expression, denote $\delta x_k(t) \triangleq x_d(t) - x_k(t)$ and $\delta u_k(t) \triangleq u_d(t) - u_k(t)$. By (2.1) and (2.3),

$$\delta x_k(t+1) = A_t \delta x_k(t) + \mathbf{b}_t \delta u_k(t) - w_k(t+1).$$

By backwardly iterating this equation, we have

$$\delta x_k(t+1) = \sum_{i=1}^{t+1} \left(\prod_{l=i}^{t} A_l\right) \mathbf{b}_{i-1}\delta u_k(i-1) - \sum_{i=0}^{t+1} \left(\prod_{l=i}^{t} A_l\right) w_k(i),$$

where $\prod_{l=i}^{j} A_l \triangleq A_j A_{j-1} \cdots A_i, j \geq i$ and $\prod_{l=i}^{j} A_l = I, j < i$. Thereby

$$y_d(t+1) - y_k(t+1) = \mathbf{c}_{t+1}\delta x_k(t+1) - v_k(t+1)$$
$$= \phi_k(t+1) + \varphi_k(t+1) - v_k(t+1),$$

where

$$\phi_k(t+1) = \mathbf{c}_{t+1} \sum_{i=1}^{t+1} \left(\prod_{l=i}^{t} A_l\right) \mathbf{b}_{i-1}\delta u_k(i-1),$$

$$\varphi_k(t+1) = \mathbf{c}_{t+1} \sum_{i=0}^{t+1} \left(\prod_{l=i}^{t} A_l\right) w_k(i).$$

By Assumptions 2.3 and 2.4 and noticing that $u_k(i) \in \mathscr{F}_{k-1}, i = 0, 1, \ldots, t$, it is clear that $\phi_k(t+1), \varphi_k(t+1)$, and $v_k(t+1)$ are mutually independent.

By using Theorem 2.8 of [1]

$$\sum_{k=1}^{n} \phi_k(t+1)(\varphi_k(t+1) - v_k(t+1))$$

$$= O\left(\left(\sum_{k=1}^{n} \|\phi_k(t+1)\|^2\right)^{\frac{1}{2}+\eta}\right), \quad a.s. \quad \forall \eta > 0,$$

$$\sum_{k=1}^{n} \varphi_k(t+1)v_k(t+1)$$

$$= O\left(\left(\sum_{k=1}^{n} \|v_k(t+1)\|^2\right)^{\frac{1}{2}+\eta}\right), \quad a.s. \quad \forall \eta > 0.$$

Consequently,

$$\limsup_{n\to\infty} \frac{1}{n} \sum_{k=1}^{n} \|y_d(t+1) - y_k(t+1)\|^2$$

$$= \limsup_{n\to\infty} \frac{1}{n} \sum_{k=1}^{n} \|\phi_k(t+1)\|^2 + \limsup_{n\to\infty} \frac{1}{n} \sum_{k=1}^{n} \|\varphi_k(t+1)\|^2$$

$$+ \limsup_{n\to\infty} \frac{1}{n} \sum_{k=1}^{n} \|v_k(t+1)\|^2$$

2.1 Problem Formulation

$$= \limsup_{n \to \infty} \frac{1}{n} \sum_{k=1}^{n} \|\phi_k(t+1)\|^2$$

$$+ \operatorname{tr}\left[\mathbf{c}_{t+1} \sum_{i=0}^{t+1} \left(\prod_{l=i}^{t} A_l\right) R_w^t \left(\prod_{l=i}^{t} A(l)\right)^T \mathbf{c}_{t+1}^T + R_v^{t+1}\right].$$

where the last term is independent of control.

Therefore, the minimum of the index for time instant $t+1$ is achieved if and only if the first term on the right side of the last equation is zero. This means that $\{u_k(i), i = 0, 1, \ldots, t, k = 1, 2, \ldots\}$ is optimal if $u_k(i) \to u_d(i)$ as $k \to \infty$, $i = 0, 1, \ldots, t$. The proof is completed.

2.2 Intermittent Update Scheme and Its Almost Sure Convergence

In the last section, the optimal control has been characterized, but it cannot be actually used, because the system information is unknown. Thus, the following P-type update law with decreasing learning gain is proposed:

$$u_{k+1}(t) = u_k(t) + a_k \mathbf{1}_{\{(t+1) \notin \mathcal{M}_k\}} \times (y_d(t+1) - y_k(t+1)), \quad (2.4)$$

where a_k is the decreasing gain such that $a_k > 0$, $a_k \to 0$, $\sum_{k=0}^{\infty} a_k = \infty$, $\sum_{k=0}^{\infty} a_k^2 < \infty$, and $a_j = a_k(1 + O(a_k))$, $\forall j = k - K + 1, \cdots, k - 1, k$ as $k \to \infty$ with K being defined in Assumption 2.2. It is clear that $a_k = \frac{1}{k+1}$ meets these requirements. Besides, $\mathbf{1}_{\{\text{event}\}}$ is an indicator function, which is equal to 1 if the event indicated in the bracket is fulfilled, and 0 if the event does not hold.

In order to analyze the convergence of the proposed update law, we first rewrite the system into super-vector form by the so-called lifting technique as follows:

$$U_k = [u_k(0), u_k(1), \ldots, u_k(N-1)]^T \in \mathbf{R}^N,$$
$$Y_k = [y_k(1), y_k(2), \ldots, y_k(N)]^T \in \mathbf{R}^N.$$

Let H be defined by (2.5),

$$H = \begin{bmatrix} \mathbf{c}_1 \mathbf{b}_0 & 0 & \cdots & 0 \\ \mathbf{c}_2 A_1 \mathbf{b}_0 & \mathbf{c}_2 \mathbf{b}_1 & \cdots & 0 \\ \vdots & \vdots & \ddots & \vdots \\ \mathbf{c}_N \prod_{l=1}^{N-1} A_l \mathbf{b}_0 & \cdots & \cdots & \mathbf{c}_N \mathbf{b}_{N-1} \end{bmatrix}, \quad (2.5)$$

then one has the following relationship between input and output:

$$Y_k = HU_k + Y_d^0 + W_k, \tag{2.6}$$

where $Y_d^0 = [(\mathbf{c}_1 A_0 x_d(0))^T, (\mathbf{c}_2 A_1 A_0 x_d(0))^T, \ldots, (\mathbf{c}_N \cdot \prod_{l=0}^{N-1} A_l \cdot x_d(0))^T]^T$ is the response to initial conditions. The stochastic noise term W_k is expressed by

$$W_k = \begin{bmatrix} \mathbf{c}_1 \sum_{j=0}^{1}(\prod_{l=j}^{0} A_l) w_k(j) + v_k(1) \\ \mathbf{c}_2 \sum_{j=0}^{2}(\prod_{l=j}^{1} A_l) w_k(j) + v_k(2) \\ \vdots \\ \mathbf{c}_N \sum_{j=0}^{N}(\prod_{l=j}^{N-1} A_l) w_k(j) + v_k(N) \end{bmatrix}. \tag{2.7}$$

By noticing Assumptions 2.3 and 2.4, the noise W_k is considered as a zero-mean Gaussian process noise with covariance matrix Q, i.e., $W_k \sim N(0, Q)$. Besides, It is noted that

$$Y_d = HU_d + Y_d^0, \tag{2.8}$$

where Y_d and U_d are defined similar to Y_k and U_k by replacing k with d.

For the lifting model (2.6), the update law could be lifted as

$$U_{k+1} = U_k + a_k \Gamma_k (Y_d - Y_k), \tag{2.9}$$

where Γ_k is defined in (2.10).

$$\Gamma_k = \begin{bmatrix} \mathbf{1}_{\{1 \notin \mathcal{M}_k\}} & 0 & \cdots & 0 \\ 0 & \mathbf{1}_{\{2 \notin \mathcal{M}_k\}} & \cdots & 0 \\ \vdots & \vdots & \ddots & \vdots \\ 0 & 0 & \cdots & \mathbf{1}_{\{N \notin \mathcal{M}_k\}} \end{bmatrix}. \tag{2.10}$$

The following theorem shows the convergence property of the proposed update law.

Theorem 2.1 *For the system (2.1) and index (2.2), assume Assumptions 2.1–2.4 hold, then the control sequence $\{u_k(t)\}$ given by ILC update law (2.4) is optimal according to Lemma 2.1. In other words, $u_k(t)$ converges to $u_d(t)$ as $k \to \infty$, a.s., for any $t \in \{0, 1, 2, \ldots, N-1\}$.*

Remark 2.1 In fact, the update law proposed is the stochastic approximation algorithm in essence [2]. However, because of the existence of random data dropout, the matrix Γ_k is no longer of full rank. Thus, the original convergence result of stochastic approximation cannot be applied here directly. More detailed analysis is needed.

Proof Instituting (2.6) and (2.8) into (2.9), we have

$$\begin{aligned} U_{k+1} &= U_k + a_k \Gamma_k (Y_d - Y_k) \\ &= U_k + a_k \Gamma_k (HU_d + Y_d^0 - HU_k - Y_d^0 - W_k). \end{aligned}$$

2.2 Intermittent Update Scheme and Its Almost Sure Convergence

Then, subtracting both sides of the above equation from U_d, we have

$$\delta U_{k+1} = (I - a_k \Gamma_k H)\delta U_k + a_k \Gamma_k W_k, \tag{2.11}$$

where $\delta U_k \triangleq U_d - U_k$.

By noticing that Γ_k is a diagonal matrix and H is a lower triangular matrix, it is clear that $\Gamma_k H$ is a lower triangular matrix, and thus the eigenvalues of $\Gamma_k H$ are its diagonal elements. By the definition of Γ_k and H, one has that all eigenvalues of $\Gamma_k H$ are either positive constants or zero. Moreover, any eigenvalue of $\Gamma_k H$ is equal to zero if and only if the corresponding packet is lost.

Now let us replace all the terms $\mathbf{1}_{\{t+1 \notin \mathcal{M}_k\}}$ with 1 in Γ_k and then the matrix becomes I. Then, it is easy to find that the eigenvalues of H are positive. Define $P = \int_0^\infty \exp\left(-H^T t\right) \exp\left(-Ht\right) dt$, then by simple calculations we have

$$PH + H^T P = I.$$

Noticing the differences between Γ_k and I, one could deduce that

$$P(\Gamma_k H) + (\Gamma_k H)^T P \geq 0.$$

By Assumption 2.2, the summation $\sum_{k=i}^{i+K-1} \Gamma_k H$ is a lower triangular matrix whose eigenvalues are positive for any $i \geq 0$, which further implies that there exists some positive constant $\beta > 0$ such that

$$P\left(\sum_{k=i}^{i+K-1} \Gamma_k H\right) + \left(\sum_{k=i}^{i+K-1} \Gamma_k H\right)^T P > \beta I, \ \forall i \geq 0. \tag{2.12}$$

Define

$$\Phi_{i,j} \triangleq (I - a_i \Gamma_i H) \cdots (I - a_j \Gamma_j H), i \geq j, \Phi_{i,i+1} \triangleq I. \tag{2.13}$$

For any $i \geq j + K$, one has

$$\Phi_{i,j}^T P \Phi_{i,j}$$
$$= \Phi_{i-1,j}^T (I - a_i(\Gamma_i H)^T) P(I - a_i(\Gamma_i H)) \Phi_{i-1,j}$$
$$= \Phi_{i-K,j}^T (I - a_{i-K+1}(\Gamma_{i-K+1} H)^T) \cdots (I - a_i(\Gamma_i H)^T) P$$
$$\times (I - a_i(\Gamma_i H)) \cdots (I - a_{i-K+1}(\Gamma_{i-K+1} H)) \Phi_{i-K,j}$$
$$= \Phi_{i-K,j}^T \left[P - \left(\sum_{k=i-K+1}^{i} a_k (\Gamma_k H)^T P \right.\right.$$
$$\left.\left. + P \sum_{k=i-K+1}^{i} a_k (\Gamma_k H) \right) + o(a_i) \right] \Phi_{i-K,j}$$

$$= \Phi_{i-K,j}^T \left\{ P - a_i \left[\sum_{k=i-K+1}^{i} (\Gamma_k H)^T P + P \sum_{k=i-K+1}^{i} (\Gamma_k H) \right] + o(a_i) \right\} \Phi_{i-K,j},$$

where for the last equality the condition $a_j = a_k(1 + O(a_k))$, $\forall j = k - K + 1, \cdots, k - 1, k$ is involved.

Noticing that $0 < a_i < 1$ for sufficiently large i, by (2.12) we have

$$\Phi_{i,j}^T P \Phi_{i,j}$$
$$\leq \Phi_{i-K,j}^T (P - a_i \beta I + o(a_i)) \Phi_{i-K,j}$$
$$\leq \Phi_{i-K,j}^T P^{\frac{1}{2}} \left(I - a_i \beta P^{-1} + o(a_i) \right) P^{\frac{1}{2}} \Phi_{i-K,j}$$
$$\leq \Phi_{i-K,j}^T P^{\frac{1}{2}} \left(I - \frac{\beta}{K} P^{-1} \sum_{l=i-K+1}^{i} a_l + o(a_i) \right) P^{\frac{1}{2}} \Phi_{i-K,j}$$
$$\leq \left(1 - \frac{\beta}{K} \lambda_{\min}(P^{-1}) \sum_{l=i-K+1}^{i} a_l + o(a_i) \right) \Phi_{i-K,j}^T P \Phi_{i-K,j}$$
$$\leq \exp\left(-c \sum_{l=i-K+1}^{i} a_l \right) \Phi_{i-K,j}^T P \Phi_{i-K,j},$$

for sufficiently large j, where c is a positive constant and $\lambda_{\min}(M)$ denotes the minimum eigenvalue of a matrix M.

Then, for sufficiently large j, say, for $j > j_0$ and $i > j + K$ we have

$$\Phi_{i,j}^T P \Phi_{i,j} \leq c_1 \exp\left(-c \sum_{l=j}^{i} a_l \right) I,$$

where $c_1 > 0$ is a suitable constant, which further, by noticing the definition of $P > 0$, implies

$$\|\Phi_{i,j}\| \leq c_2 \exp\left(-\frac{c}{2} \sum_{l=j}^{i} a_l \right),$$

where $c_2 > 0$ is another suitable constant. Then, for $\forall i > j_0 + K$, $\forall j > 0$, we have

$$\|\Phi_{i,j}\| = \|\Phi_{i,j_0}\| \|\Phi_{j_0-1,j}\| \leq c_0 \exp\left(-\frac{c}{2} \sum_{l=j}^{i} a_l \right), \qquad (2.14)$$

2.2 Intermittent Update Scheme and Its Almost Sure Convergence

for some $c_0 >$ by noticing $\Phi_{j,j+1} \triangleq I$.

From (2.11), we have

$$\delta U_{k+1} = \Phi_{k,0}\delta U_0 + \sum_{i=0}^{k}\Phi_{k,i+1}a_i\Gamma_i W_i. \tag{2.15}$$

Comparing with Lemma A.3 in Appendix, it is found that (2.14) and (2.15) correspond to (A.11) and (A.17) of that lemma, respectively. Thus, the rest steps can be carried out along the lines of the proof of Lemma A.3 in Appendix. The proof of the theorem is completed.

Remark 2.2 As one can see, the initial condition Assumption 2.4 is formed as a normal distributed random variable. On the other hand, if the initial state is asymptotically available, in other words, $x_k(0) \to x_d(0)$ as $k \to \infty$, Theorem 2.1 is still valid with a similar proof.

2.3 Extension to Arbitrary Relative Degree Case with Mean Square Convergence

In this section, we will extend the SISO system in Sect. 2.1 to the multi-input multi-output (MIMO) system. Moreover, we also consider the arbitrary relative degree. In particular, consider the following linear time-varying system:

$$\begin{aligned} x_k(t+1) &= A_t x_k(t) + B_t u_k(t) + w_k(t+1), \\ y_k(t) &= C_t x_k(t) + v_k(t), \end{aligned} \tag{2.16}$$

where k is the iteration number, $k = 1, 2, \ldots$, t is the time instant, $t = 0, 1, \ldots, N$, and N is the iteration length. The variables $x_k(t) \in \mathbf{R}^n$, $u_k(t) \in \mathbf{R}^p$, and $y_k(t) \in \mathbf{R}^q$ are the system state, input, and output, respectively. The notations $w_k(t) \in \mathbf{R}^n$ and $v_k(t) \in \mathbf{R}^q$ are the system and measurement noises, respectively. In addition, A_t, B_t, and C_t are system matrices with appropriate dimensions.

In this section, we assume the system relative degree is τ, $\tau \geq 1$; that is, for any $t \geq \tau$,

$$C_t A_{t+1-i}^{t-1} B_{t-i} = 0, \quad 1 \leq i \leq \tau - 1, \tag{2.17}$$

$$C_t A_{t+1-\tau}^{t-1} B_{t-\tau} \neq 0, \tag{2.18}$$

where $A_i^j \triangleq A_j A_{j-1} \cdots A_i, j \geq i$, and $A_{i+1}^i \triangleq I_n$.

Remark 2.3 The relative degree implies the smallest structure delay of the input effect on its corresponding output. For example, if the relative degree $\tau = 1$, the input at time instant t would have an effect on the output at time instant $t + 1$ but no effect on the output at time instant t. The relative degree is an intrinsic property of the system and thus is usually time-invariant. Moreover, assuming the relative degree to be τ and starting the operation from the time instant $t = 0$, we find that the first controllable output appears at time instant $t = \tau$, which is driven by $u_k(0)$. In other words, the outputs at time $t = 0$ up to $t = \tau - 1$ are uncontrollable in such situation. As a consequence, these outputs would be formulated in the initialization condition. In addition, considering the MIMO system formulation, the relative degree may vary for different dimensions of the output vector; that is, different dimensions of the output vector have different relative degree values. It is straightforward to extend the following derivations to this case. Therefore, we omit the tedious extensions to make a concise layout.

Denote the desired reference as $y_d(t)$, $t \in \{0, 1, \ldots, N\}$. Without loss of any generality, we assume the reference is achievable; that is, with suitable initial value of $x_d(0)$, there exists a unique input $u_d(t)$ such that

$$\begin{aligned} x_d(t+1) &= A_t x_d(t) + B_t u_d(t), \\ y_d(t) &= C_t x_d(t). \end{aligned} \quad (2.19)$$

Denote the tracking error as $e_k(t) \triangleq y_d(t) - y_k(t)$, $t \in \{0, 1, \ldots, N\}$.

Remark 2.4 Note that the system relative degree is τ, implying that the output at time instant $t = 0$ up to $t = \tau - 1$ cannot be affected by the input. Therefore, the actual tracking reference is $y_d(t)$, $\tau \le t \le N$, while the initial τ outputs from $t = 0$ up to $\tau - 1$ is regulated by the initialization condition. Moreover, the uniqueness of the desired input $u_d(t)$ can be guaranteed if the matrix $C_t A_{t+1-\tau}^{t-1} B_{t-\tau}$ is of full-column rank. That is, the input $u_d(t)$ can be recursively computed from the nominal model (2.19) for $t \ge \tau$ as follows:

$$\begin{aligned} u_d(t - \tau) =& \left[\left(C_t A_{t+1-\tau}^{t-1} B_{t-\tau} \right)^T \left(C_t A_{t+1-\tau}^{t-1} B_{t-\tau} \right) \right]^{-1} \left(C_t A_{t+1-\tau}^{t-1} B_{t-\tau} \right)^T \\ & \times \left(y_d(t) - C_t A_{t-\tau}^{t-1} x_d(t - \tau) \right). \end{aligned} \quad (2.20)$$

The special case of (2.20) with τ being 1 has been explicitly given in many existing papers such as [3]. It should be emphasized that the full-column rank requirement is not strict as it has been proved necessary for the perfect tracking [4, 5]. As a consequence, the formulation (2.19) is a mild assumption for the system. When the coupling matrix is of full-row rank rather than full-column rank, which usually implies that the dimension of the input is greater than that of the output, it is found that only the asymptotical convergence of the tracking error is ensured in existing literature (e.g., [6]).

Without loss of generality (see the analysis after Assumption 2.4), we restrict the initial state condition to be the desired one so that the additional initial term is omitted for brevity. In other words, we use the following assumption.

2.3 Extension to Arbitrary Relative Degree Case with Mean Square Convergence

Assumption 2.5 The system initial value satisfies that $x_k(0) = x_d(0)$, where $x_d(0)$ is consistent with the desired reference $y_d(0)$ in the sense that $y_d(0) = C_0 x_d(0)$.

This initialization condition is critical for ensuring the accurate tracking performance of noise-free systems and thus is an important issue in the ILC field. Assumption 2.5 is the well-known identical initialization condition. This condition is a basic requirement for both time and space resetting of the system operation and thus is widely used in ILC papers. Moreover, many scholars contributed to relax this condition by introducing initial rectifying or learning mechanisms; however, either additional system information or tracking information is required when using the initial learning mechanisms [7, 8]. In Assumption 2.4, the initial state is assumed to be a random variable with expectation being the desired state, which actually is an equivalent form of Assumption 2.5 as we have reformulated the difference as a random variable.

Moreover, the assumption on stochastic noises can be relaxed as follows.

Define the σ-algebra $\mathscr{F}_k = \sigma\{x_i(t), u_i(t), y_i(t), w_i(t), v_i(t), 1 \leq i \leq k, 0 \leq t \leq N\}$ (i.e., the set of all events induced by these random variables) for $k \geq 1$.

Assumption 2.6 The stochastic noises $\{w_k(t)\}$ and $\{v_k(t)\}$ are martingale difference sequences along the iteration axis with finite conditional second moments. That is, for $t \in \{0, 1, \ldots, N\}$, $\mathbb{E}\{w_{k+1}(t) \mid \mathscr{F}_k\} = 0$, $\sup_k \mathbb{E}\{\|w_{k+1}(t)\|^2 \mid \mathscr{F}_k\} < \infty$, $\mathbb{E}\{v_{k+1}(t) \mid \mathscr{F}_k\} = 0$, $\sup_k \mathbb{E}\{\|v_{k+1}(t)\|^2 \mid \mathscr{F}_k\} < \infty$.

The system for which the ILC method is applicable should be repeated so that the tracking performance can be gradually improved along the iteration axis. Consequently, the stochastic noises are usually independent along the iteration axis (which is described in Assumption 2.3), whence Assumption 2.6 is mild and widely satisfied in practical applications. It is evident that the classical zero-mean white noise satisfies this assumption.

To facilitate the analysis in the following subsections, we give the lifting forms of the system (2.16). To this end, define the super-vectors

$$U_k = \left[u_k^T(0), u_k^T(1), \ldots, u_k^T(N-\tau) \right]^T, \tag{2.21}$$

$$Y_k = \left[y_k^T(\tau), y_k^T(\tau+1), \ldots, y_k^T(N) \right]^T. \tag{2.22}$$

Similarly, U_d and Y_d can be defined by replacing the subscript k in the above equations with d. The associated transfer matrix $\mathbf{H}$ can be formulated as

$$\mathbf{H} = \begin{bmatrix} C_\tau A_1^{\tau-1} B_0 & \mathbf{0}_{q \times p} & \cdots & \mathbf{0}_{q \times p} \\ C_{\tau+1} A_1^\tau B_0 & C_{\tau+1} A_2^\tau B_1 & \cdots & \mathbf{0}_{q \times p} \\ \vdots & \vdots & \ddots & \vdots \\ C_N A_1^{N-1} B_0 & C_N A_2^{N-1} B_1 & \cdots & C_N A_{N-\tau+1}^{N-1} B_{N-\tau} \end{bmatrix}. \tag{2.23}$$

Therefore, we have the following relationship between the input and output:

$$Y_k = \mathbf{H}U_k + Mx_k(0) + \xi_k \qquad (2.24)$$

and

$$Y_d = \mathbf{H}U_d + Mx_d(0), \qquad (2.25)$$

where $M = [(C_\tau A_0^{\tau-1})^T, \ldots, (C_N A_0^{N-1})^T]^T$ and

$$\xi_k = \left[\left(\sum_{i=1}^{\tau} C_\tau A_i^{\tau-1} w_k(i) + v_k(\tau)\right)^T, \left(\sum_{i=1}^{\tau+1} C_{\tau+1} A_i^\tau w_k(i) + v_k(\tau+1)\right)^T, \ldots,\right.$$
$$\left.\left(\sum_{i=1}^{N} C_N A_i^{N-1} w_k(i) + v_k(N)\right)^T\right]^T. \qquad (2.26)$$

Recalling the tracking error $e_k(t) = y_d(t) - y_k(t)$, we denote the lifted tracking error $E_k \triangleq Y_d - Y_k$. Then, it is evident that

$$E_k = Y_d - Y_k = \mathbf{H}(U_d - U_k) - \xi_k, \qquad (2.27)$$

where Assumption 2.5 is applied. These formulations will be used in the convergence analysis only.

2.3.1 Noise-Free System Case

In this subsection, we consider the case that the stochastic noises are absent in (2.16). That is, we consider the noise-free system

$$\begin{aligned} x_k(t+1) &= A_t x_k(t) + B_t u_k(t), \\ y_k(t) &= C_t x_k(t). \end{aligned} \qquad (2.28)$$

For such system, the randomness is only resulted from the data dropouts, which provides us a concise view to address the influences of data dropouts and stochastic noises.

The P-type ILC update law is designed as follows:

$$u_{k+1}(t) = u_k(t) + \rho \gamma_k(t+\tau) L_t e_k(t+\tau), \qquad (2.29)$$

for $t = 0, \ldots, N - \tau$, where ρ is a positive constant to be specified later and $L_t \in \mathbf{R}^{p \times q}$ is the learning gain matrix for regulating the control direction. Moreover, $\gamma_k(t+\tau) \triangleq \mathbf{1}_{\{t+\tau \notin \mathcal{M}_k\}}$. That is, $\gamma_k(t) = 1$ means the corresponding data packet is successfully transmitted, and $\gamma_k(t) = 0$ means the corresponding data packet is lost.

2.3 Extension to Arbitrary Relative Degree Case with Mean Square Convergence

Remark 2.5 First, we emphasize again that the ILC update law is not limited to the classical P-type law although we mainly focus on such type in this chapter to make a concise expression. Second, it is evident that the design of the positive constant ρ can be blended into the design of L_t. However, here we provide the separated design procedure to elaborate a clear design principle in the following analysis of this subsection as well as to provide a comparison with the design for the stochastic system case in the next subsection.

Now lift the input along the time axis as (2.21). The update law (2.29) can be rewritten as follows:
$$U_{k+1} = U_k + \rho \Gamma_k \mathbf{L} E_k, \tag{2.30}$$

where $E_k = Y_d - Y_k$, defined as (2.27) with $\xi_k = 0$, is the stacked vector of the tracking errors for $t = \tau, \ldots, N$, and Γ_k and $\mathbf{L}$ are defined by

$$\Gamma_k = \begin{bmatrix} \gamma_k(\tau) I_q & & & \\ & \gamma_k(\tau+1) I_q & & \\ & & \ddots & \\ & & & \gamma_k(N) I_q \end{bmatrix}, \tag{2.31}$$

$$\mathbf{L} = \begin{bmatrix} L_0 & & & \\ & L_1 & & \\ & & \ddots & \\ & & & L_{N-\tau} \end{bmatrix}. \tag{2.32}$$

Obviously, $\Gamma_k = \text{diag}\{\gamma_k(\tau), \gamma_k(\tau+1), \ldots, \gamma_k(N)\} \otimes I_q$.

Recalling that $E_k = \mathbf{H}(U_d - U_k)$ and substituting this equation into (2.30), we have
$$U_{k+1} = U_k + \rho \Gamma_k \mathbf{L} \mathbf{H}(U_d - U_k). \tag{2.33}$$

We define $\Lambda_k \triangleq \Gamma_k \mathbf{L}\mathbf{H}$ and

$$\mathbf{LH} = \begin{bmatrix} L_0 C_\tau A_1^{\tau-1} B_0 & \mathbf{0}_p & \cdots & \mathbf{0}_p \\ L_1 C_{\tau+1} A_1^\tau B_0 & L_1 C_{\tau+1} A_2^\tau B_1 & \cdots & \mathbf{0}_p \\ \vdots & \vdots & \ddots & \vdots \\ L_{N-\tau} C_N A_1^{N-1} B_0 & L_{N-\tau} C_N A_2^{N-1} B_1 & \cdots & L_{N-\tau} C_N A_{N-\tau+1}^{N-1} B_{N-\tau} \end{bmatrix}. \tag{2.34}$$

Since **LH** is a block lower triangular matrix, it is clear that the eigenvalue set of **LH** is a combination of the eigenvalue sets of $L_t C_{t+\tau} A_{t+1}^{t+\tau-1} B_t$, $t = 0, \ldots, N - \tau$. Moreover, Γ_k is a block diagonal matrix, and thus $\Lambda_k = \Gamma_k \mathbf{L}\mathbf{H}$ is also a block lower triangular matrix with all eigenvalues being the eigenvalues of its diagonal blocks. Specifically, the eigenvalue of Λ_k is either equal to the eigenvalue of **LH** or equal to zero, depending on whether the corresponding variable $\gamma_k(t)$ is 1 or 0, respectively.

Note that each $\gamma_k(t)$ has two possible values, i.e., 1 and 0, corresponding to the data that is successfully transmitted and not; thus, Γ_k have $\kappa \triangleq 2^{N+1-\tau}$ possible outcomes due to the independence of $\gamma_k(t)$ for different time instants. As a consequence, the newly defined $\Lambda_k = \Gamma_k \mathbf{LH}$ also has κ possible outcomes. Denote the set of all possible outcomes as $\mathfrak{S} = \{\Lambda^{(1)}, \ldots, \Lambda^{(\kappa)}\}$. Without loss of generality, we denote $\Lambda^{(1)} = \mathbf{LH}$ and $\Lambda^{(\kappa)} = \mathbf{0}_{(N+1-\tau)p}$, corresponding to the cases that all $\gamma_k(t)$ equal to 1 and 0, respectively. The other $\kappa - 2$ alternatives are also block lower triangular matrices similar to $\mathbf{LH}$ but with one or more block rows of $\mathbf{LH}$ are zero rows, corresponding to the time instants at which the packets are lost during the transmission. In other words, for a matrix Λ_k, if the data packet at time instant t is lost during the transmission, $t \geq \tau$, then the $(t+1-\tau)$th block row of Λ_k is a zero block row.

Now, we give the design condition of the learning gain matrix L_t, $0 \leq t \leq N - \tau$.

Learning gain matrix condition: In order to ensure the convergence of the P-type update law (2.29), the learning gain matrix L_t should satisfy that $-L_t C_{t+\tau} A_{t+1}^{t+\tau-1} B_t$ is a Hurwitz matrix, where a square matrix M is called Hurwitz if all the eigenvalues of M are with negative real parts.

Recalling the formulation of $\mathbf{LH}$ in (2.34), we have that all eigenvalues of $-\mathbf{LH}$ are with negative real parts if $-L_t C_{t+\tau} A_{t+1}^{t+\tau-1} B_t$ is a Hurwitz matrix for $t = 0, \ldots, N - \tau$. By the Lyapunov theorem, for any negative definite matrix $\mathbf{S}$ with appropriate dimension, there is a positive definite matrix $\mathbf{Q}$ such that $-\left((\mathbf{LH})^T \mathbf{Q} + \mathbf{Q}\mathbf{LH}\right) = \mathbf{S}$. In the following, to facilitate the analysis, we let $\mathbf{S} = -I$. That is, there exists a positive matrix $\mathbf{Q}$ such that

$$(\mathbf{LH})^T \mathbf{Q} + \mathbf{Q}\mathbf{LH} = I. \quad (2.35)$$

Noting the difference between $\Lambda^{(i)}$ and $\mathbf{LH}$, we have

$$\left(\Lambda^{(i)}\right)^T \mathbf{Q} + \mathbf{Q}\Lambda^{(i)} \geq 0, \quad (2.36)$$

for $i = 2, \ldots, \kappa - 1$.

Define $\delta U_k \triangleq U_d - U_k$. Subtracting both sides of (2.33) from U_d yields that

$$\delta U_{k+1} = \left(I_{(N+1-\tau)p} - \rho \Lambda_k\right)\delta U_k. \quad (2.37)$$

When considering the RSM case, no statistical property of the data dropout can be accessed and used; however, the bounded length assumption of successive data dropouts ensures a somewhat deterministic way for convergence analysis.

To make a clear insight of the influence of the RSM case of data dropouts, we rewrite (2.37) as follows:

$$\delta U_{k+K} = \left(I - \rho \Lambda_{k+K-1}\right) \cdots \left(I - \rho \Lambda_k\right)\delta U_k. \quad (2.38)$$

Denote

$$\Phi_{m,n} = \left(I - \rho \Lambda_m\right) \cdots \left(I - \rho \Lambda_n\right), \quad m \geq n. \quad (2.39)$$

Now, we give an estimate of $\Phi_{k+K-1,k}$ in the following lemma.

2.3 Extension to Arbitrary Relative Degree Case with Mean Square Convergence

Lemma 2.2 *Consider the matrix product (2.39). If the learning matrix L_t satisfies that $-L_t C_{t+\tau} A_{t+1}^{t+\tau-1} B_t$ is a Hurwitz matrix and ρ is small enough, then there exists a positive definite matrix $\mathbf{Q}$ such that*

$$\Phi_{k+K-1,k}^T \mathbf{Q} \Phi_{k+K-1,k} \leq \eta \mathbf{Q}, \quad 0 < \eta < 1, \quad \forall k. \tag{2.40}$$

Proof As has been explained above, all Λ_k are block lower triangular matrices, thus the summation of Λ_k is also a block lower triangular matrix. In other words, $\sum_{i=0}^{K-1} \Lambda_{k+i}$ is a block lower triangular matrix. Moreover, the RSM assumption of data dropouts implies that all the diagonal blocks of $\sum_{i=0}^{K-1} \Lambda_{k+i}$ are with positive real parts in their eigenvalues for $k \geq 0$, which further implies that there exists some positive constant $c_1 > 0$ such that

$$\left(\sum_{i=0}^{K-1} \Lambda_{k+i}\right)^T \mathbf{Q} + \mathbf{Q}\left(\sum_{i=0}^{K-1} \Lambda_{k+i}\right) \geq c_1 I, \quad \forall k \geq 0. \tag{2.41}$$

Now revisit the recursion of $\Phi_{k+K-1,k}$ and we have

$$\Phi_{k+K-1,k}^T \mathbf{Q} \Phi_{k+K-1,k}$$
$$= (I - \rho \Lambda_k^T) \cdots (I - \rho \Lambda_{k+K-1}^T) \mathbf{Q} (I - \rho \Lambda_{k+K-1}) \cdots (I - \rho \Lambda_k)$$
$$\leq \mathbf{Q} - \rho \left[\left(\sum_{i=0}^{K-1} \Lambda_{k+i}\right)^T \mathbf{Q} + \mathbf{Q}\left(\sum_{i=0}^{K-1} \Lambda_{k+i}\right)\right]$$
$$+ \rho^2 \left[\sum_{k \leq i,j \leq k+K-1} \Lambda_i^T \mathbf{Q} \Lambda_j + \sum_{k \leq i < j \leq k+K-1} (\mathbf{Q} \Lambda_i \Lambda_j + \Lambda_j^T \Lambda_i^T \mathbf{Q}) + \cdots \right].$$

Note that $\|\Lambda_i\| \leq \|\mathbf{LH}\|$ and the possible combinations are finite due to the boundedness of K, and thus there exists a constant $c_2 > 0$ such that the last term on the right-hand side of the last inequality over ρ^2 is bounded by $c_2 I$. Moreover, $\mathbf{Q}$ is a positive definite matrix, and thus there is a suitable constant $c_3 > 0$ such that $c_3 \mathbf{Q} \leq I$. Then, we have

$$\Phi_{k+K-1,k}^T \mathbf{Q} \Phi_{k+K-1,k} \leq \mathbf{Q} - \rho c_1 I + \rho^2 c_2 I \leq \mathbf{Q} - [\rho c_1 - \rho^2 c_2] c_3 \mathbf{Q}$$

as long as ρ is small enough such that $\rho c_1 - \rho^2 c_2 > 0$ and $\rho c_1 c_3 < 1$. In such case, denote $\eta = 1 - [\rho c_1 - \rho^2 c_2] c_3$ and it is clear

$$\Phi_{k+K-1,k}^T \mathbf{Q} \Phi_{k+K-1,k} \leq \eta \mathbf{Q}, \quad 0 < \eta < 1, \quad \forall k. \tag{2.42}$$

The proof is completed.

Remark 2.6 It is worth pointing out that the proof of Lemma 2.2 is quite technical; however, the inherent principle is not so complicated. Specifically, the introduction of the positive definite matrix $\mathbf{Q}$ is to make well-defined expressions in the analysis. The contract effect of $\Phi_{k+K-1,k}$ can be interpreted as follows. Λ_k is a block lower triangular matrix and then $I - \rho \Lambda_k$ is a block lower triangular matrix with its eigenvalues being $1 - \rho \gamma_k(t+\tau)\lambda_{t,i}$, $1 \leq i \leq p$, $0 \leq t \leq N-\tau$, where $\lambda_{t,i}$ denotes the eigenvalue of $L_t C_{t+\tau} A_{t+1}^{t+\tau-1} B_t$. Therefore, when $\gamma_k(t+\tau) = 0$, the corresponding eigenvalue of $I - \rho \Lambda_k$ is 1, implying that no contraction occurs but neither any expansion occurs; when $\gamma_k(t+\tau) = 1$, the corresponding eigenvalue of $I - \rho \Lambda_k$ is less than 1 provided that the eigenvalues $\lambda_{t,i}$ are positive and ρ is small enough, implying a contraction. The bounded length assumption on successive data dropouts actually guarantees the infinitely often contractions along the iteration axis.

Remark 2.7 From the technical viewpoint, the parameter ρ can be solved from the relationship $\rho c_1 - \rho^2 c_2 > 0$ and $\rho c_1 c_3 < 1$, i.e., $\rho < \min\{c_1 c_2^{-1}, (c_1 c_3)^{-1}\}$. Apparently, ρ should be small enough when little information of the system is known. However, a small ρ would render a large value of η, which limit the contraction effect. Thus, there is a trade-off in selecting the parameter ρ. In addition, the proof provides a rather conservative estimation of η while the actual contract influence is usually more efficient.

With the help of Lemma 2.2, we can give the convergence for the input sequence now.

Theorem 2.2 *Consider the noise-free linear system (2.28) and the ILC update law (2.29), where the random data dropouts follow the RSM case. Assume Assumption 2.5 hold. Then, the input sequence $\{u_k(t)\}$, $t = 0, \ldots, N-\tau$, achieves mean square convergence to the desired input $u_d(t)$, $t = 0, \ldots, N-\tau$, if the learning gain matrix L_t satisfies that $-L_t C_{t+\tau} A_{t+1}^{t+\tau-1} B_t$ is a Hurwitz matrix and ρ is small enough.*

Proof The proof is carried out based on the inherent convergence principle that there exists at least once contraction during any K successive iterations. To this end, we can group the iteration number by modulo operator with respect to K; that is, all iterations are divided into K subsets, $\{iK+j, i \geq 0\}$, $0 \leq j \leq K-1$. Then, we show the strict contraction mapping of the input sequence with the subscripts valued in each subset given above.

Define a weighted norm of δU_k as $V_k = \|\delta U_k\|_{\mathbf{Q}} \triangleq (\delta U_k)^T \mathbf{Q} \delta U_k$, which can be regarded as a Lyapunov function. Then, we have $\forall 0 \leq j \leq K-1$,

$$\begin{aligned} V_{iK+j} &= (\delta U_{iK+j})^T \mathbf{Q} \delta U_{iK+j} \\ &= (\Phi_{iK+j-1,(i-1)K+j} \delta U_{(i-1)K+j})^T \mathbf{Q} \Phi_{iK+j-1,(i-1)K+j} \delta U_{(i-1)K+j} \\ &= (\delta U_{(i-1)K+j})^T \Phi_{iK+j-1,(i-1)K+j}^T \mathbf{Q} \Phi_{iK+j-1,(i-1)K+j} \delta U_{(i-1)K+j} \\ &\leq \eta (\delta U_{(i-1)K+j})^T \mathbf{Q} \delta U_{(i-1)K+j} \\ &= \eta V_{(i-1)K+j}, \quad i \geq 1, \end{aligned}$$

2.3 Extension to Arbitrary Relative Degree Case with Mean Square Convergence

where (2.37) and Lemma 2.2 are used. Consequently, we have

$$\mathbb{E}\|\delta U_{iK+j}\|_{\mathbf{Q}} \leq \eta \mathbb{E}\|\delta U_{(i-1)K+j}\|_{\mathbf{Q}}, \quad \forall 0 \leq j \leq K-1, \quad i \geq 1. \tag{2.43}$$

Then, it directly leads to that

$$\mathbb{E}\|\delta U_{iK+j}\|_{\mathbf{Q}} \leq \eta^i \mathbb{E}\|\delta U_j\|_{\mathbf{Q}}, \quad 0 \leq j \leq K-1. \tag{2.44}$$

Meanwhile, following the same idea in Lemma 2.2, the weighted norms of the inputs for the first K iterations, i.e., δU_j with $0 \leq j \leq K-1$, are bounded by the initial one. That is, $\forall 0 \leq j \leq K-1$,

$$\begin{aligned} V_j &= (\delta U_j)^T \mathbf{Q} \delta U_j \\ &= (\delta U_0)^T (I - \rho \Lambda_0^T) \cdots (I - \rho \Lambda_{j-1}^T) \mathbf{Q} (I - \rho \Lambda_{j-1}) \cdots (I - \rho \Lambda_0) \delta U_0 \\ &\leq (\delta U_0)^T \mathbf{Q} \delta U_0 = \|\delta U_0\|_{\mathbf{Q}}, \end{aligned}$$

where U_0 denotes the initial input. Incorporating with (2.44), we are evident to derive that

$$\mathbb{E}\|\delta U_{iK+j}\|_{\mathbf{Q}} \xrightarrow[i \to \infty]{} 0, \quad 0 \leq j \leq K-1. \tag{2.45}$$

Note that $\mathbf{Q}$ is a fixed positive definite matrix; therefore, a direct corollary of (2.45) is that $\lim_{k \to \infty} \mathbb{E}\|\delta U_k\|^2 = 0$. In other words, the mean square convergence of the update law is established.

Remark 2.8 In fact, based on the above derivations of the mean square convergence, we can derive the almost sure convergence by employing the Borel–Cantelli lemma. In particular, recalling the inequality (2.44) and noting that $\mathbf{Q}$ is a positive definite matrix, we have

$$\mathbb{E}\|\delta U_{iK+j}\|^2 \leq \lambda_{\min}^{-1}(\mathbf{Q}) \eta^i \mathbb{E}\|\delta U_j\|_{\mathbf{Q}}, \quad 0 \leq j \leq K-1, \tag{2.46}$$

where $\lambda_{\min}(\cdot)$ denotes the minimum eigenvalue of its indicated matrix. It follows that

$$\begin{aligned} \sum_{i=0}^{\infty} \mathbb{E}\|\delta U_{iK+j}\| &\leq \sum_{i=0}^{\infty} \lambda_{\min}^{-1/2}(\mathbf{Q}) \eta^{i/2} \left(\mathbb{E}\|\delta U_j\|_{\mathbf{Q}}\right)^{1/2} \\ &\leq \lambda_{\min}^{-1/2}(\mathbf{Q}) \left(\mathbb{E}\|\delta U_0\|_{\mathbf{Q}}\right)^{1/2} \sum_{i=0}^{\infty} \eta^{i/2} \\ &= \lambda_{\min}^{-1/2}(\mathbf{Q}) \left(\mathbb{E}\|\delta U_0\|_{\mathbf{Q}}\right)^{1/2} \frac{1}{1 - \eta^{1/2}} < \infty, \end{aligned}$$

which further yields

$$\sum_{k=0}^{\infty} \mathbb{E}\|\delta U_k\| = \sum_{j=0}^{K-1}\sum_{i=0}^{\infty} \mathbb{E}\|\delta U_{iK+j}\| < \infty.$$

Then, by the Markov inequality, for any $\varepsilon > 0$, we have

$$\sum_{k=1}^{\infty} \mathbb{P}(\|\delta U_k\| > \varepsilon) \leq \sum_{k=1}^{\infty} \frac{\mathbb{E}\|\delta U_k\|}{\varepsilon} < \infty.$$

This fact leads to $\mathbb{P}(\|\delta U_k\| > \varepsilon, \text{i.o.}) = 0$ by the Borel–Cantelli lemma, $\forall \varepsilon > 0$, where "i.o." is short for "infinitely often". That is, $\mathbb{P}(\lim_{k\to\infty} \|\delta U_k\| = 0) = 1$. In other words, δU_k converges to zero almost surely.

In this subsection, the noise-free system is taken into account; therefore, the precise convergence of the input sequence ensures that the system output $y_k(t)$ can precisely track the desired reference $y_d(t)$, $\forall t$, with the help of Assumption 2.5. Moreover, it is noticed from (2.44) that the update law (2.29) for the noise-free system ensures an exponential convergence speed. Meanwhile, such exponential convergence speed enables us to establish the almost sure convergence based on the Borel–Cantelli lemma.

2.3.2 Stochastic System Case

In this subsection, we consider the stochastic linear system (2.16). Due to the existence of stochastic noises, the ILC update law (2.29) cannot guarantee a stable convergence of the input sequence. Take the lifted form (2.30) for intuitive understanding of this limitation. If the input sequence $\{U_k\}$ exists a stable convergence limitation, then taking limitation to both sides of (2.30) leads to $\lim_{k\to\infty} U_{k+1} = \lim_{k\to\infty} U_k + \lim_{k\to\infty} \rho \Gamma_k L E_k$. We can derive a simple corollary that $\lim_{k\to\infty} E_k = 0$. This corollary contradicts with the randomness of E_k in (2.27). That is, the tracking error E_k consists of two parts, $\mathbf{H}(U_d - U_k)$ and ξ_k; thus, it is impossible to derive $\lim_{k\to\infty} E_k = 0$. Moreover, by Assumption 2.6, the stochastic noises cannot be predicted and eliminated by any algorithm; thus, we have to impose an additional mechanism to reduce the effect of noises along the iteration axis. As a matter of fact, it is well known that an appropriate decreasing gain for the correction term in updating processes is a necessary requirement to ensure convergence in the recursive computation for optimization, identification, and tracking of stochastic systems. Inspired by this recognition, we replace the design parameter ρ in (2.29) with a decreasing sequence to cope with the stochastic noises. Specifically, the ILC update law for the stochastic system is modified as follows:

$$u_{k+1}(t) = u_k(t) + a_k \gamma_k(t+\tau) L_t e_k(t+\tau), \tag{2.47}$$

2.3 Extension to Arbitrary Relative Degree Case with Mean Square Convergence

where the learning step size $\{a_k\}$ is a decreasing sequence satisfying that

$$a_k \in (0, 1), \quad a_k \to 0, \quad \sum_{k=1}^{\infty} a_k = \infty, \quad \sum_{k=1}^{\infty} a_k^2 < \infty, \quad \frac{1}{a_{k+1}} - \frac{1}{a_k} \to \chi > 0. \tag{2.48}$$

Remark 2.9 As has been shown by many results in stochastic control and optimization, the introduction of a decreasing step size would slow down the convergence speed. This fact is due to that the suppression effect of a_k is not only imposed on the stochastic noises but also on the correction information. In fact, it is a classic trade-off between the stable zero-error convergence and convergence speed for stochastic control. Roughly speaking, the exponential convergence speed for the noise-free case is no longer guaranteed. We can only ensure an asymptotical convergence for stochastic systems.

Similarly to the noise-free case, we lift the input along the time axis. The update law (2.47) is rewritten as follows:

$$U_{k+1} = U_k + a_k \Gamma_k \mathbf{L} E_k, \tag{2.49}$$

where Γ_k and $\mathbf{L}$ is given in (2.31) and (2.32). Subtracting both sides of (2.49) from U_d, substituting the definition of $E_k = \mathbf{H}(U_d - U_k) - \xi_k$ (see (2.27)), and using the notation $\delta U_k = U_d - U_k$ leads to

$$\delta U_{k+1} = (I - a_k \Lambda_k)\delta U_k + a_k \xi_k, \tag{2.50}$$

where $\Lambda_k = \Gamma_k \mathbf{L} \mathbf{H}$ is specified in the last section.

Before proceeding to the detailed convergence analysis, we need to declare that the design condition for the learning gain matrix L_t retains the same to the noise-free case. That is, the learning gain matrix L_t should satisfy that $-L_t C_{t+\tau} A_{t+1}^{t+\tau-1} B_t$ is a Hurwitz matrix.

Similar to the noise-free case, we first give a decreasing property for the multiple products of $I - a_k \Lambda_k$ and then show the convergence with the help of such technical lemma.

Denote

$$\Psi_{m,n} = (I - a_m \Lambda_m) \cdots (I - a_n \Lambda_n), \quad m \geq n \tag{2.51}$$

and $\Psi_{m,m+1} \triangleq I$. Then, the estimate of $\Psi_{m,n}$ is given in the following lemma.

Lemma 2.3 *Consider the matrix product (2.51). If the learning gain matrix L_t satisfies that $-L_t C_{t+\tau} A_{t+1}^{t+\tau-1} B_t$ is a Hurwitz matrix, $\forall t$, then there exist constants $c_4 > 0$ and $c_5 > 0$ such that, for $m > n + K$,*

$$\|\Psi_{m,n}\| \leq c_4 \exp\left(-c_5 \sum_{i=n}^{m} a_i\right), \quad \forall n \geq 1. \tag{2.52}$$

Proof First, we recall that $(\mathbf{LH})^T\mathbf{Q} + \mathbf{QLH} = I$ and $(\Lambda^{(i)})^T\mathbf{Q} + \mathbf{Q}\Lambda^{(i)} \geq 0$ for $i = 2, \ldots, \kappa$ (see (2.35) and (2.36)). The RSM of data dropouts results in that

$$\left(\sum_{i=0}^{K-1} \Lambda_{k+i}\right)^T \mathbf{Q} + \mathbf{Q}\left(\sum_{i=0}^{K-1} \Lambda_{k+i}\right) \geq c_1 I, \quad \forall k \geq 0. \tag{2.53}$$

Moreover, from (2.48), we have, for $1 \leq i \leq K$,

$$\frac{a_{m-i}}{a_m} - 1 = a_{m-i}\left(\frac{1}{a_m} - \frac{1}{a_{m-i}}\right) = O(a_m). \tag{2.54}$$

For any $m \geq n + K - 1$, we have

$$\begin{aligned}\Psi_{m,n}^T \mathbf{Q} \Psi_{m,n} &= \Psi_{m-1,n}^T (I - a_m \Lambda_m)^T \mathbf{Q} (I - a_m \Lambda_m) \Psi_{m-1,n} \\ &= \Psi_{m-K,n}^T (I - a_{m-K+1}\Lambda_{m-K+1})^T \cdots (I - a_m\Lambda_m)^T \mathbf{Q} \\ &\quad \times (-a_m\Lambda_m) \cdots (I - a_{m-K+1}\Lambda_{m-K+1}) \Psi_{m-K,n} \\ &= \Psi_{m-K,n}^T \left[\mathbf{Q} - \left(\sum_{i=m-K+1}^{m} a_i \Lambda_i^T \mathbf{Q} + \sum_{i=m-K+1}^{m} a_i \mathbf{Q}\Lambda_i\right) + o(a_m)\right] \Psi_{m-K,n} \\ &= \Psi_{m-K,n}^T \left\{\mathbf{Q} - a_m\left[\left(\sum_{i=m-K+1}^{m} \Lambda_i^T\right)\mathbf{Q}\right.\right. \\ &\quad \left.\left. + \mathbf{Q}\left(\sum_{i=m-K+1}^{m} \Lambda_i\right)\right] + o(a_m)\right\} \Psi_{m-K,n}, \end{aligned} \tag{2.55}$$

where for the equality (2.54) is invoked.

Noticing $0 < a_m < 1$ for large enough m and using (2.53), we have

$$\begin{aligned}\Psi_{m,n}^T \mathbf{Q} \Psi_{m,n} &\leq \Psi_{m-K,n}^T (\mathbf{Q} - a_m c_1 I + o(a_m))\Psi_{m-K,n} \\ &\leq \Psi_{m-K,n}^T \mathbf{Q}^{\frac{1}{2}} (I - a_m c_1 \mathbf{Q}^{-1} + o(a_m))\mathbf{Q}^{\frac{1}{2}} \Psi_{m-K,n} \\ &\leq \Psi_{m-K,n}^T \mathbf{Q}^{\frac{1}{2}} \left(I - \frac{c_1}{K}\mathbf{Q}^{-1}\sum_{i=m-K+1}^{m} a_i + o(a_m)\right)\mathbf{Q}^{\frac{1}{2}} \Psi_{m-K,n} \\ &\leq \left(1 - \frac{c_1}{K}\lambda_{\min}(\mathbf{Q}^{-1})\sum_{i=m-K+1}^{m} a_i + o(a_m)\right) \Psi_{m-K,n}^T \mathbf{Q}\Psi_{m-K,n} \\ &\leq \exp\left(-c_6 \sum_{i=m-K+1}^{m} a_i\right) \Psi_{m-K,n}^T \mathbf{Q}\Psi_{m-K,n} \end{aligned} \tag{2.56}$$

for sufficiently large n, where c_6 is a positive constant.

Therefore, for sufficiently large n, say, for $n \geq n_0$ and $m \geq n + K$, we have

2.3 Extension to Arbitrary Relative Degree Case with Mean Square Convergence

$$\Psi_{m,n}^T \mathbf{Q} \Psi_{m,n} \leq c_7 \exp\left(-c_6 \sum_{i=n}^{m} a_i\right) I \quad \text{with} \quad c_7 > 0, \tag{2.57}$$

which, by noticing the definition of $\mathbf{Q} > 0$, implies

$$\|\Psi_{m,n}\| \leq c_8 \exp\left(-\frac{c_6}{2} \sum_{i=n}^{m} a_i\right) \quad \text{with} \quad c_8 > 0. \tag{2.58}$$

Consequently, for $\forall n \geq n_0 + K$, $\forall n > 0$, by (2.58) and the definition $\Psi_{m,m+1} \triangleq I$, we have

$$\|\Psi_{m,n}\| = \|\Psi_{m,n_0}\| \cdot \|\Psi_{n_0-1,n}\| \leq c_4 \exp\left(-c_5 \sum_{i=n}^{m} a_i\right), \tag{2.59}$$

where c_4 is a suitable constant and $c_5 = c_6/2$. The proof is completed.

Remark 2.10 Comparing the estimations of the corresponding product (2.40) for the noise-free case and (2.52) for the noise case, we can have a clear understanding of the difference between the fixed step size ρ and the decreasing step size a_k. Specifically, these two estimations are consistent as if we replace the decreasing step size a_k with the fixed but small enough ρ, the estimation (2.52) actually turns into (2.40). In other words, (2.40) can be regarded as a special case of (2.52).

Now we can move to show the convergence of stochastic systems.

Theorem 2.3 *Consider the stochastic linear system (2.16) and the ILC update law (2.47), where the random data dropouts follow the RSM case. Assume Assumptions 2.5 and 2.6 hold. Then, the input sequence $\{u_k(t)\}$, $t = 0, \ldots, N - \tau$, achieves mean square convergence to the desired input $u_d(t)$, $t = 0, \ldots, N - \tau$, if the learning gain matrix L_t satisfies that $-L_t C_{t+\tau} A_{t+1}^{t+\tau-1} B_t$ is a Hurwitz matrix.*

Proof The proof is carried out through grouping the iterations by modulo operator with respect to K. To this end, all iterations are divided into K subsets, $\{iK+j, i \geq 0\}$, $0 \leq j \leq K - 1$. Now we check the contraction for successive K iterations, that is, we check the convergence for each subset.

From (2.50), it follows, $\forall 0 \leq j \leq K - 1$,

$$\delta U_{iK+j} = \Psi_{iK+j-1,(i-1)K+j} \delta U_{(i-1)K+j}$$
$$+ \sum_{l=0}^{K-1} \Psi_{iK+j-1,(i-1)K+j+l+1} a_{(i-1)K+j+l} \xi_{(i-1)K+j+l}. \tag{2.60}$$

Apply the weighted norm $V_k = \|\delta U_k\|_{\mathbf{Q}} = \delta U_k^T \mathbf{Q} \delta U_k$. We have that

$$V_{iK+j} = \delta U_{iK+j}^T \mathbf{Q} \delta U_{iK+j}$$
$$= \left(\Psi_{iK+j-1,(i-1)K+j}\delta U_{(i-1)K+j}\right)^T \mathbf{Q} \Psi_{iK+j-1,(i-1)K+j}\delta U_{(i-1)K+j}$$
$$+ 2\left(\Psi_{iK+j-1,(i-1)K+j}\delta U_{(i-1)K+j}\right)^T \mathbf{Q}\phi_* + \phi_*^T \mathbf{Q}\phi_*, \quad (2.61)$$

where

$$\phi_* \triangleq \sum_{l=0}^{K-1} \Psi_{iK+j-1,(i-1)K+j+l+1} a_{(i-1)K+j+l} \xi_{(i-1)K+j+l}. \quad (2.62)$$

From the proof of Lemma 2.3, it follows that

$$\Psi_{iK+j-1,(i-1)K+j}^T \mathbf{Q} \Psi_{iK+j-1,(i-1)K+j} \leq (1 - c_9 a_{iK+j-1} + c_{10} a_{iK+j-1}^2)\mathbf{Q}, \quad (2.63)$$

which implies that

$$\left(\Psi_{iK+j-1,(i-1)K+j}\delta U_{(i-1)K+j}\right)^T \mathbf{Q}\Psi_{iK+j-1,(i-1)K+j}\delta U_{(i-1)K+j}$$
$$\leq (1 - c_9 a_{iK+j-1} + c_{10} a_{iK+j-1}^2)\|\delta U_{(i-1)K+j}\|_{\mathbf{Q}}. \quad (2.64)$$

Noticing that ϕ_* is a sum of random noises and the noises are independent of the data dropout variables, we have

$$\mathbb{E}\left(\Psi_{iK+j-1,(i-1)K+j}\delta U_{(i-1)K+j}\right)^T \mathbf{Q}\phi_*$$
$$= \left(\mathbb{E}\Psi_{iK+j-1,(i-1)K+j}\delta U_{(i-1)K+j}\right)^T \mathbf{Q}(\mathbb{E}\phi_*)$$
$$= \left(\mathbb{E}\Psi_{iK+j-1,(i-1)K+j}\delta U_{(i-1)K+j}\right)^T \mathbf{Q}\mathbb{E}\left(\mathbb{E}(\phi_* \mid \mathscr{F}'_{(i-1)K+j-1})\right)$$
$$= 0, \quad (2.65)$$

where the σ-algebra $\mathscr{F}'_k$ is augmented from $\mathscr{F}_k$ as $\mathscr{F}'_k = \sigma\{x_i(t), u_i(t), y_i(t), w_i(t), v_i(t), \gamma_i(t), 1 \leq i \leq k, 0 \leq t \leq N\}$. Moreover, by Assumption 2.6, the stochastic noises are conditionally independent along the iteration axis, thus it follows that

$$\mathbb{E}\phi_*^T \mathbf{Q}\phi_*$$
$$= \mathbb{E}\left(\sum_{l=0}^{K-1} \Psi_{iK+j-1,(i-1)K+j+l+1} a_{(i-1)K+j+l} \xi_{(i-1)K+j+l}\right)^T$$
$$\times \mathbf{Q}\left(\sum_{l=0}^{K-1} \Psi_{iK+j-1,(i-1)K+j+l+1} a_{(i-1)K+j+l} \xi_{(i-1)K+j+l}\right)$$
$$= \mathbb{E}\left(\sum_{l=0}^{K-1} a_{(i-1)K+j+l}^2 \xi_{(i-1)K+j+l}^T \Psi_{iK+j-1,(i-1)K+j+l+1}^T\right.$$
$$\left.\times \mathbf{Q}\Psi_{iK+j-1,(i-1)K+j+l+1} \xi_{(i-1)K+j+l}\right)$$

2.3 Extension to Arbitrary Relative Degree Case with Mean Square Convergence

$$\leq \sum_{l=0}^{K-1} a_{(i-1)K+j+l}^2 c_4^2 \exp\left(-2c_5 \sum_{i=(i-1)K+j+l+1}^{iK+j-1}\right) \mathbb{E}\|\xi_{(i-1)K+j+l}\|^2$$

$$\leq a_{iK+j-1}^2 c_{11}, \tag{2.66}$$

where c_{11} is a suitable constant such that

$$c_{11} \geq c_4^2 \sup_k \mathbb{E}\|\xi_k\|^2 \sum_{l=0}^{K-1} \left(a_{(i-1)K+j+l}^2/a_{iK+j-1}^2\right).$$

Taking mathematical expectation to both sides of (2.61) and substituting (2.63)–(2.66), we have

$$\mathbb{E}V_{iK+j} \leq (1 - c_9 a_{iK+j-1})\mathbb{E}V_{(i-1)K+j} + c_{10} a_{iK+j-1}^2 (\mathbb{E}V_{(i-1)K+j} + c_{11}/c_{10}),$$
$$\forall 0 \leq j \leq K-1. \tag{2.67}$$

Comparing (2.67) with (A.1) in Lemma A.1, it is found that $\mathbb{E}V_{iK+j}$, a_{iK+j-1} (with respect to recursive index i), c_9, c_{10}, and c_{11}/c_{10} correspond to ϑ_{k+1}, a_k (with respect to recursive index k), d_1, d_2, and d_3, respectively. Then by Lemma A.1, we have that $\lim_{i\to\infty} \mathbb{E}V_{iK+j} = 0, \forall 0 \leq j \leq K-1$. Moreover, incorporating with the fact that $\mathbf{Q}$ is a positive definite matrix, the mean square convergence is established for each subset of iteration number $\{iK+j, i \geq 0\}$, i.e., $\lim_{i\to\infty} \mathbb{E}\|\delta U_{iK+j}\|^2 = 0, \forall 0 \leq j \leq K-1$. The mean square convergence of the input sequence $\{U_k, k \geq 1\}$ to the desired input U_d is thus obvious.

Remark 2.11 For the stochastic system case, we can also prove the almost sure convergence based on the above mean square convergence with the help of a technical lemma given in Appendix (Lemma A.2). In particular, taking a conditional expectation to (2.61) with respect to σ-algebra $\mathscr{F}'_{(i-1)K+j-1}$ it follows

$$\mathbb{E}\left(V_{iK+j} \mid \mathscr{F}'_{(i-1)K+j-1}\right) \leq V_{(i-1)K+j} + c_{10} a_{iK+j-1}^2 (V_{(i-1)K+j} + c_{11}/c_{10}),$$
$$\forall 0 \leq j \leq K-1. \tag{2.68}$$

Note that the two terms on the right-hand side of the last inequality, i.e., $V_{(i-1)K+j}$ and $c_{10} a_{iK+j-1}^2 (V_{(i-1)K+j} + c_{11}/c_{10})$, correspond to $X(n)$ and $Z(n)$ in Lemma A.2, respectively. Moreover, it has been shown that $\mathbb{E}V_{(i-1)K+j}$ converges to zero as $i \to \infty$, and thus it is evident

$$\sum_{i=0}^{\infty} \mathbb{E}\left[c_{10} a_{iK+j-1}^2 (V_{(i-1)K+j} + c_{11}/c_{10})\right] \leq \left(c_{10} \sup_i \mathbb{E}V_{(i-1)K+j} + c_{11}\right) \sum_{i=0}^{\infty} a_{iK+j-1}^2 < \infty. \tag{2.69}$$

In other words, the conditions in Lemma A.2 are fulfilled. Therefore, it follows that V_{iK+j} converges almost surely as $i \to \infty$, $\forall j$. On the other hand, we have show that δU_{iK+j} converges to zero in mean square sense. Then, the almost surely convergent limitation of δU_{iK+j} should also be zero.

2.4 Illustrative Simulations

Example 2.1 Consider the following second-order linear system:

$$x_k(t+1) = \begin{bmatrix} 1 & 0.1 \\ 0.05 & 1.1 \end{bmatrix} x_k(t) + \begin{bmatrix} 0.45 \\ 0.6 \end{bmatrix} u_k(t) + w_k(t+1),$$

$$y_k(t) = \begin{bmatrix} 1 & 1 \end{bmatrix} x_k(t) + v_k(t).$$

For simple illustration, let $N = 10$, and the noises are assumed zero-Gaussian distributed, i.e., $w_k(t) \sim N(0, 0.05^2 I_2)$, $v_k(t) \sim N(0, 0.1^2)$.

In order to simulate measurement loss, let us fix $K = 4$. Iteration steps are separated into groups of four successive iterations, i.e., $\{1, 2, 3, 4\}, \{5, 6, 7, 8\}, \ldots$, and randomly select one iteration from each group for each t, say, 1, 6, 9, ... for example. For these selected steps, the control is not updated. Figure 2.2 shows an illustration of data dropout step for the first 100 iterations, where 0 denotes that the packet is lost and 1 means that the packet is successfully transmitted.

The reference trajectory is $y_d(t) = 2t$. The initial control action is simply given as $u_0(t) = 0$, $\forall t$. The learning gain chooses $a_k = \frac{1}{k+1}$. The algorithm has run 300

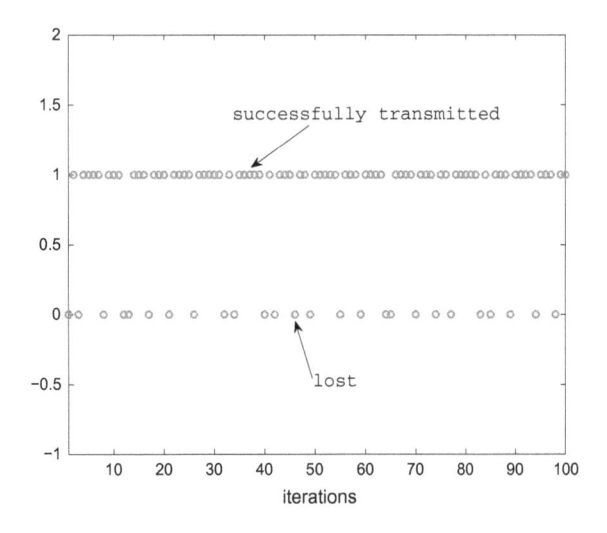

Fig. 2.2 Illustration of packet loss

2.4 Illustrative Simulations

Fig. 2.3 $y_{300}(t)$ versus $y_d(t)$

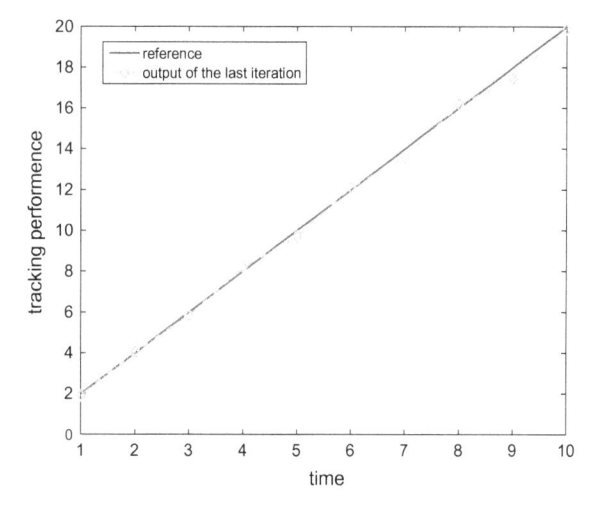

Fig. 2.4 Tracking errors $e_k(t)$ at $t = 3$ and $t = 5$

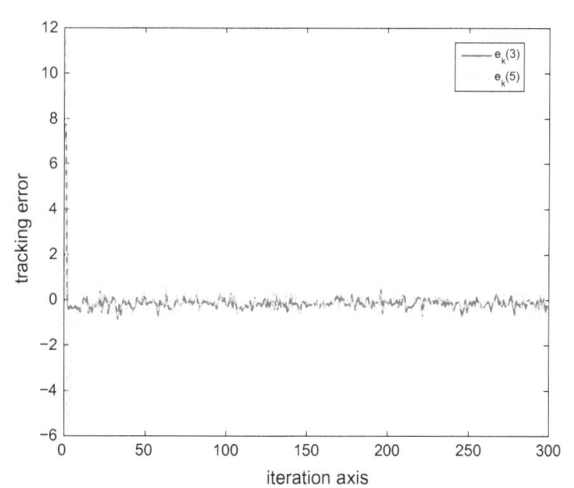

iterations. The output of the 300th iteration is shown in Fig. 2.3, where the solid line is the reference signals and the dotted-dashed line with diamonds denotes the output $y_{300}(t)$. As one could see, the actual output could track the desired reference trajectory with quite small tracking errors, which are mainly involved by the stochastic noises. These observations verify the effectiveness and precision of the proposed ILC algorithm under random data dropouts and stochastic noises.

The tracking errors at $t = 3$ and $t = 5$ as examples are demonstrated in Fig. 2.4. The tracking errors along the iteration axis at the other time instants are similar to those in Fig. 2.4. As is seen, the tracking errors decay to zero in few iterations and then fluctuate around zero in a very narrow range. The fluctuations are caused by the stochastic noises at each iteration. That is to say, if the stochastic noises are eliminated from the actual tracking error, the tracking error would decrease to zero.

Example 2.2 Consider a time-varying linear system (A_t, B_t, C_t) where

$$A_t = \begin{bmatrix} 0.2\exp(-t/100) & -0.6 & 0 \\ 0 & 0 & 0.50\sin(t) \\ 0 & 0 & 0.7 \end{bmatrix},$$

$$B_t = \begin{bmatrix} 0 & 0.3\sin(t) & 1 \end{bmatrix}^T,$$

$$C_t = \begin{bmatrix} 0 & 0.1 & 0.8 \end{bmatrix}.$$

The iteration length is set to be $N = 100$. The tracking reference is $y_d(t) = 0.5\sin\left(\frac{t}{20}\pi\right) + 0.25\sin\left(\frac{t}{10}\pi\right)$. The initial state for all the iterations is set $x_k(0) = x_d(0) = 0$. The algorithm is run 150 iterations for each case.

We consider five cases. To simulate the data missing, we first separate the iterations into groups of M successive iterations, $M = 2, \ldots, 6$. That is, the iterations are separated as $\{kM+1, kM+2, \ldots, (k+1)M\}, k = 0, 1, 2, \ldots$, and randomly select one iteration from each group denoting the one whose data is dropped during the transmission. For example, take $M = 3$, then the iterations are separated as $\{1, 2, 3\}$, $\{4, 5, 6\}, \ldots$, and from each group one iteration is selected randomly. Therefore, the DDR for the above five cases is equal to $1/2, 1/3, 1/4, 1/5$, and $1/6$, respectively.

We first check the noise-free system case. In this case, we set $\rho = 0.4$ and $L_t = 1$. The maximal tracking error for each iteration is defined as $\max_{1 \le j \le N} |e_k(j)|$. The maximal tracking error profiles along the iteration axis are plotted in Fig. 2.5. The figure turns out two observations. One is that the convergence speed slows down as the DDR increases; that is, a larger DDR would result in a slower convergence speed. The other one is that the maximal tracking error profiles approximate straight lines

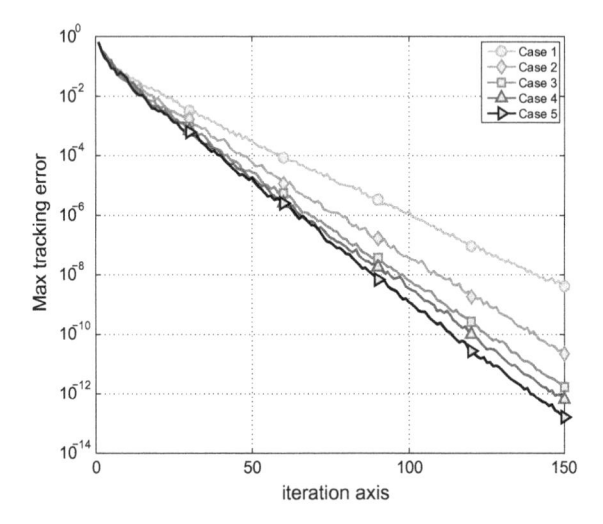

Fig. 2.5 Maximal tracking error profiles for the noise-free system along the iteration axis, where Cases 1 to 5 correspond to DDR being 1/2, 1/3, 1/4, 1/5, and 1/6, respectively

Fig. 2.6 Maximal tracking error profiles for the noised system along the iteration axis, where Cases 1 to 5 correspond to DDR being 1/2, 1/3, 1/4, 1/5, and 1/6, respectively

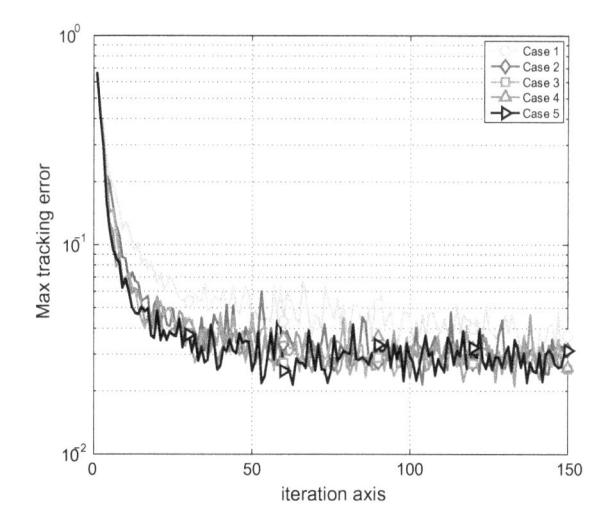

in the logarithm axis, which demonstrates that the convergence is exponential when no noise is involved in the system.

When the system is involved with random noises, an additional decreasing learning sequence should be introduced to the ILC rule to guarantee a stable convergence of the proposed algorithms. The tracking performance is shown in Fig. 2.6, where the random noise is assumed to be a white Gaussian noise, namely, subject to $N(0, \sigma^2)$ with $\sigma = 0.1$. In the simulation, the learning gain is set as $L_t = 1.5$ and the decreasing sequence selects $a_k = \frac{1}{k+1}$. We have some observations from Fig. 2.6. First of all, due to the existence of random noises, the final tracking error cannot reduce to zero as the iteration number increases and the maximal tracking error profiles would fluctuate heavily. Moreover, the introduction of $\{a_k\}$ makes the convergence speed much slower than the noise-free case. However, it is a natural requirement for the control of stochastic systems.

2.5 Summary

In this chapter, we consider ILC for linear systems with data dropouts, which are described by the random sequence model. That is, none statistical property is imposed to the random data dropouts, but we require that the packet at arbitrary time instant should be successfully transmitted at least once for arbitrary K successive iterations. This model cannot be covered by the conventional Bernoulli variable model and vice versa. Under this data dropout model, we employ the intermittent update scheme and establish the almost sure convergence results. Moreover, the mean square convergence is also established for MIMO systems with arbitrary relative degree. The corresponding derivations for nonlinear systems will be given in the next chapter. The results in this chapter are based on [9, 10].

References

1. Chen, H.F., Guo, L.: Identification and Stochastic Adaptive Control. Birkhäuser Boston (1991)
2. Chen, H.F.: Stochastic Approximation and Its Applications. Kluwer (2002)
3. Saab, S.S.: A discrete-time stochastic learning control algorithm. IEEE Trans. Autom. Control **46**(6), 877–887 (2001)
4. Huang, S.N., Tan, K.K., Lee, T.H.: Necessary and sufficient condition for convergence of iterative learning algorithm. Automatica **38**(7), 1257–1260 (2002)
5. Meng, D., Jia, Y., Du, J., Yu, F.: Necessary and sufficient stability condition of LTV iterative learning control systems using a 2-D approach. Asian J. Control **13**(1), 25–37 (2011)
6. Saab, S.S.: Selection of the learning gain matrix of an iterative learning control algorithm in presence of measurement noise. IEEE Trans. Autom. Control **50**(11), 1761–1774 (2005)
7. Chen, Y., Wen, C., Gong, Z., Sun, M.: An iterative learning controller with initial state learning. IEEE Trans. Autom. Control **44**(2), 371–376 (1999)
8. Sun, M., Wang, D.: Initial shift issues on discrete-time iterative learning control with system relative degree. IEEE Trans. Autom. Control **48**(1), 144–148 (2003)
9. Shen, D., Wang, Y.: Iterative learning control for networked stochastic systems with random packet losses. Int. J. Control **88**(5), 959–968 (2015)
10. Shen, D., Xu, J.-X.: A framework of iterative learning control under random data dropouts: mean square and almost sure convergence. Int. J. Adap. Control Signal Process. **31**(12), 1825–1852 (2017)

Chapter 3
Random Sequence Model for Nonlinear Systems

3.1 Problem Formulation

Consider the following affine nonlinear systems with measurement noises:

$$\begin{aligned} x_k(t+1) &= f(t, x_k(t)) + \mathbf{b}(t, x_k(t))u_k(t), \\ y_k(t) &= \mathbf{c}(t)x_k(t) + v_k(t), \end{aligned} \quad (3.1)$$

where $t \in \{0, 1, \ldots, N\}$ denotes the time instant in an iteration of the process, and $k = 1, 2, \ldots$ labels different iterations. $u_k(t) \in \mathbf{R}$, $x_k(t) \in \mathbf{R}^n$, and $y_k(t) \in \mathbf{R}$ denote the input, state, and output, respectively. Besides, $f(t, x_k(t))$ and $\mathbf{b}(t, x_k(t))$ are continuous functions and $\mathbf{c}(t)$ denotes output coupling coefficients. $v_k(t)$ is the stochastic measurement noise.

Similar to Chap. 2, the setup of the control system is illustrated in Fig. 3.1. For convenience, only data dropout at the measurement side is considered in this chapter. As shown in Fig. 3.1, the measurement signals are transmitted back through a lossy channel. The occurrence of data dropout can be regarded as a random switch between opening and closing states. Denote $\mathcal{M}_k$ as the random set of time locations at which measurements are lost in the kth iteration. In other words, $t_0 \in \mathcal{M}_k$ if $y_k(t_0)$ is lost.

Let $\mathscr{F}_k \triangleq \sigma\{y_j(t), x_j(t), v_j(t), 0 \leq j \leq k, t \in \{0, \ldots, N\}\}$ be the σ-algebra generated by $y_j(t), x_j(t), v_j(t), 0 \leq t \leq N, 0 \leq j \leq k$. Then, the set of admissible controls is defined as

$$U = \{u_{k+1}(t) \in \mathscr{F}_k, \sup_k \|u_k(t)\| < \infty, \text{a.s.}$$
$$t = 0, \ldots, N-1, \quad k = 0, 1, 2, \ldots\}.$$

The control purpose is to find a control sequence $\{u_k(t), k = 0, 1, 2, \ldots\} \subset U$ under random data dropout environments to minimize the following averaged-tracking index, $\forall t \in \{0, 1, \ldots, N\}$

$$V(t) = \limsup_{n \to \infty} \frac{1}{n} \sum_{k=1}^{n} \|y_k(t) - y_d(t)\|^2, \quad (3.2)$$

Fig. 3.1 Block diagram of networked control system with measurement data dropout

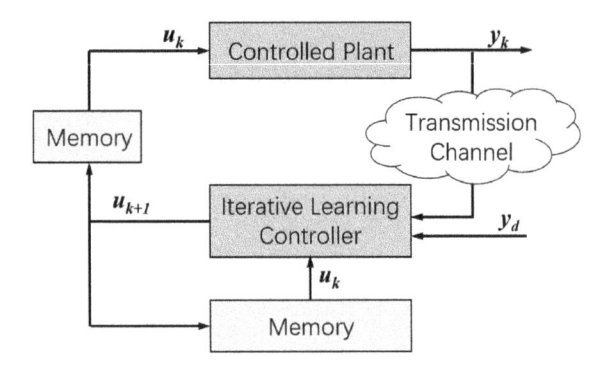

where $y_d(t), t \in \{0, 1, \ldots, N\}$ is the tracking target. The following assumptions are used.

Assumption 3.1 The tracking target $y_d(t)$ is realizable in the sense that there exist $u_d(t)$ and $x_d(0)$ such that

$$\begin{aligned} x_d(t+1) &= f(t, x_d(t)) + \mathbf{b}(t, x_d(t))u_d(t), \\ y_d(t) &= \mathbf{c}(t)x_d(t). \end{aligned} \quad (3.3)$$

Assumption 3.2 The functions $f(\cdot, \cdot)$ and $\mathbf{b}(\cdot, \cdot)$ are continuous with respect to the second argument.

Assumption 3.3 The real number $\mathbf{c}(t+1)\mathbf{b}(t, x)$ coupling the input and output is unknown, but its sign, characterizing the control direction, is assumed known. Without loss of generality, it is assumed that $\mathbf{c}(t+1)\mathbf{b}(t, x) > 0$.

Assumption 3.4 The initial values can be precisely reset asymptotically, i.e., $x_k(0) \to x_d(0)$ as $k \to \infty$.

Assumption 3.5 For each time instant t, the measurement noise $\{v_k(t)\}$ is a sequence of independent and identically distributed (i.i.d.) random variables with $\mathbb{E}v_k(t) = 0$, $\sup_k \mathbb{E}v_k^2(t) < \infty$, and $\lim_{n \to \infty} \frac{1}{n} \sum_{k=1}^n v_k^2(t) = R_v^t$, a.s., where R_v^t is unknown.

Assumption 3.6 For each time instant t, the measurement data dropout is random without obeying any certain probability distribution, but there exists a sufficiently large number K such that during any successive K iterations, at least in one iteration the measurement is successfully sent back.

Remark 3.1 The number K is not necessary to be known prior. In other words, only the existence of such number is required. Thus, this condition means that the measurements should not be lost too much to guarantee the convergence in almost sure sense. As a matter of fact, $K - 1$ represents the worst case of successive iteration number of data dropout along the iteration axis.

3.1 Problem Formulation

For simplicity of writing let us set $f_k(t) = f(t, x_k(t))$, $f_d(t) = f(t, x_d(t))$, $\mathbf{b}_k(t) = \mathbf{b}(t, x_k(t))$, $\mathbf{b}_d(t) = \mathbf{b}(t, x_d(t))$, $\delta u_k(t) = u_d(t) - u_k(t)$, $\delta f_k(t) = f_d(t) - f_k(t)$, $\delta \mathbf{b}_k(t) = \mathbf{b}_d(t) - \mathbf{b}_k(t)$, $\mathbf{c}^+\mathbf{b}_k(t) = \mathbf{c}(t+1)\mathbf{b}(t, x_k(t))$.

Lemma 3.1 *Assume Assumptions 3.1–3.4 hold for system (3.1). If $\lim_{k \to \infty} \delta u_k(s) = 0$, $s = 0, 1, \ldots, t$, then at time instant $t+1$, $\|\delta x_k(t+1)\| \xrightarrow[k \to \infty]{} 0$, $\|\delta f_k(t+1)\| \xrightarrow[k \to \infty]{} 0$, $\|\delta \mathbf{b}_k(t+1)\| \xrightarrow[k \to \infty]{} 0$.*

Proof The proof of this lemma can be carried out by mathematical induction along the time axis t. By (3.1) and Assumption 3.1,

$$\begin{aligned}\delta x_k(t+1) &= f_d(t) - f_k(t) + \mathbf{b}_d(t)u_d(t) - \mathbf{b}_k(t)u_k(t) \\ &= \delta f_k(t) + \delta \mathbf{b}_k(t)u_d(t) + \mathbf{b}_k(t)\delta u_k(t).\end{aligned} \quad (3.4)$$

Initial Step: For $t = 0$, noticing Assumptions 3.2 and 3.4, we have

$$\delta f_k(0) = f_d(0) - f_k(0) \xrightarrow[k \to \infty]{} 0,$$
$$\delta \mathbf{b}_k(0) = \mathbf{b}_d(0) - \mathbf{b}_k(0) \xrightarrow[k \to \infty]{} 0,$$

which imply that the first two terms at the right-hand side (RHS) of (3.4) tend to zero as $k \to \infty$. Since

$$\|\mathbf{b}_k(0)\| \leq \|\mathbf{b}_d(0)\| + \|\delta \mathbf{b}_k(0)\|,$$

it follows that $\mathbf{b}_k(0)$ is bounded. Thus if $\delta u_k(0) \xrightarrow[k \to \infty]{} 0$, then the third term at the RHS of (3.4) also tends to zero. It further implies that $\delta x_k(1) \xrightarrow[k \to \infty]{} 0$ and then by Assumption 3.2 again, $\delta f_k(1) \xrightarrow[k \to \infty]{} 0$ and $\delta \mathbf{b}_k(1) \xrightarrow[k \to \infty]{} 0$. That is, the conclusions are valid for $t = 0$.

Inductive Step: We now assume the conclusions of the lemma are true for $s = 0, 1, \ldots, t-1$. It suffices to show that the conclusions also hold for time instant t, i.e., $\|\delta x_k(t+1)\| \xrightarrow[k \to \infty]{} 0$, $\|\delta f_k(t+1)\| \xrightarrow[k \to \infty]{} 0$, $\|\delta \mathbf{b}_k(t+1)\| \xrightarrow[k \to \infty]{} 0$. This could be done following the same procedure as above. This completes the proof.

Based on Lemma 3.1, the conclusion of Lemma 2.1 also holds for nonlinear system (3.1) with Assumptions 2.1–2.4 replaced by Assumptions 3.1–3.5, by using similar steps of Lemma 2.1. Thus, we only provide the lemma without proof.

Lemma 3.2 *Consider stochastic system (3.1) and tracking objective $y_d(t+1)$, and assume Assumptions 3.1–3.5 hold, then for any arbitrary time instant $t+1$, the index (3.2) will be minimized if the control sequence $\{u_k(t)\}$ is admissible and satisfies $u_k(i) \xrightarrow[k \to \infty]{} u_d(i)$, $i = 0, 1, \ldots, t$. In this case, $\{u_k(t)\}$ is called the optimal control sequence.*

3.2 Intermittent Update Scheme and Its Convergence

In this section, we apply the following intermittent update scheme for the nonlinear system (3.1):

$$u_{k+1}(t) = u_k(t) + a_k \mathbf{1}_{\{(t+1) \notin \mathcal{M}_k\}} \times (y_d(t+1) - y_k(t+1)). \tag{3.5}$$

For any fixed time instant t from (3.5), we have

$$\delta u_{k+1}(t) = \delta u_k(t) - a_k \mathbf{1}_{\{(t+1) \notin \mathcal{M}_k\}} [\mathbf{c}^+ \mathbf{b}_k(t) \delta u_k(t) + \varphi_k(t) - v_k(t+1)], \tag{3.6}$$

where

$$\varphi_k(t) = \mathbf{c}^+ \delta f_k(t) + \mathbf{c}^+ \delta \mathbf{b}_k(t) u_d(t). \tag{3.7}$$

We have the following convergence results.

Theorem 3.1 *For system* (3.1) *and index* (3.2), *assume Assumptions 3.1–3.6 hold, then the control sequence* $\{u_k(t)\}$ *given by the ILC update law* (3.5) *is optimal. In other words,* $u_k(t)$ *converges to* $u_d(t)$ *a.s. as* $k \to \infty$ *for any* $t \in \{0, 1, \ldots, N-1\}$.

Proof Due to the existence of nonlinear functions $f_k(t)$ and $\mathbf{b}_k(t)$, it is hard to formulate the super-vector form of affine nonlinear system (3.1) just as the linear system case. Thus, the proof in Chap. 2 cannot be directly applied here. Instead, the proof of this theorem is carried out by mathematical induction along the time axis t with similar techniques of the proof of Theorem 2.1.

For $t = 0$ the algorithm (3.5) is written as

$$\delta u_{k+1}(0) = (1 - a_k \mathbf{1}_{\{1 \notin \mathcal{M}_k\}} \mathbf{c}^+ \mathbf{b}_k(0)) \delta u_k(0) - a_k \mathbf{1}_{\{1 \notin \mathcal{M}_k\}} \varphi_k(0) + a_k \mathbf{1}_{\{1 \notin \mathcal{M}_k\}} v_k(1). \tag{3.8}$$

Since $\mathbf{b}_k(0)$ is continuous in the initial state by Assumption 3.2, we have $\mathbf{b}_k(0) \xrightarrow[k \to \infty]{} \mathbf{b}_d(0)$ by Assumption 3.4 and $\mathbf{c}^+ \mathbf{b}_k(0)$ converges to a positive constant by Assumption 3.3. Therefore, by Assumption 3.6 it follows

$$\sum_{k=i}^{i+K-1} \left(-\mathbf{1}_{\{1 \notin \mathcal{M}_k\}} \mathbf{c}^+ \mathbf{b}_k(0) \right) < -\gamma, \quad \gamma > 0, \tag{3.9}$$

for all sufficiently large i.

Set $\phi_{i,j} \triangleq (1 - a_i \mathbf{1}_{\{1 \notin \mathcal{M}_i\}} \mathbf{c}^+ \mathbf{b}_i(0)) \cdots (1 - a_j \mathbf{1}_{\{1 \notin \mathcal{M}_j\}} \mathbf{c}^+ \mathbf{b}_j(0)), i \geq j, \phi_{i,i+1} \triangleq 1$. It is clear that $1 - a_j \mathbf{1}_{\{1 \notin \mathcal{M}_j\}} \mathbf{c}^+ \mathbf{b}_j(0) > 0$ for all sufficiently large j, say, $j \geq j_0$. Then for any $i \geq j + K, j \geq j_0$ by (3.8) and (3.9), we have that

3.2 Intermittent Update Scheme and Its Convergence

$$\phi_{i,j} = \phi_{i-K,j}\left(1 - a_i \sum_{k=i-K+1}^{i} \mathbf{1}_{\{1 \notin \mathcal{M}_k\}} \mathbf{c}^+ \mathbf{b}_k(0) + o(a_i)\right)$$

$$\leq \phi_{i-K,j}(1 - \gamma a_i + o(a_i))$$

$$= \phi_{i-K,j}\left(1 - \frac{\gamma}{K} \sum_{k=i-K+1}^{i} a_k + o(a_i)\right)$$

$$\leq \exp\left(-c \sum_{k=i-K+1}^{i} a_k\right) \phi_{i-K,j} \quad \text{with } c > 0.$$

It follows from this equation that $\phi_{i,j} \leq c_3 \exp\left(-\frac{c}{2}\sum_{k=j}^{i} a_k\right)$, $\forall j \geq j_0$ for some $c_3 > 0$, and hence there is a c_4 such that

$$|\phi_{i,j}| \leq c_4 \exp\left(-\frac{c}{2} \sum_{k=j}^{i} a_k\right), \quad \forall i \geq j+K, j \geq j_0.$$

Therefore, for $\forall i \geq j_0 + K, \forall j \geq 0$ we have

$$|\phi_{i,j}| \leq |\phi_{i,j_0}||\phi_{j_0-1,j}| \leq c_5 \exp\left(-\frac{c}{2} \sum_{k=j}^{i} a_k\right), \tag{3.10}$$

for some $c_5 > 0$.

From (3.8), it follows that

$$\delta u_{k+1}(0) = \phi_{k,0} \delta u_0(0) - \sum_{j=0}^{k} \phi_{k,j+1} a_k \mathbf{1}_{\{1 \notin \mathcal{M}_k\}} \varphi_k(0) + \sum_{j=0}^{k} \phi_{k,j+1} a_k \mathbf{1}_{\{1 \notin \mathcal{M}_k\}} v_k(1) \tag{3.11}$$

where the first term at the RHS of the above equation tends to zero as $k \to \infty$ because of (3.10). By Assumptions 3.2 and 3.4, it is clear that $\varphi_k(0) \xrightarrow[k\to\infty]{} 0$. By Assumption 3.5, it follows that

$$\sum_{k=1}^{\infty} a_k \mathbf{1}_{\{1 \notin \mathcal{M}_k\}} v_k(1) < 0.$$

Thus, the last two terms at the RHS of (3.11) also tend to zero as $k \to \infty$ using similar proof steps to Lemma A.3 in Appendix [1]. Thus, the optimality of the control sequence $\delta u_k(0)$ has been proved.

Inductively, assume that the optimality holds for $t = 0, 1, \ldots, s-1$. It is sufficient to show the validity for $t = s$.

By the inductive assumption, we have $\delta u_k(t) \xrightarrow[k\to\infty]{} 0, t = 0, 1, \ldots, s-1$, and by Lemma 3.1,

$$\delta x_k(s) \xrightarrow[k\to\infty]{} 0, \delta f_k(s) \xrightarrow[k\to\infty]{} 0, \delta b_k(s) \xrightarrow[k\to\infty]{} 0.$$

Then, it leads to $\varphi_k(s) \xrightarrow[k\to\infty]{} 0$. Similar to the case $t = 0$, the same treatment leads to $u_k(s) \xrightarrow[k\to\infty]{} u_d(s)$. This proves optimality of control at $t = s$ and completes the proof.

3.3 Successive Update Scheme and Its Convergence

In the last section, an intermittent update scheme was used there. However, it is noted that the algorithm would not update if the corresponding measurement packet is lost. In other words, if the tracking data is available, the algorithm uses such data for updating; otherwise, if the tracking data is dropped, then the algorithm would stop updating until the arrival of new data. This is a simple mechanism to deal with random data dropouts. However, as can be observed in practical applications, if the corresponding packet is lost in the kth iteration, the last available packet at the same time instant from the previous iteration could be used for updating. In this section, we are interested in how the algorithm behaves if the latest available packet is used for successive updating.

To this end, let $e_k(t) \triangleq y_d(t) - y_k(t)$ be the tracking error,

$$e_k(t+1) = \mathbf{c}^+ \mathbf{b}_k(t)\delta u_k(t) + \varphi_k(t) - v_k(t+1), \tag{3.12}$$

where

$$\varphi_k(t) = \mathbf{c}^+ \delta f_k(t) + \mathbf{c}^+ \delta b_k(t) u_d(t). \tag{3.13}$$

Then, define the updating error used in ILC update law as

$$e_k^*(t) = \begin{cases} e_k(t), & t \notin \mathcal{M}_k \\ e_{k-1}^*(t), & \text{otherwise} \end{cases} \tag{3.14}$$

It should be pointed out that the data dropout may happen in successive iterations randomly, and thus the tracking error $e_k^*(t)$ is recursively defined along the iteration axis. That is, a tracking error $e_k(t)$ may be used for several successive iterations if sequential data dropouts occur.

Then, the update law of the control sequence is given as

$$u_{k+1}(t) = u_k(t) + a_k e_k^*(t+1), \tag{3.15}$$

3.3 Successive Update Scheme and Its Convergence

where a_k is a decreasing gain such that $a_k > 0$, $a_k \to 0$, $\sum_{k=0}^{\infty} a_k = \infty$, $\sum_{k=0}^{\infty} a_k^2 < \infty$. It is clear that $a_k = \frac{1}{k+1}$ meets these requirements.

For any fixed time instant t, from (3.5) and (3.14), we have

$$u_{k+1}(t) = u_k(t) + a_k \mathbf{1}_{\{(t+1) \notin \mathcal{M}_k\}} e_k(t+1) \\ + a_k \mathbf{1}_{\{(t+1) \in \mathcal{M}_k\}} e_{k-1}^*(t+1), \quad (3.16)$$

where $\mathbf{1}_{\{\text{event}\}}$ is an indicator function meaning that it is equal to 1 if the event indicated in the bracket is fulfilled, and 0 if the event does not hold.

Now, we have the following convergence results.

Theorem 3.2 *Consider system (3.1) and index (3.2), and assume Assumptions 3.1–3.6 hold, then the control sequence $\{u_k(t)\}$ generated by the ILC update scheme (3.14) and (3.15) is optimal. In other words, $u_k(t)$ converges to $u_d(t)$ almost surely as the iteration number k goes to infinity, $\forall t$.*

Proof The proof is carried out by mathematical induction along the time axis t.

Initial Step: Consider the case for $t = 0$.

It is noted from (3.16) that the control signal is updated with the latest available tracking error. To be specific, the updating information comes from the last iteration if no data dropout happens, and from previous but non-adjacent iteration otherwise. In order to clarify it, let us introduce a stopping time sequence $\tau_0 < \tau_1 < \tau_2 < \cdots$ to denote all the iterations of which the measurements at time instant $t = 1$ are successfully transmitted. That is, $1 \in \mathcal{M}_{\tau_i}$. Without loss of generality, let $\tau_0 = 1$. In the light of Assumption 3.6, it is obvious that $\tau_i - \tau_{i-1} < K$ and $i \leq \tau_i \leq Ki$, $\forall i \in \mathbb{N}$.

As shown in Fig. 3.2, the tracking error would be transmitted successfully at the τ_i iteration while lost at the other iterations. As a result, the updating error $e_k^*(1)$ with $k \in [\tau_i + 1, \tau_{i+1}]$ actually is $e_{\tau_i}(1)$.

From (3.14), (3.16), and the above explanations, we have

$$u_{\tau_i+1}(0) = u_{\tau_i}(0) + a_{\tau_i} e_{\tau_i}(1), \quad (3.17)$$

and for $\tau_i < k \leq \tau_{i+1} - 1$,

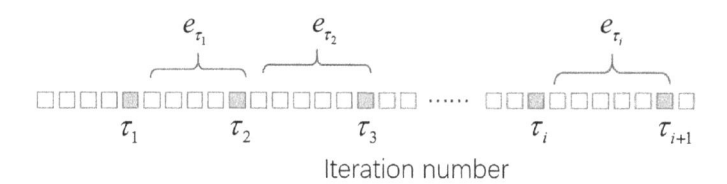

Fig. 3.2 Illustration of stopping times τ_i and updating error e_k^*

$$u_{k+1}(0) = u_k(0) + a_k e_k^*(1)$$
$$= u_k(0) + a_k e_{k-1}^*(1)$$
$$= u_k(0) + a_k e_{\tau_i}(1)$$
$$= u_{k-1}(0) + a_{k-1} e_{\tau_i}(1) + a_k e_{\tau_i}(1)$$
$$= u_{\tau_i}(0) + (a_{\tau_i} + a_{\tau_i+1} + \cdots + a_k) e_{\tau_i}(1).$$

It follows that

$$u_{\tau_{i+1}}(0) = u_{\tau_i}(0) + (a_{\tau_i} + a_{\tau_i+1} + \cdots + a_{\tau_{i+1}-1}) e_{\tau_i}(1)$$
$$= u_{\tau_i}(0) + \left(\sum_{k=\tau_i}^{\tau_{i+1}-1} a_k \right) e_{\tau_i}(1). \qquad (3.18)$$

Next, we first show the convergence of $u_{\tau_i}(0)$ and then extend it to all $u_k(0)$. Subtracting both sides of (3.18) from $u_d(0)$ and noticing (3.12), we have

$$\delta u_{\tau_{i+1}}(0) = \delta u_{\tau_i}(0) - \left(\sum_{k=\tau_i}^{\tau_{i+1}-1} a_k \right) e_{\tau_i}(1)$$
$$= \left[1 - \left(\sum_{k=\tau_i}^{\tau_{i+1}-1} a_k \right) \mathbf{c}^+ \mathbf{b}_k(0) \right] \delta u_{\tau_i}(0)$$
$$- \left(\sum_{k=\tau_i}^{\tau_{i+1}-1} a_k \right) \varphi_{\tau_i}(0) + \left(\sum_{k=\tau_i}^{\tau_{i+1}-1} a_k \right) v_{\tau_i}(1). \qquad (3.19)$$

Denote

$$\Phi_{i,j} \triangleq \left(1 - \left(\sum_{k=\tau_i}^{\tau_{i+1}-1} a_k \right) \mathbf{c}^+ \mathbf{b}_{\tau_i}(0) \right) \times \cdots$$
$$\times \left(1 - \left(\sum_{k=\tau_j}^{\tau_{j+1}-1} a_k \right) \mathbf{c}^+ \mathbf{b}_{\tau_j}(0) \right)$$

for $i \geq j$ and $\Phi_{i,i+1} = 1$.

Since $\mathbf{b}_k(0)$ is continuous with respect to the initial state by Assumption 3.2, we have $\mathbf{b}_k(0) \xrightarrow[k \to \infty]{} \mathbf{b}_d(0)$ by Assumption 3.4 and then $\mathbf{c}^+ \mathbf{b}_k(0)$ converges to a positive constant by Assumption 3.3. Therefore, for sufficiently large j, say $j \geq j_0$,

3.3 Successive Update Scheme and Its Convergence

$$1 - \left(\sum_{k=\tau_j}^{\tau_{j+1}-1} a_k \right) \mathbf{c}^+ \mathbf{b}_{\tau_j}(0) \geq 1 - K a_{\tau_j} \mathbf{c}^+ \mathbf{b}_{\tau_j}(0) > 0. \qquad (3.20)$$

Then

$$\begin{aligned}\Phi_{i,j} &= \left(1 - \left(\sum_{k=\tau_i}^{\tau_{i+1}-1} a_k \right) \mathbf{c}^+ \mathbf{b}_{\tau_i}(0) \right) \Phi_{i-1,j} \\ &\leq \exp\left(-c \sum_{k=\tau_i}^{\tau_{i+1}-1} a_k \right) \Phi_{i-1,j},\end{aligned}$$

where c is a suitable constant and the elementary inequality $1 - x \leq e^{-x}$ is used.

From the above estimation, it is clear that

$$\Phi_{i,j} = c_1 \exp\left(-c \sum_{k=\tau_j}^{\tau_{i+1}-1} a_k \right), \quad \forall j \geq j_0, i \geq j, \qquad (3.21)$$

where $c_1 > 0$ is a suitable constant. Thus, $\forall i \geq j, j \geq 0$, we have

$$\begin{aligned}|\Phi_{i,j}| &\leq |\Phi_{i,j_0}| \cdot |\Phi_{j_0,j}| \\ &\leq c_2 \exp\left(-c \sum_{k=\tau_j}^{\tau_{i+1}-1} a_k \right),\end{aligned}$$

for some $c_2 > 0$.

Then, (3.19) yields that

$$\begin{aligned}\delta u_{\tau_{i+1}}(0) = \Phi_{i,0} \delta u_{\tau_0}(0) &- \sum_{j=0}^{i} \Phi_{i,j+1} \left(\sum_{k=\tau_j}^{\tau_{j+1}-1} a_k \right) \varphi_{\tau_j}(0) \\ &+ \sum_{j=0}^{i} \Phi_{i,j+1} \left(\sum_{k=\tau_j}^{\tau_{j+1}-1} a_k \right) v_{\tau_j}(1),\end{aligned} \qquad (3.22)$$

where the first term at the RHS of the last equation tends to zero as k goes to infinity almost surely.

According to Assumption 3.6 and $j \leq \tau_j \leq Kj$, $\sum_{k=\tau_j}^{\tau_{j+1}-1} a_k \leq K a_{\tau_j} \leq K a_j$ and $\sum_{k=\tau_j}^{\tau_{j+1}-1} a_k \geq a_{\tau_j} \geq a_{Kj}$. Thus, it is clear to derive $\sum_{k=\tau_j}^{\tau_{j+1}-1} a_k \to 0$. Besides,

$$\sum_{j=1}^{\infty}\left(\sum_{k=\tau_j}^{\tau_{j+1}-1} a_k\right) = \sum_{j=1}^{\infty} a_j = \infty,$$

$$\sum_{j=1}^{\infty}\left(\sum_{k=\tau_j}^{\tau_{j+1}-1} a_k\right)^2 \leq \sum_{j=1}^{\infty} K \left(\sum_{k=\tau_j}^{\tau_{j+1}-1} a_k^2\right) = K \sum_{j=1}^{\infty} a_j^2 < \infty.$$

Assumption 3.5 leads to

$$\sum_{j=1}^{\infty}\left(\sum_{k=\tau_j}^{\tau_{j+1}-1} a_k\right) v_{\tau_j}(1) < \infty.$$

Besides, by Assumptions 3.2 and 3.4, it is clear that $\varphi_{\tau_j}(0) \xrightarrow[j \to \infty]{} 0$. Thus, the last two terms at the RHS of (3.22) also tend to zero as $j \to \infty$ by using similar proof steps to Lemma A.3 in Appendix. Thus, we have proved $\delta u_{\tau_i}(0) \to 0$ as i goes to infinity.

Next, we proceed to prove the zero convergence of $\delta u_k(0)$, $\tau_i < k < \tau_{i+1} - 1$.

$$\delta u_{k+1}(0) = \left(1 - \left(\sum_{j=\tau_i}^{k} a_j\right) c^+ b_{\tau_i}(0)\right) \delta u_{\tau_i}(0)$$

$$- \left(\sum_{j=\tau_i}^{k} a_j\right) \varphi_{\tau_i}(0) + \left(\sum_{j=\tau_i}^{k} a_j\right) v_{\tau_i}(1).$$

Thus, the optimality of the control sequence $\delta u_k(0)$ has been proved.

Inductive Step: Assume that the optimality holds for $t = 0, 1, \ldots, s - 1$. It is sufficient to show the validity for $t = s$.

The inductive assumption implies $\delta u_k(t) \xrightarrow[k \to \infty]{} 0$, $t = 0, 1, \ldots, s - 1$. By Lemma 3.1, we have

$$\delta x_k(s) \xrightarrow[k \to \infty]{} 0, \delta f_k(s) \xrightarrow[k \to \infty]{} 0, \delta b_k(s) \xrightarrow[k \to \infty]{} 0.$$

Then leads to $\varphi_k(s) \xrightarrow[k \to \infty]{} 0$. Similar to the case $t = 0$, the same treatment leads to $u_k(s) \xrightarrow[k \to \infty]{} u_d(s)$. This proves optimality of control at $t = s$ and completes the proof of the theorem.

3.4 Illustrative Simulations

Consider the following system:

$$x_k^{(1)}(t+1) = 0.9x_k^{(1)}(t) + 0.3\sin(x_k^{(2)}(t)) + 0.45u_k(t),$$
$$x_k^{(2)}(t+1) = 0.2\cos(x_k^{(1)}(t)) + 0.95x_k^{(2)}(t) + 0.5u_k(t),$$
$$y_k(t) = x_k^{(1)}(t) + x_k^{(2)}(t) + v_k(t),$$

where $x_k^{(1)}(t)$ and $x_k^{(2)}(t)$ denote the first and second dimension of $x_k(t)$, respectively.

For simple illustration, let $N = 40$, and the measurement noises are assumed zero-Gaussian distributed, i.e., $v_k(t) \sim N(0, 0.1^2)$. The reference trajectory is $y_d(t) = 20\sin(\frac{t}{20}\pi)$. The initial control action is simply given as $u_0(t) = 0, \forall t$. The learning gain chooses $a_k = \frac{1}{k+1}$. The algorithm has run 160 iterations.

It is worth mentioning that K denotes the worst level of data dropout, and data dropout rate (DDR) denotes the average loss level. Thus, two scenarios are provided. One is with $K = 3$ and DDR $= 33\%$, illustrated in Fig. 3.3 and called Case-I, and the other is with $K = 5$ and DDR $= 66.7\%$, illustrated in Fig. 3.4 and called Case-II. The two cases correspond to low loss level and high loss level, respectively. In Figs. 3.3 and 3.4, the value 1 denotes successful transmission and 0 means data dropout.

In order to make a comparison, both intermittent and successive update schemes are simulated under the same conditions. In the intermittent update scheme, the control signal would remain unchanged when data dropout occurs until the arrival of a new measurement. Figure 3.5 shows the tracking performance of last iteration under data dropout Case-I, where solid line, dashed line with cycles, and dot-dash line represent the desired trajectory, the output generated by successive update scheme,

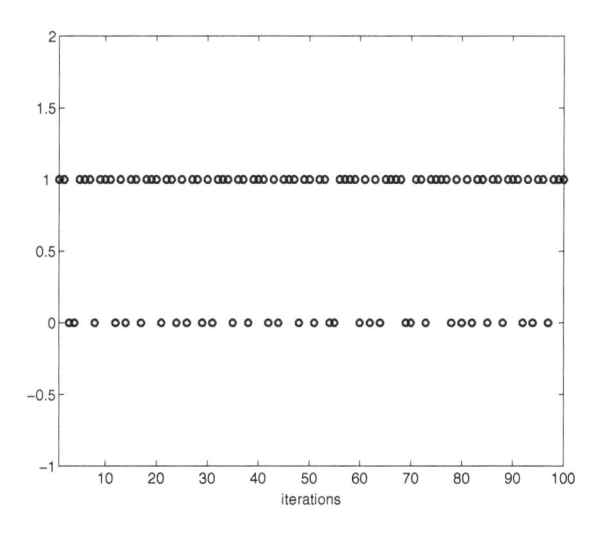

Fig. 3.3 Illustrative data dropout of Case-I

Fig. 3.4 Illustrative data dropout of Case-II

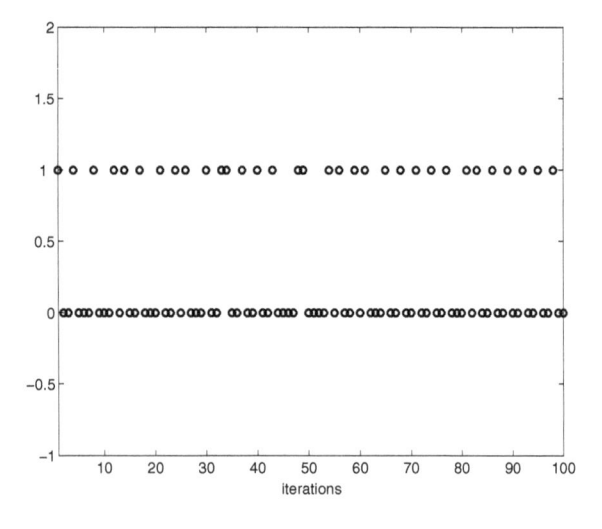

Fig. 3.5 Tracking performance for last iteration: Case-I

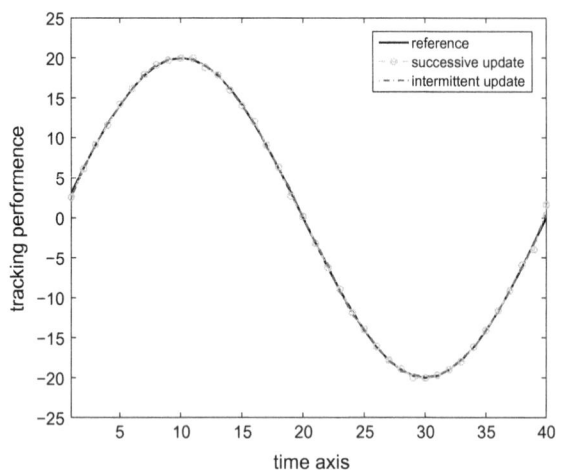

and the output generated by intermittent update scheme, respectively. As one could see, these three curves are in mutual coincidence, which means that both intermittent and successive update schemes could achieve good tracking performance under data dropouts. This performance is also demonstrated by Fig. 3.6, where the absolute tracking error at time instant $t = 12$ is presented as an illustration. The solid line and the dashed line denote successive and intermittent update schemes, respectively. The sawtooth of the curves is mainly caused by stochastic noises. As is seen, both update schemes behave similarly.

Then, both update schemes are simulated for data dropout Case-II and the results are shown in Figs. 3.7 and 3.8, where the former is the tracking performance of last iteration and the latter is absolute tracking error at time instant $t = 20$ as an

3.4 Illustrative Simulations

Fig. 3.6 Absolute error at time instant $t = 20$: Case-I

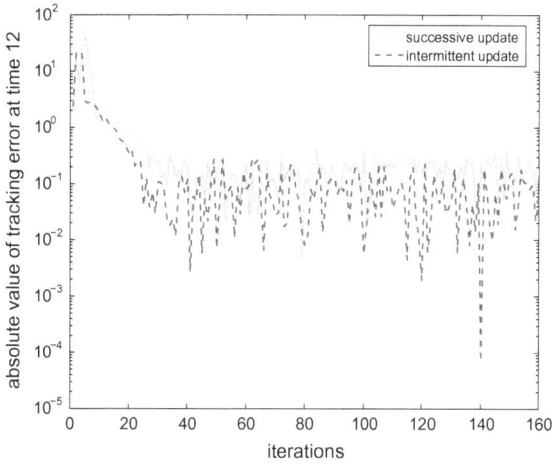

Fig. 3.7 Tracking performance for last iteration: Case-II

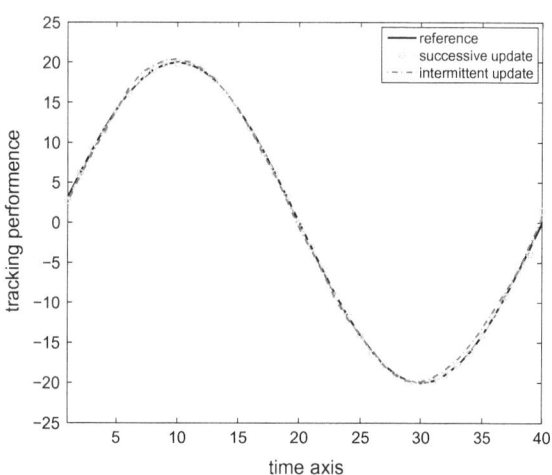

illustration. It is seen from Fig. 3.7 that the final tracking performance of successive update scheme behaves a little better than that of intermittent update scheme. The reason we believe is that successive update scheme could continually improve the tracking performance after enough updating, whereas intermittent update scheme would need more iterations to achieve similar performance. This is also displayed in Fig. 3.8, where the tracking error of successive update scheme, denoted by solid line, is smaller than that of intermittent update scheme, denoted by dashed line, after about 50 iterations. However, as could be seen from Fig. 3.8, successive update scheme results in a worse performance for the first serval iterations, which is an interesting phenomenon.

Fig. 3.8 Absolute error at time instant $t = 20$: Case-II

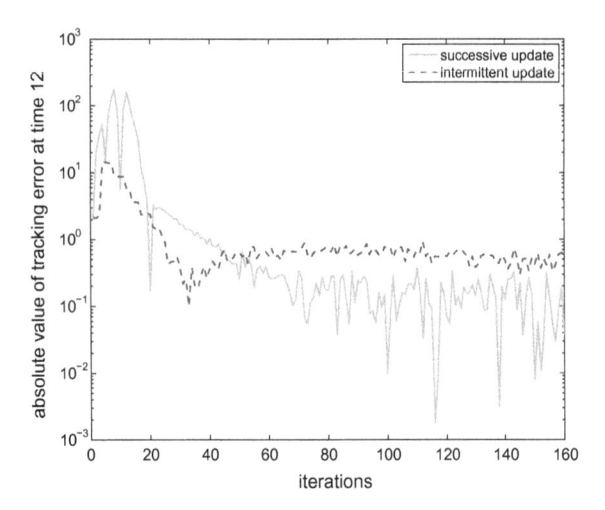

3.5 Summary

In this chapter, we address ILC problem for affine nonlinear systems with observation noises, where the networks from the measured output to the controller suffer data dropouts. For the random sequence model of data dropouts, both intermittent and successive update schemes are investigated and the almost sure convergence is strictly proved. However, in these results, we require the prior knowledge of the control direction, i.e., the sign of the input/output coupling value. A natural and interesting problem is how to deal with the unknown control direction case. This issue will be given in the next chapter. The results in this chapter are mainly based on [2, 3].

References

1. Chen, H.F.: Stochastic Approximation and Its Applications. Kluwer, Dordrecht (2002)
2. Shen, D., Wang, Y.: Iterative learning control for networked stochastic systems with random packet losses. Int. J. Control **88**(5), 959–968 (2015)
3. Shen, D., Xu, Y.: Iterative learning control for networked nonlinear systems using latest information. In: The 34th Chinese Control Conference, pp. 3079–3084 (2015)

Chapter 4
Random Sequence Model for Nonlinear Systems with Unknown Control Direction

4.1 Problem Formulation

Consider the following SISO affine nonlinear discrete-time system:

$$\begin{aligned} x_k(t+1) &= f(t, x_k(t)) + \mathbf{b}(t, x_k(t))u_k(t), \\ y_k(t) &= \mathbf{c}(t)x_k(t) + w_k(t), \end{aligned} \quad (4.1)$$

where $t \in \{0, 1, \ldots, N\}$ denotes the time instant in an iteration of the process, and $k = 1, 2, \ldots$ labels different iterations. $u_k(t) \in \mathbf{R}$, $x_k(t) \in \mathbf{R}^n$, and $y_k(t) \in \mathbf{R}$ denote the input, state, and output, respectively. $f(t, x_k(t))$, $\mathbf{b}(t, x_k(t))$, and $\mathbf{c}(t)$ denote unknown system information. $w_k(t)$ is the measurement noise.

The setup of the control system is illustrated in Fig. 4.1. For convenience, only data dropout at the measurement side is taken into account in this chapter, which can be extended to the case that both measurement and control signals are lossy using techniques in Part II. As is shown in Fig. 4.1, the measurement signals are transmitted back through a lossy channel, which could be regarded as a switch that opens and closes in a random manner. The rest of the block diagram follows the conventional ILC framework.

Similar to the previous two chapters, in this chapter, the random sequence model of the random data dropout is employed. To this end, denote $\mathcal{M}_k$ as the random set of time locations at which measurements are lost in the kth iteration. In other words, $t_0 \in \mathcal{M}_k$ if $y_k(t_0)$ is lost.

Let $\mathcal{F}_k \triangleq \sigma\{y_j(t), x_j(t), w_j(t), v_j(t), 0 \leq j \leq k, t \in \{0, \ldots, N\}\}$ be the σ-algebra generated by $y_j(t), x_j(t), w_j(t), v_j(t), 0 \leq t \leq N, 0 \leq j \leq k$. Then, the set of admissible controls is defined as

$$U = \{u_{k+1}(t) \in \mathcal{F}_k, \sup_k \|u_k(t)\| < \infty, \text{a.s. } t = 0, \ldots, N-1, k = 0, 1, 2, \ldots\}.$$

Fig. 4.1 Block diagram of networked control system with measurement packet loss

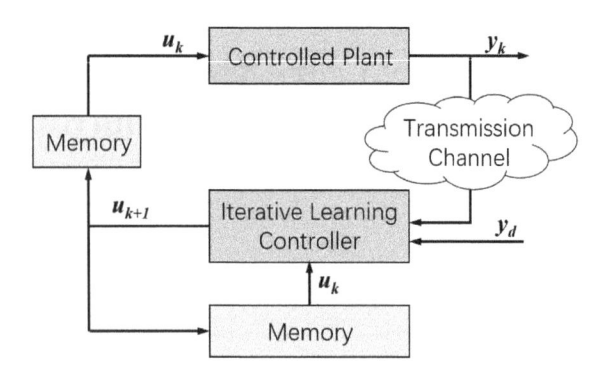

The control objective is to find a control sequence $\{u_k(t), k = 0, 1, 2, \ldots\} \subset U$ under random data dropout environments to minimize the limitation of the averaged tracking errors, $\forall t \in \{0, 1, \ldots, N\}$,

$$V(t) = \limsup_{n \to \infty} \frac{1}{n} \sum_{k=1}^{n} \|y_k(t) - y_d(t)\|^2, \qquad (4.2)$$

where $y_d(t), t \in \{0, 1, \ldots, N\}$ is the tracking target.

The following assumptions are used.

Assumption 4.1 The tracking target $y_d(t)$ is realizable in the sense that there exist $u_d(t)$ and $x_d(0)$ such that

$$\begin{aligned} x_d(t+1) &= f(t, x_d(t)) + \mathbf{b}(t, x_d(t))u_d(t), \\ y_d(t) &= \mathbf{c}(t)x_d(t). \end{aligned} \qquad (4.3)$$

Assumption 4.2 The functions $f(\cdot, \cdot)$ and $\mathbf{b}(\cdot, \cdot)$ are continuous with respect to the second argument.

Assumption 4.2 could be relaxed to the case that the functions $f(t, x)$ and $\mathbf{b}(t, x)$ have discontinuities with respect to x away from $x_d(t)$. Since $x = x_d(t)$ is unknown generally, Assumption 4.2 is assumed for simplicity.

Assumption 4.3 The input/output coupling value $\mathbf{c}(t+1)\mathbf{b}(t, x_k(t))$ is unknown, but it is nonzero and does not change its sign. Without loss of any generality, it is assumed $\mathbf{c}(t+1)\mathbf{b}(t, x_k(t)) > 0$. This sign information is unknown to the controller.

Assumption 4.3 is the assumption on control direction. As the SISO model is taken into account, the control direction actually is the sign of the coupling value. That is, the control direction is either $+1$ or -1. However, this case can be extended to the MIMO case by using similar techniques of [1].

Assumption 4.4 The initial values can be asymptotically reset in the sense that $x_k(0) \to x_d(0)$ as $k \to \infty$.

4.1 Problem Formulation

Assumption 4.4 is the re-initialization condition. In many papers, the initial state is usually required to be $x_d(0)$ [2, 3], whereas in this chapter it is only required that the accurate initial state is asymptotically achievable.

Assumption 4.5 For each time instant t, the measurement noise $\{w_k(t)\}$ is a sequence of independent and identically distributed random variables with $\mathbb{E} w_k(t) = 0$, $\sup_k \mathbb{E} w_k^2(t) < \infty$, and $\lim_{n \to \infty} \frac{1}{n} \sum_{k=1}^n w_k^2(t) = R_w^t$, a.s., where R_w^t is unknown.

In Assumption 4.5, the noise condition is made according to the iteration axis rather than the time axis. This requirement is not rigorous as the process would be performed repeatedly.

Assumption 4.6 For each time instant t, the data dropout at the measurement side is random without obeying any certain probability distribution, but there is a number K such that during any successive K iterations, at least in one iteration the measurement is successfully transmitted back.

For simplicity of writing, let us set $f_k(t) = f(t, x_k(t))$, $f_d(t) = f(t, x_d(t))$, $\mathbf{b}_k(t) = \mathbf{b}(t, x_k(t))$, $\mathbf{b}_d(t) = \mathbf{b}(t, x_d(t))$, $e_k(t) = y_d(t) - y_k(t)$, $\delta x_k(t) = x_d(t) - x_k(t)$, $\delta u_k(t) = u_d(t) - u_k(t)$, $\delta f_k(t) = f_d(t) - f_k(t)$, $\delta \mathbf{b}_k(t) = \mathbf{b}_d(t) - \mathbf{b}_k(t)$, $\mathbf{c}^+ f_k(t) = \mathbf{c}(t+1) f(t, x_k(t))$, $\mathbf{c}^+ \mathbf{b}_k(t) = \mathbf{c}(t+1) \mathbf{b}(t, x_k(t))$.

Now the following lemma on the continuity property of nonlinear functions is given, whose proof is provided in Chap. 3.

Lemma 4.1 *Assume Assumptions 4.1–4.4 hold for system (4.1). If* $\lim_{k \to \infty} \delta u_k(s) = 0$, $s = 0, 1, \ldots, t$, *then at time instant* $t+1$, $\|\delta x_k(t+1)\| \xrightarrow[k \to \infty]{} 0$, $\|\delta f_k(t+1)\| \xrightarrow[k \to \infty]{} 0$, $\|\delta \mathbf{b}_k(t+1)\| \xrightarrow[k \to \infty]{} 0$.

Theorem 4.1 *For the nonlinear system (4.1) and tracking objective* $y_d(t)$, *assume Assumptions 4.1–4.6 hold, then the index (4.2) will be minimized for any arbitrary time instant* t *if the control sequence* $\{u_k(s)\}$ *is admissible and satisfies* $u_k(s) \xrightarrow[k \to \infty]{} u_d(s)$, $s = 0, 1, \ldots, t-1$. *In this case,* $\{u_k(s)\}$ *is called the optimal control sequence.*

Proof According to Assumption 4.5 and the definition of $\mathscr{F}_k$, it follows that $\mathscr{F}_k$ is independent of $\{w_l(t), l = k+i, i = 1, 2, \ldots, \forall t\}$, thus $\{w_k(t), \mathscr{F}_k\}$ is a martingale difference sequence. Meanwhile, the input, output, and state vectors are all adapted to $\mathscr{F}_k$. Therefore, by (4.1) and Assumption 4.1,

$$\limsup_{n \to \infty} \frac{1}{n} \sum_{k=1}^n \|y_k(t) - y_d(t)\|^2$$

$$= \limsup_{n \to \infty} \frac{1}{n} \sum_{k=1}^n \|\mathbf{c}(t)(x_k(t) - x_d(t)) + w_k(t)\|^2$$

$$= \limsup_{n \to \infty} \frac{1}{n} \sum_{k=1}^n \|\mathbf{c}(t) \delta x_k(t)\|^2 (1 + o(1))$$

$$+ \limsup_{n\to\infty} \frac{1}{n} \sum_{k=1}^{n} \|w_k(t)\|^2$$

$$\geq \limsup_{n\to\infty} \frac{1}{n} \sum_{k=1}^{n} \|w_k(t)\|^2 = R_w^t.$$

The sufficient and necessary condition to achieve the minimum is

$$\limsup_{n\to\infty} \frac{1}{n} \sum_{k=1}^{n} \|\mathbf{c}(t)\delta x_k(t)\|^2 = 0,$$

which is true when $\mathbf{c}(t)\delta x_k(t) \xrightarrow[k\to\infty]{} 0$. While the latter holds if $\delta u_k(s) \xrightarrow[k\to\infty]{} 0$, $s = 0, 1, \ldots, t-1$ by Lemma 4.1. The proof is completed.

4.2 Intermittent Update Scheme and Its Almost Sure Convergence

Now we define the ILC algorithms leading the performance index (4.2) to its minimum. As the control direction is unknown, we have to introduce a switch mechanism to make the algorithm find the correct control direction adaptively. Intuitively, the algorithm should track well enough if it chooses the correct control direction and behave badly if it chooses the wrong control direction. This motivates us to define the following update law by using the tracking performance for searching the control direction, which is inspired from [4]. Remind that the correct control direction is "+1" in this chapter by Assumption 4.3.

$$\bar{u}_{k+1}(t) = u_k(t) + a_k(-1)^{\sigma_k(t)}$$
$$\times \mathbf{1}_{\{(t+1)\notin \mathcal{M}_k\}}(y_d(t+1) - y_k(t+1)), \quad (4.4)$$

$$u_{k+1} = \bar{u}_{k+1}\mathbf{1}_{\{|\bar{u}_{k+1}|<M_{\sigma_k(t)}\}}, \quad (4.5)$$

$$\sigma_k(t) = \sum_{j=1}^{k-1} \mathbf{1}_{\{|\bar{u}_{j+1}|>M_{\sigma_j(t)}\}}, \quad \sigma_0(t) = 0, \quad (4.6)$$

where $\{M_k\}$ is a sequence of positive real numbers such that $M_{k+1} > M_k$, $\forall k$ and $M_k \xrightarrow[k\to\infty]{} \infty$. Besides, a_k is the decreasing gain such that $a_k > 0$, $a_k \to 0$, $\sum_{k=0}^{\infty} a_k = \infty$, $\sum_{k=0}^{\infty} a_k^2 < \infty$, and $\mathbf{1}_{\{\text{event}\}}$ is an indicator function which is equal to 1 if the event indicated in the bracket is fulfilled, and 0 if the event does not hold. The initial input $u_0(t)$ is simply valued zero.

Remark 4.1 The number of truncations $\sigma_k(t)$ is used to regulate the control direction in the algorithm. In particular, the term $(-1)^{\sigma_k(t)}$ is the regulating term, and "+1" is

4.2 Intermittent Update Scheme and Its Almost Sure Convergence

the correct direction. If the wrong direction "-1" is applied in the algorithm, it will be proved that the algorithm diverges and forces the truncation mechanism to work (see Lemma 4.4 below for details). The underlining reason is the formulation of the learning step-length a_k. The divergence of the algorithm with wrong direction could be seen from the numerical simulations. As a result, the direction shifts to the correct "$+1$".

Remark 4.2 In this chapter, noticing the term $\mathbf{1}_{\{(t+1)\notin \mathcal{M}_k\}}$ in (4.4), the algorithm would not update if the corresponding measurement packet is lost. However, in practice, if the corresponding packet is lost in the kth iteration, the latest available packet at the same time instant from the previous iteration could be used for updating. This is another kind of data selection method. It should be pointed out that the data dropout may happen in successive iterations randomly. This will result in multiple updates of the input signal with one packet of tracking information. The algorithm would remain convergent if Assumption 4.6 holds. The proof is similar to the followings but more complex. Therefore, a simple data selection mechanism is adopted in this chapter. In other words, if the tracking data is available, the algorithms use this data for updating; while if the tracking data is dropped, then the algorithms would stop updating until the arrival of the new data.

The following theorem shows the convergence results of the proposed algorithm.

Theorem 4.2 *For the nonlinear system (4.1) and index (4.2), assume Assumptions 4.1–4.6 hold, then (a) the number of truncations of the algorithm (4.4)–(4.6) is finite; (b) the algorithm (4.4)–(4.6) shifts the correct direction when stopping switching; and (c) the input sequence $\{u_k(t)\}$ generated by (4.4)–(4.6) is optimal.*

In order to make the proof of the theorem concise and clear, some key steps of the proof are made into auxiliary lemmas as follows. The proofs of these lemmas are given in the next section. Readers who are not interested in complex derivations may skip the next technical section.

Lemma 4.2 *Consider the algorithm (4.4)–(4.6) without truncations and remove the regulating term $(-1)^{\sigma_k(t)}$, then the input sequence generated by the modified algorithm converges to the desired input in (4.3) for time instant $t = 0$.*

Lemma 4.3 *Consider the algorithm (4.4)–(4.6) without truncations for time instant $t = 0$, remove the regulating term $(-1)^{\sigma_k(0)}$, and replace a_k by a_{k+m}, then $\delta u_k(0)$ generated by the modified algorithm with any fixed initial value is uniformly bounded for any integer $m > 0$.*

Lemma 4.4 *Consider the algorithm (4.4)–(4.6) for time instant $t = 0$, then the algorithm stops at the correct direction and converges to the optimal control whenever the algorithm stops truncation.*

Based on the above lemmas, the proof of Theorem 4.2 can be given now.

Proof The proof is carried by mathematical induction along the time axis t.
Base step: In this step, the validity of the theorem for $t = 0$ is proved.
Subtracting the algorithm without truncations

$$u_{k+1}(0) = u_k(0) + a_k \mathbf{1}_{\{1 \notin \mathcal{M}_k\}}(y_d(1) - y_k(1)) \tag{4.7}$$

from the term $u_d(0)$ leads to

$$\begin{aligned}
\delta u_{k+1}(0) &= \delta u_k(0) - a_k \mathbf{1}_{\{1 \notin \mathcal{M}_k\}}(y_d(1) - y_k(1)) \\
&= \delta u_k(0) - a_k \mathbf{1}_{\{1 \notin \mathcal{M}_k\}} \mathbf{c}^+ \mathbf{b}_k(0) \delta u_k(0) \\
&\quad - a_k \mathbf{1}_{\{1 \notin \mathcal{M}_k\}}[\varphi_k(0) - w_k(1)],
\end{aligned} \tag{4.8}$$

where $\varphi_k(t)$ is defined by

$$\varphi_k(t) = \mathbf{c}^+ \delta f_k(t) + \mathbf{c}^+ \delta \mathbf{b}_k(t) u_d(t), \tag{4.9}$$

and $\varphi_k(0)$ is the value of (4.9) at $t = 0$. Here, $\mathbf{c}^+ \delta f_k(t) = \mathbf{c}(t+1) \delta f(t, x_k(t))$, $\mathbf{c}^+ \delta \mathbf{b}_k(t) = \mathbf{c}(t+1) \delta \mathbf{b}(t, x_k(t))$.

By Lemma 4.2, we have $\delta u_k(0) \xrightarrow[k \to \infty]{} 0$ for (4.8) with any initial value. Therefore, if the algorithm (4.8) is with initial value $\delta u_0(0) = u_d(0)$, then there is an $L > 0$ such that $\|\delta u_k(0)\| < L$ for $\{\delta u_k(0)\}$ generated by the algorithm. Then, by Lemma 4.3, it is found that $\{\delta u_k(0)\}$ produced by the following algorithm with any integer $m > 0$ is bounded if the initial value is taken as $\delta u_0(0) = u_d(0)$:

$$\begin{aligned}
\delta u_{k+1}(0) &= \delta u_k(0) - a_{k+m} \mathbf{1}_{\{1 \notin \mathcal{M}_k\}} \mathbf{c}^+ \mathbf{b}_k(0) \delta u_k(0) \\
&\quad - a_{k+m} \mathbf{1}_{\{1 \notin \mathcal{M}_k\}}[\varphi_k(0) - w_k(1)].
\end{aligned} \tag{4.10}$$

Without loss of any generality, denote the boundary as L. In other words, for any integer $m > 0$ by taking the initial value $\delta u_0(0) = u_d(0)$, $\{\delta u_k(0)\}$ generated by (4.10) satisfies $\|\delta u_k(0)\| < L$. After each truncation, the algorithm (4.4)–(4.6) is pulled back to zero, and hence the control sequence has an upper bound $L' \triangleq L + \|u_d(0)\|$, whenever the algorithm has switched on the correct direction.

If the number of truncations was infinite, then the algorithm would be switched to the correct direction infinitely many times. Let the algorithm be with the correct direction and let k be sufficiently large so that $M_{\sigma_k(0)} > L'$. This implies that the algorithm (4.4)–(4.6) will no longer be truncated since then provided that the correct direction is switched on, which contradicts the assumption of infinitely many truncations. Therefore, the algorithm (4.4)–(4.6) may have only finite number of truncations, and hence the algorithm is bounded. Thus, Theorem 4.2(a) is proved for $t = 0$.

4.2 Intermittent Update Scheme and Its Almost Sure Convergence

By Lemma 4.4, the impossibility of stopping at the wrong direction is verified, thus the algorithm stops at the correct direction. Then, by Lemma 4.2, the input sequence generated by the algorithm converges to the desired control, implying that Theorem 4.2(b) and (c) hold for $t = 0$.

Inductive step: The cases for the other time instants are proved in this step.

Now assume that the conclusion of the theorem is valid for $0, 1, \ldots, t-1$, we proceed to prove that it is also true for t. By the inductive assumptions, it is seen that $u_k(0), u_k(1), \ldots, u_k(t-1)$ are optimal, i.e., $\delta u_k(s) \xrightarrow[k\to\infty]{} 0, s = 0, 1, \ldots, t-1$, and then $\varphi_k(s) \xrightarrow[k\to\infty]{} 0$ by Lemma 4.1. Then, completely the same argument as that used for $t = 0$ leads to the desired conclusion. The proof of the theorem is completed.

Remark 4.3 For continuous-time systems, Nussbaum-type gain is a major technique for searching the control direction. However, for discrete-time systems, it is difficult to establish a similar gain. As a matter of fact, the intuitive idea for direction switching is that the algorithm would diverge if the wrong direction is selected and the algorithm would converge if the correct direction is selected. The truncation used in this chapter is a simple implementation of this idea.

4.3 Proofs of Lemmas

Proof of Lemma 4.2

Proof If no truncation is considered in (4.4) and (4.5) for $t = 0$, it reduces to (4.7), which further leads to

$$\delta u_{k+1}(0) = (1 - a_k \mathbf{1}_{\{1 \notin \mathcal{M}_k\}} \mathbf{c}^+ \mathbf{b}_k(0)) \delta u_k(0) \\ - a_k \mathbf{1}_{\{1 \notin \mathcal{M}_k\}} \varphi_k(0) + a_k \mathbf{1}_{\{1 \notin \mathcal{M}_k\}} w_k(1). \qquad (4.11)$$

Since $\mathbf{b}_k(0)$ is continuous with respect to the initial state by Assumption 4.2, we have $\mathbf{b}_k(0) \xrightarrow[k\to\infty]{} \mathbf{b}_d(0)$ by Assumption 4.4 and $\mathbf{c}^+ \mathbf{b}_k(0)$ converges to a positive constant by Assumption 4.3. Therefore, by Assumption 4.6, it follows

$$\sum_{k=i}^{i+K-1} \left(-\mathbf{1}_{\{1 \notin \mathcal{M}_k\}} \mathbf{c}^+ \mathbf{b}_k(0)\right) < -\gamma, \quad \gamma > 0, \qquad (4.12)$$

for all sufficiently large i.

Set $\phi_{i,j} \triangleq (1 - a_i \mathbf{1}_{\{1 \notin \mathcal{M}_i\}} \mathbf{c}^+ \mathbf{b}_i(0)) \cdots (1 - a_j \mathbf{1}_{\{1 \notin \mathcal{M}_j\}} \mathbf{c}^+ \mathbf{b}_j(0)), i \geq j, \phi_{i,i+1} \triangleq 1$. It is clear that $1 - a_j \mathbf{1}_{\{1 \notin \mathcal{M}_j\}} \mathbf{c}^+ \mathbf{b}_j(0) > 0$ for all sufficiently large j, say, $j \geq j_0$. Then, for any $i \geq j + K, j \geq j_0$ by (4.11) and (4.12), we have that

$$\phi_{i,j} = \phi_{i-K,j}\left(1 - a_i \sum_{k=i-K+1}^{i} \mathbf{1}_{\{1 \notin \mathcal{M}_k\}}\mathbf{c}^+\mathbf{b}_k(0) + o(a_i)\right)$$

$$\leq \phi_{i-K,j}(1 - \gamma a_i + o(a_i))$$

$$\leq \phi_{i-K,j}\left(1 - \frac{\gamma}{K}\sum_{k=i-K+1}^{i} a_k + o(a_i)\right)$$

$$\leq \exp\left(-c\sum_{k=i-K+1}^{i} a_k\right)\phi_{i-K,j} \quad \text{with } c > 0.$$

It follows from here that $\phi_{i,j} \leq c_1 \exp\left(-\frac{c}{2}\sum_{k=j}^{i} a_k\right)$, $\forall j \geq j_0$ for some $c_1 > 0$, and hence there is a c_2 such that

$$|\phi_{i,j}| \leq c_2 \exp\left(-\frac{c}{2}\sum_{k=j}^{i} a_k\right), \quad \forall i \geq j + K, j \geq j_0.$$

Therefore, for $\forall i \geq j_0 + K, \forall j \geq 0$, we have

$$|\phi_{i,j}| \leq |\phi_{i,j_0}||\phi_{j_0-1,j}| \leq c_3 \exp\left(-\frac{c}{2}\sum_{k=j}^{i} a_k\right), \tag{4.13}$$

for some $c_3 > 0$.

From (4.11), it follows that

$$\delta u_{k+1}(0) = \phi_{k,0}\delta u_0(0) - \sum_{j=0}^{k}\phi_{k,j+1}a_j\mathbf{1}_{\{1\notin\mathcal{M}_j\}}\varphi_j(0) + \sum_{j=0}^{k}\phi_{k,j+1}a_j\mathbf{1}_{\{1\notin\mathcal{M}_j\}}w_j(1), \tag{4.14}$$

where the first term at the right-hand side (RHS) of the above equation tends to zero as $k \to \infty$ because of (4.13). By Assumptions 4.2 and 4.4, it is clear that $\varphi_k(0) \xrightarrow{k\to\infty} 0$. By Assumption 4.5, it follows that

$$\sum_{k=1}^{\infty} a_k\mathbf{1}_{\{1\notin\mathcal{M}_k\}}w_k(1) < 0.$$

Thus, the last two terms at the RHS of (4.14) also tend to zero as $k \to \infty$ by using similar proof steps to Lemma A.3 in Appendix. Thus, it concludes that $\delta u_k(0) \to 0$. In other words, the convergence of control sequence $u_k(0)$, $u_k(0) \xrightarrow{k\to\infty} u_d(0)$, has been proved.

4.3 Proofs of Lemmas

Proof of Lemma 4.3

Proof If no truncation is considered in (4.4) and (4.5) for $t=0$ and a_k is replaced by a_{k+m}, $\forall m$, it leads to

$$\delta u_{k+1}(0) = \delta u_k(0) - a_{k+m}\mathbf{1}_{\{1\notin\mathscr{M}_k\}}\mathbf{c}^+\mathbf{b}_k(0)\delta u_k(0)$$
$$- a_{k+m}\mathbf{1}_{\{1\notin\mathscr{M}_k\}}[\varphi_k(0) - w_k(1)].$$

Set

$$h_k = a_{k+m},$$
$$\psi_{i,j} = (1 - h_i\mathbf{1}_{\{1\notin\mathscr{M}_i\}}\mathbf{c}^+\mathbf{b}_i(0))\cdots(1 - h_j\mathbf{1}_{\{1\notin\mathscr{M}_j\}}\mathbf{c}^+\mathbf{b}_j(0)).$$

Then, we have

$$\delta u_{k+1}(0) = \psi_{k,0}\delta u_0(0) - \sum_{j=0}^{k}\psi_{k,j+1}h_j\mathbf{1}_{\{1\notin\mathscr{M}_j\}}\phi_j(0) + \sum_{j=0}^{k}\psi_{k,j+1}h_j\mathbf{1}_{\{1\notin\mathscr{M}_j\}}w_j(1).$$
(4.15)

To prove the uniform boundedness of $\delta u_k(0)$ for any $m > 0$ and fixed initial value $\delta u_0(0)$, it suffices to show that all the three terms at the RHS of (4.15) are uniformly bounded with respect to m.

Similar to the proof of Lemma 4.2, we have

$$|\psi_{i,j}| \leq c_4 \exp\left(-c\sum_{k=j}^{i}h_k\right), \forall i \geq j, \forall j \geq 0,$$
(4.16)

for some constants $c_4 > 0$ and $c > 0$.

Then, it is obvious that $|\psi_{k,0}| \leq c_4$, $\forall m > 0$, and hence

$$|\psi_{k,0}\delta u_0(0)| \leq c_4|\delta u_0(0)|.$$
(4.17)

This shows the uniform boundedness of the first term at the RHS of (4.15).

Since $\varphi_k(0) \xrightarrow[k\to\infty]{} 0$, $\sup_k |\varphi_k(0)| < \infty$. By noticing that $h_k \to 0$, there exists k_1 such that

$$h_k \leq 2\left(h_k - \frac{ch_k^2}{2}\right), \text{ and } 0 < ch_k < 1, \forall m > 0,$$

$\forall k \geq k_1$. Hence, it follows that

$$\left| \sum_{j=k_1}^{k} \psi_{k,j+1} h_j \mathbf{1}_{\{1 \notin \mathcal{M}_j\}} \varphi_j(0) \right|$$

$$\leq c_4 \sup_k |\varphi_k(0)| \sum_{j=k_1}^{k} \left[\exp\left(-c \sum_{i=j+1}^{k} h_i\right) h_j \right]$$

$$\leq 2c_4 \sup_k |\varphi_k(0)| \sum_{j=k_1}^{k} \left(h_j - \frac{ch_j^2}{2} \right) \exp\left(-c \sum_{i=j+1}^{k} h_i\right)$$

$$\leq \frac{2c_4}{c} \sup_k |\varphi_k(0)| \sum_{j=k_1}^{k} (1 - e^{-ch_j}) \exp\left(-c \sum_{i=j+1}^{k} h_i\right)$$

$$= \frac{2c_4}{c} \sup_k |\varphi_k(0)| \sum_{j=k_1}^{k} \left[\exp\left(-c \sum_{i=j+1}^{k} h_i\right) - \exp\left(-c \sum_{i=j}^{k} h_i\right) \right]$$

$$\leq \frac{2c_4}{c} \sup_k |\varphi_k(0)|.$$

Besides, it is clear that $|\sum_{j=0}^{k_1} \psi_{k,j+1} h_j \mathbf{1}_{\{1 \notin \mathcal{M}_j\}} \varphi_j(0)|$ is bounded. Thus, the uniform boundedness for the second term at the RHS of (4.15) is proved.

Now check the last term $\sum_{j=0}^{k} \psi_{k,j+1} h_j \mathbf{1}_{\{1 \notin \mathcal{M}_j\}} w_j(1)$. By the definition of h_k, it is clear that $\sum_{k=1}^{\infty} h_k = \infty$ and $\sum_{k=1}^{\infty} h_k^2 < \infty$. Let

$$s_k = \sum_{j=1}^{k} h_j \mathbf{1}_{\{1 \notin \mathcal{M}_j\}} w_j(1), \quad s_{-1} = 0. \tag{4.18}$$

By Assumption 4.5, it is clear that $s_k \to s < \infty$. Hence, for any $\varepsilon > 0$, there is k_2 such that $\|s_j - s\| \leq \varepsilon, \forall j \geq k_2$. By a partial summation

$$\sum_{j=0}^{k} \psi_{k,j+1} h_j \mathbf{1}_{\{1 \notin \mathcal{M}_j\}} w_j(1) = \sum_{j=0}^{k} \psi_{k,j+1}(s_j - s_{j-1})$$

$$= s_k - \sum_{j=0}^{k} (\psi_{k,j+1} - \psi_{k,j}) s_{j-1}$$

$$= s_k - \sum_{j=0}^{k} (\psi_{k,j+1} - \psi_{k,j}) s - \sum_{j=0}^{k} (\psi_{k,j+1} - \psi_{k,j})(s_{j-1} - s)$$

$$= s_k - s + \psi_{k,0} s - \sum_{j=0}^{k_2} (\psi_{k,j+1} - \psi_{k,j})(s_{j-1} - s)$$

$$- \sum_{j=k_2}^{k} \psi_{k,j+1} h_j \mathbf{1}_{\{1 \notin \mathcal{M}_j\}} \mathbf{c}^{\top} \mathbf{b}_k(0)(s_{j-1} - s),$$

4.3 Proofs of Lemmas

where all terms tend to zero as $k \to \infty$ except the last one. By noticing $\|s_j - s\| \leq \varepsilon$, $\forall j \geq k_2$, we have

$$\|\sum_{j=k_2}^{k} \psi_{k,j+1} h_j \mathbf{1}_{\{1 \notin \mathcal{M}_j\}} \mathbf{c}^+ \mathbf{b}_k(0)(s_{j-1} - s)\|$$

$$\leq \varepsilon \sup_{k} |\mathbf{c}^+ \mathbf{b}_k(0)| \cdot \|\sum_{j=k_2}^{k} \psi_{k,j+1} h_j\|,$$

which tends to zero as $k \to \infty$ and $\varepsilon \to 0$. Thus, the third term at the RHS of (4.15) is also uniformly bounded. The proof of this lemma is completed.

Proof of Lemma 4.4

Proof It is observed from Lemmas 4.2, 4.3, and first part proof of Theorem 4.2 that the algorithm (4.4)–(4.6) may have only finite number of truncations for $t = 0$.

If the algorithm stops at the correct direction, then by Lemma 4.2 the algorithm converges to the optimal control.

For the impossibility of stopping at the wrong direction, one may consider the following regression:

$$u_{k+1}(0) = u_k(0) - a_k \mathbf{1}_{\{1 \notin \mathcal{M}_k\}} \mathbf{c}^+ \mathbf{b}_k(0) \delta u_k(0)$$
$$- a_k \mathbf{1}_{\{1 \notin \mathcal{M}_k\}} \varphi_k(0) + a_k \mathbf{1}_{\{1 \notin \mathcal{M}_k\}} w_k(1). \tag{4.19}$$

The proof consists of two steps. The first is to show that $u_k(0)$ defined by (4.19) converges and the second is to show that $u_k(0)$ cannot converge to a point $u'(0) \neq u_d(0)$.

First, assume that $u_k(0)$ defined by (4.19) is not convergent. Take a Lyapunov function $\zeta(x) \triangleq (u_d(0) - x)^2$, then we have

$$0 \leq \liminf_{k \to \infty} \zeta(u_k(0)) < \limsup_{k \to \infty} \zeta(u_k(0)) < \infty. \tag{4.20}$$

We say that $\zeta(x_{n_k}), \ldots, \zeta(x_{m_k})$ down-cross the interval $[\delta_1, \delta_2]$, if $\zeta(x_{n_k}) \geq \delta_2$, $\zeta(x_{m_k}) \leq \delta_1$, and $\delta_1 < \zeta(x_j) < \delta_2, \forall j : n_k < j < m_k$. From (4.20), it follows that $\zeta(u_k(0))$ down-cross a nonempty interval $[\delta_1, \delta_2]$ infinitely many times. Without loss of generality, it is assumed $d([\delta_1, \delta_2], \zeta(u_d(0))) > 0$, where $d(\cdot, \cdot)$ denotes the distance.

Let $\zeta(u_{n_k}(0)), \ldots, \zeta(u_{m_k}(0)), k = 1, 2, \ldots$, be down-crossings.

By the boundedness of $u_k(0)$, without loss of generality, one may assume $u_{n_k}(0) \xrightarrow[k \to \infty]{} \bar{u}$.

Set $m(k, T) \triangleq \max\{m : \sum_{i=k}^{m} a_i \mathbf{1}_{\{1 \notin \mathcal{M}_i\}} \leq T\}$. For sufficiently large k and small enough T, we have

$$|u_{m+1}(0) - u_{n_k}(0)| \leq |\sum_{j=n_k}^{m} a_j \mathbf{1}_{\{1 \notin \mathscr{M}_j\}} \mathbf{c}^+ \mathbf{b}_j(0) \delta u_j(0)|$$

$$+ |\sum_{j=n_k}^{m} a_j \mathbf{1}_{\{1 \notin \mathscr{M}_j\}} \mathbf{c}^+ \mathbf{b}_j(0) \varphi_j(0)|$$

$$+ |\sum_{j=n_k}^{m} a_j \mathbf{1}_{\{1 \notin \mathscr{M}_j\}} \mathbf{c}^+ \mathbf{b}_j(0) w_j(1)|$$

$$\leq \sup_j |\mathbf{c}^+ \mathbf{b}_j(0) \delta u_j(0)| \sum_{j=n_k}^{m} a_j \mathbf{1}_{\{1 \notin \mathscr{M}_j\}} + o(\tau)$$

$$\leq c_5 \tau \quad \forall m : n_k \leq m \leq m(n_k, \tau), \forall \tau \in [0, T], \tag{4.21}$$

where c_5 is a positive constant and $o(\cdot)$ denotes infinitesimal of high order.

Set $\tau = a_{n_k} \mathbf{1}_{\{1 \notin \mathscr{M}_{n_k}\}}$ in the above expression, we have

$$|u_{n_k+1}(0) - u_{n_k}(0)| \leq c_5 a_{n_k} \mathbf{1}_{\{1 \notin \mathscr{M}_{n_k}\}} \xrightarrow[k \to \infty]{} 0. \tag{4.22}$$

By the definition of down-crossing $\zeta(u_{n_k}) \geq \delta_2 > \zeta(u_{n_k+1})$, it follows that

$$\zeta(u_{n_k}(0)) \xrightarrow[k \to \infty]{} \delta_2 = \zeta(\bar{u}), \quad d(\bar{u}, u_d(0)) \triangleq \delta > 0. \tag{4.23}$$

For small enough τ and sufficiently large k, we have

$$d(u_m(0), u_d(0)) \geq \frac{\delta}{2}, \quad \forall m : n_k \leq m \leq m(n_k, \tau). \tag{4.24}$$

Consequently, for sufficiently large k, it follows

$$\zeta(u_{m(n_k,\tau)+1}(0)) - \zeta(u_{n_k}(0))$$

$$= - \sum_{j=n_k}^{m(n_k,\tau)} a_j \mathbf{1}_{\{1 \notin \mathscr{M}_j\}} (y_d(1) - y_j(1)) \zeta_u(\bar{u}) + o(\tau)$$

$$= - \sum_{j=n_k}^{m(n_k,\tau)} a_j \mathbf{1}_{\{1 \notin \mathscr{M}_j\}} \mathbf{c}^+ \mathbf{b}_j(0) \delta u_j(0) \zeta_u(u_j(0))$$

$$+ \sum_{j=n_k}^{m(n_k,\tau)} a_j \mathbf{1}_{\{1 \notin \mathscr{M}_j\}} \mathbf{c}^+ \mathbf{b}_j(0) \delta u_j(0) (\zeta_u(u_j(0)) - \zeta_u(\bar{u}))$$

$$- \sum_{j=n_k}^{m(n_k,\tau)} a_j \mathbf{1}_{\{1 \notin \mathscr{M}_j\}} [\varphi_j(0) - w_j(1)] \zeta_u(\bar{u}) + o(\tau).$$

4.3 Proofs of Lemmas

Since $\varphi_k(0) \to 0$, we have that

$$\limsup_{k \to \infty} \left| \zeta_u(\bar{u}) \sum_{j=n_k}^{m(n_k,\tau)} a_j \mathbf{1}_{\{1 \notin \mathcal{M}_j\}}[\varphi_j(0) - w_j(1)] \right| = o(\tau), \tag{4.25}$$

and

$$\left| \sum_{j=n_k}^{m(n_k,\tau)} a_j \mathbf{1}_{\{1 \notin \mathcal{M}_j\}} \mathbf{c}^+ \mathbf{b}_j(0) \delta u_j(0)(\zeta_u(u_j(0)) - \zeta_u(\bar{u})) \right| = o(\tau). \tag{4.26}$$

By (4.24) and Assumption 4.3, there exists an $\alpha > 0$ such that

$$-\mathbf{c}^+ \mathbf{b}_j(0) \delta u_j(0) \zeta_u(u_j(0)) = 2\mathbf{c}^+ \mathbf{b}_j(0) [\delta u_j(0)]^2 > \alpha,$$
$$\forall j : n_k \leq j \leq m(n_k, \tau). \tag{4.27}$$

Combining the last four formulas leads to

$$\zeta(u_{m(n_k,\tau)+1}(0)) - \zeta(u_{n_k}(0)) \geq \beta\tau, \tag{4.28}$$

for some $\beta > 0$ and $\tau > 0$, if k is sufficiently large.

From (4.23) and (4.28), it follows that

$$\liminf_{k \to \infty} \zeta(u_{m(n_k,\tau)+1}(0)) \geq \delta_2 + \beta\tau. \tag{4.29}$$

On the other hand, from (4.21), it is observed that

$$\lim_{\tau \to 0} \max_{n_k \leq j \leq m(n_k,\tau)} |\zeta(u_{j+1}(0)) - \zeta(u_{n_k}(0))| = 0. \tag{4.30}$$

This means that $m(n_k, \tau) + 1 < m_k$ for small enough τ, and hence $\zeta(u_{m(n_k,\tau)+1}(0)) \in (\delta_1, \delta_2]$. This contradicts with (4.29). Hence, $\{\zeta(u_k(0))\}$ is impossible to have infinitely many down-crossings, and $\{u_k(0)\}$ is convergent.

Then, it comes to show that $\{u_k(0)\}$ cannot converge to a point $u'(0) \neq u_d(0)$. To show this, assume the converse: $u_k(0) \xrightarrow[k \to \infty]{} u'(0) \neq u_d(0)$.

From (4.19), it follows

$$u_{k_0+n}(0) = u_{k_0}(0) - \sum_{j=k_0}^{k_0+n-1} a_j \mathbf{1}_{\{1 \notin \mathcal{M}_j\}}[\mathbf{c}^+ \mathbf{b}_j(0) \delta u_j(0) + \varphi_j(0)]$$

$$+ \sum_{j=k_0}^{k_0+n-1} a_j \mathbf{1}_{\{1 \notin \mathcal{M}_j\}} w_j(1), \quad \forall n. \tag{4.31}$$

It is clear that

$$\sum_{j=k_0}^{\infty} a_j \mathbf{1}_{\{1 \notin \mathcal{M}_j\}} w_j(1) < \infty, \qquad (4.32)$$

and by Assumptions 4.2 and 4.4, $\varphi_j(0) \xrightarrow[j \to \infty]{} 0$. Since $u'(0) \neq u_d(0)$, $\mathbf{c}^+ \mathbf{b}_j(0) \delta u_j(0)$ would converge to a nonzero constant. Then, we have

$$\sum_{j=k_0}^{k_0+n-1} a_j \mathbf{1}_{\{1 \notin \mathcal{M}_j\}} [\mathbf{c}^+ \mathbf{b}_j(0) \delta u_j(0) + \varphi_j(0)] \xrightarrow[n \to \infty]{} \infty, \qquad (4.33)$$

which implies $u_{k_0+n}(0) \xrightarrow[n \to \infty]{} \infty$. This, however, contradicts with the boundedness of $\{u_k(0)\}$. Thus, the proof is completed.

4.4 Illustrative Simulations

Consider the following affine nonlinear system as an example, where the state is of two dimensions:

$$x_k^{(1)}(t+1) = 0.8 x_k^{(1)}(t) + 0.3 \sin(x_k^{(2)}(t)) + 0.23 u_k(t),$$
$$x_k^{(2)}(t+1) = 0.4 \cos(x_k^{(1)}(t)) + 0.85 x_k^{(2)}(t) + 0.33 u_k(t),$$
$$y_k(t) = x_k^{(1)}(t) + x_k^{(2)}(t) + w_k(t),$$

where $x_k^{(1)}(t)$ and $x_k^{(2)}(t)$ denote the first and second dimension of $x_k(t)$, respectively. It is easy to see that $\mathbf{c}^+ \mathbf{b}(t) = 0.23 \times 1 + 0.33 \times 1 = 0.56 > 0$, and thus $+1$ is the correct control direction.

For simple illustration, let $N = 40$, and the measurement noise $w_k(t)$ is assumed zero-Gaussian distributed, i.e., $w_k(t) \sim N(0, 0.1^2)$.

In order to simulate data dropouts, we first separate iteration steps into groups of four successive iterations, i.e., $\{1, 2, 3, 4\}$, $\{5, 6, 7, 8\}$, …, and then randomly select one iteration from each group for each t, say, 1, 6, 9, … for example. For these selected iterations, the control is not updated, which means the corresponding measurements are lost. Figure 4.2 shows an arbitrary illustration of data dropout iterations for the first 100 iterations, where 0 denotes that the packet is lost while 1 means that the packet is successfully transmitted.

The reference trajectory is $y_d(t) = 20 \sin(\frac{t}{20} \pi)$. The initial control action is simply given as $u_0(t) = 0, \forall t$. The learning gain chooses $a_k = \frac{1}{k+1}$ and the parameter used in the algorithm is $M_k = 4^k$. The algorithm has run 300 iterations.

The control direction regulation is demonstrated for time instant $t = 4$ in Fig. 4.3 as an illustration. For the example given here, the correct control direction is $+1$, and from the figure it is seen that the algorithm switches to the wrong control direction at

4.4 Illustrative Simulations

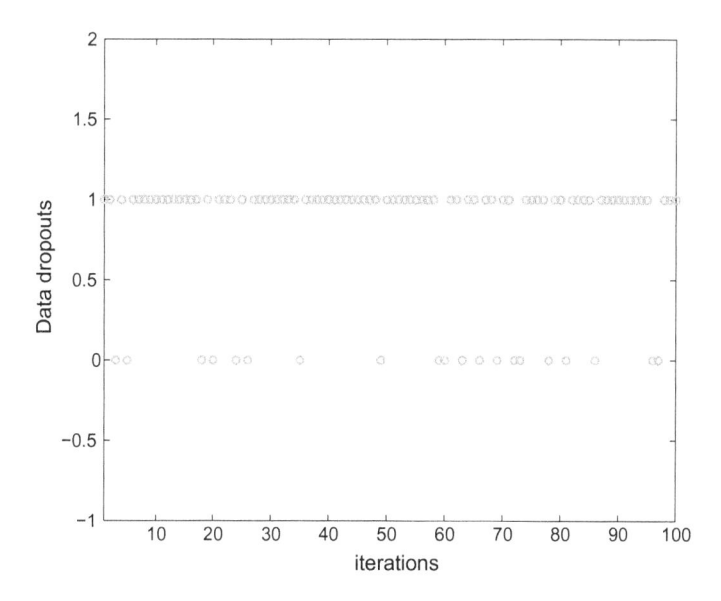

Fig. 4.2 Illustration of data dropout in the stochastic sequence case for an arbitrary time instant

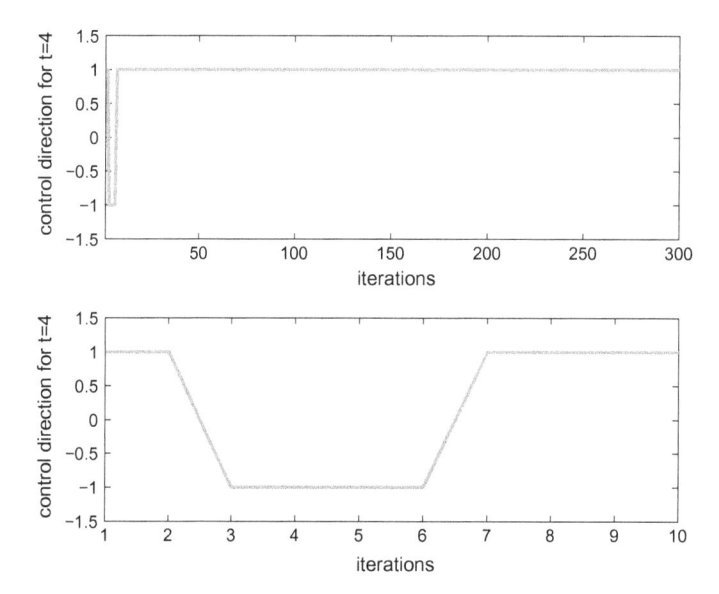

Fig. 4.3 Control direction regulation at time instant $t = 4$

the 3rd, 4th, 5th, and 6th iterations, and switches to the correct control direction since then. This shows that the algorithm could find its correct control direction adaptively. In Fig. 4.3, the upper figure shows the control direction for the whole 300 cycles, while the lower figure shows the first 10 cycles with enlarged scale.

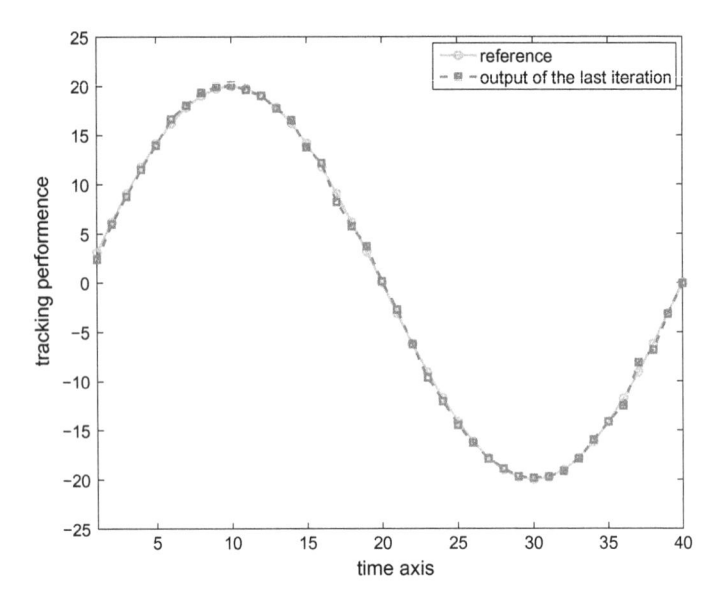

Fig. 4.4 $y_{300}(t)$ versus $y_d(t)$

The tracking result of the 300th iteration is shown in Fig. 4.4, where the solid line with circle is the reference signal and the dashed line with square denotes the actual output $y_{300}(t)$. In our study, the discrete-time case is considered and thus the reference trajectory actually is a set of values rather than a curve. It is noticed that the output could track the desired positions effectively, which shows the convergence and effectiveness of the proposed algorithms. Besides, the deviations in the figure are caused by stochastic measurement noises, which cannot be canceled by any learning algorithms since the noise is completely unpredictable.

Besides, the averaged absolute tracking error for each iteration is defined as $\sqrt{\frac{\sum_{t=1}^{N} \|e_k(t)\|^2}{N}}$. Noticing that stochastic noise is involved in the above index, thus the defined error could not decrease to zero as the number of iterations goes to infinity. Figure 4.5 shows that trajectory behaviors of the defined tracking error.

From the implementation of data dropout and Fig. 4.2, one could find that the maximum number of successive dropping iterations is 2, and thus in this case it is noted $K = 3$. In other words, during successive three iterations, at least in one iteration the measurement is successfully sent back. Besides, it is easy to compute that the data dropout data (DDR) is 25%, since only one iteration out of every four iterations misses its measurement data. One may be interested in the influence of K and the DDR, thus more cases of data dropouts should be simulated. To be specific, three more cases are considered: $K = 3$, DDR = 10%; $K = 3$, DDR = 50%; and $K = 7$, DDR = 75%. The averaged absolute tracking errors $\sqrt{\frac{\sum_{t=1}^{N} \|e_k(t)\|^2}{N}}$ for each case are shown in Fig. 4.6. It is worth mentioning that K denotes the worst level of

4.4 Illustrative Simulations   81

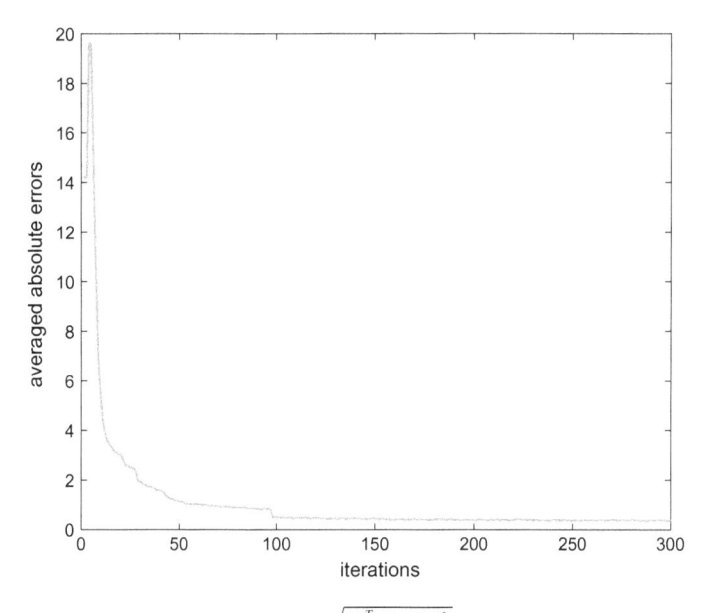

Fig. 4.5 The average absolute tracking error $\sqrt{\frac{\sum_{t=1}^{T}\|e_k(t)\|^2}{T}}$

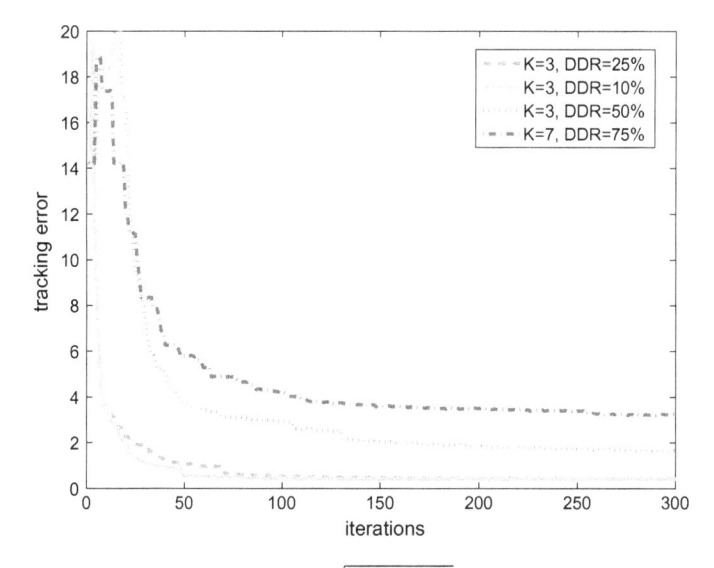

Fig. 4.6 The average absolute tracking error $\sqrt{\frac{\sum_{t=1}^{T}\|e_k(t)\|^2}{T}}$ for four different cases of K and DDR

data dropout, and DDR denotes the average loss level. As one could see, larger K and/or larger DDR would result in larger tracking errors.

At the end of this section, some remarks are listed as follows. The DDR has a great impact on the tracking performance. To be specific, the lower the DDR is, the better the tracking performance behaves, and the sooner the convergence speed is. If the rate is zero, which means no packet is lost, then the update algorithm becomes a common ILC algorithm. If the rate is 100%, which means no tracking information is fed back, then the update algorithm does not work anymore. For saving space, the simulation verification is omitted.

4.5 Summary

The ILC is considered for networked nonlinear systems with random measurement losses and unknown control direction. In this chapter and previous two chapters, the random data dropout is modeled by a random sequence model with bounded length requirement, which is different from many previous publications where the random data dropout is modeled by a binary Bernoulli random variable. Besides, the control direction, which plays an important role in the control design, is assumed unknown and a novel direction regulating approach is introduced in this chapter. Based on this direction regulating approach, the P-type update algorithm is proposed for SISO affine nonlinear system with stochastic measurement noises and the convergence in almost sure sense is strictly proved. The results of this chapter could be extended to the MIMO case by minor modifications. The results in this chapter are mainly based on [5].

References

1. Jiang, P., Chen, H., Bamforth, C.A.: A universal iterative learning stabilizer for a class of MIMO systems. Automatica **42**(6), 973–981 (2006)
2. Ahn, H.-S., Chen, Y.Q., Moore, K.L.: Iterative learning control: Survey and categorization from 1998 to 2004. IEEE Trans. Syst. Man Cybern. Part C **37**(6), 1099–1121 (2007)
3. Xu, J.-X., Yan, R.: On initial conditions in iterative learning control. IEEE Trans. Autom. Control **50**(9), 1349–1354 (2005)
4. Shen, D.: Iterative Learning Control for Nonlinear Stochastic Systems. Ph.D. thesis, Chinese Academy of Sciences (2010)
5. Shen, D., Wang, Y.: ILC for networked nonlinear systems with unknown control direction through random lossy channel. Syst. Control Lett. **77**, 30–39 (2015)

Chapter 5
Bernoulli Variable Model for Linear Systems

5.1 Problem Formulation

Consider the following discrete stochastic system:

$$\begin{aligned} x_k(t+1) &= A(t)x_k(t) + B(t)u_k(t) + w_k(t+1), \\ y_k(t) &= C(t)x_k(t) + v_k(t), \end{aligned} \quad (5.1)$$

where $k = 1, 2, \ldots$ are the different iteration numbers, $t = 0, 1, \ldots, N$ are the different time instances in an iteration, and N is the length of each iteration. $x_k(t) \in \mathbf{R}^n$, $u_k(t) \in \mathbf{R}^p$, and $y_k(t) \in \mathbf{R}^q$ are the state, input, and output of the system, respectively. $A(t)$, $B(t)$, and $C(t)$ are system matrices with appropriate dimensions. The random variables $w_k(t)$ and $v_k(t)$ are system noises and measurement noises, respectively.

Let $y_d(t)$, $t = 0, 1, \ldots, N$ be the tracking reference.

The following mild assumptions are given for the system (5.1).

Assumption 5.1 The input–output coupling matrix $C(t+1)B(t)$ is assumed to have a full-column rank for all t.

Assumption 5.2 For each time instant t, the independent and identically distributed (i.i.d.) sequence $\{w_k(t), k = 0, 1, \ldots\}$ is independent of the i.i.d. sequence $\{v_k(t), k = 0, 1, \ldots\}$ with $\mathbb{E}w_k(t) = 0$, $\mathbb{E}v_k(t) = 0$, $\sup_k \mathbb{E}\|w_k(t)\|^2 < \infty$, $\sup_k \mathbb{E}\|v_k(t)\|^2 < \infty$, $\lim_{n\to\infty} \frac{1}{n}\sum_{k=1}^n w_k(t)w_k^T(t) = R_w^t$, and $\lim_{n\to\infty} \frac{1}{n}\sum_{k=1}^n v_k(t)v_k^T(t) = R_v^t$, a.s., where R_w^t and R_v^t are unknown matrices.

Assumption 5.3 The initial state sequence $\{x_k(0)\}$ is i.i.d. with $\mathbb{E}x_k(0) = x_d(0)$, $\sup_k \mathbb{E}\|x_k(0)\|^2 < \infty$, and $\lim_{n\to\infty} \frac{1}{n}\sum_{k=1}^n x_k(0)x_k^T(0) = R_0$. Further, the sequences $\{x_k(0), k = 0, 1, \ldots\}$, $\{w_k(t), k = 0, 1, \ldots\}$, and $\{v_k(t), k = 0, 1, \ldots\}$ are mutually independent.

Remark 5.1 For any given initial value $x_d(0)$, the following expression of $u_d(t)$ can be computed recursively from the nominal model:

$$u_d(t) = [(C^+B(t))^T(C^+B(t))]^{-1}$$
$$\times (C^+B(t))^T(y_d(t+1) - C(t+1)A(t)x_d(t)),$$

where Assumption 5.1 is used and $C^+B(t) \triangleq C(t+1)B(t) \in \mathbf{R}^{q \times p}$. Evidently, the following equations are fulfilled:

$$\begin{aligned} x_d(t+1) &= A(t)x_d(t) + B(t)u_d(t), \\ y_d(t) &= C(t)x_d(t). \end{aligned} \quad (5.2)$$

Moreover, Assumption 5.1 further implies that the relative degree is one and the dimension of input is not larger than the dimension of output, i.e., $p \leq q$. Assumption 5.1 is used to guarantee the existence of the desired control that generates the desired tracking reference from the nominal model. Assumption 5.2 is a common condition of unknown random noises in stochastic control. The independence condition in Assumption 5.2 is required along the iteration axis, and thus, it is rational for practical applications because the process is repeatable. The initial resetting condition Assumption 5.3 enables a random initial shift around the desired initial state. The classical precise resetting condition of the initial state can be regarded as a special case of Assumption 5.3. To facilitate the expression, denote $w_k(0) = x_k(0) - x_d(0)$. Then, defining $\lim_{n \to \infty} \frac{1}{n} \sum_{k=1}^{n} w_k(0)w_k^T(0) = R_w^0$ is easy to satisfy the formulation of Assumption 5.2. In other words, Assumption 5.3 can be compressed into Assumption 5.2. Thus, all the assumptions are mild.

The setup of the control system is illustrated in Fig. 5.1, where the plant and learning controller are located separately and communicate via networks. The data may be dropped out through the networks because of network congestion, linkage interruption, and transmission error. However, to make the expression concise and without loss of any generality, data dropout is only considered for the output side. That is, random data dropouts only occur in the network from the measurement output to the memory, and the network from the learning controller to the control plant is assumed to work well.

Fig. 5.1 Block diagram of the networked iterative learning control

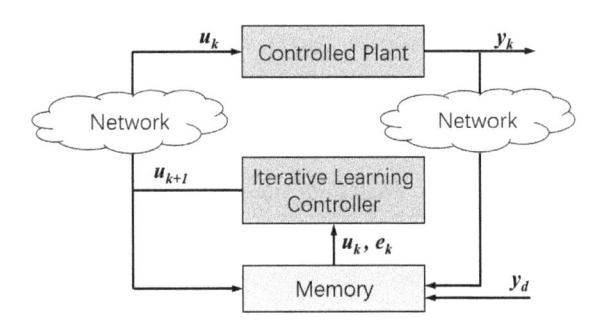

5.1 Problem Formulation

Similar to [1–5], we adopt a Bernoulli random variable to model the random data dropouts. Specifically, a random variable $\gamma_k(t)$ is introduced to indicate whether or not the measurement packet $y_k(t)$ is successfully transmitted,

$$\gamma_k(t) = \begin{cases} 1, & y_k(t) \text{ is successfully transmitted} \\ 0, & \text{otherwise} \end{cases} \tag{5.3}$$

and without loss of any generality,

$$\mathbb{P}(\gamma_k(t) = 1) = \rho, \quad \mathbb{P}(\gamma_k(t) = 0) = 1 - \rho, \tag{5.4}$$

where $0 < \rho < 1$. That is, the probability that the measurement $y_k(t)$ is successfully transmitted is ρ for all t and k.

Remark 5.2 The Bernoulli random variable has been used in many publications to describe random data dropouts. In the field of ILC, papers [1–5] are typical illustrations. However, in most cases, only convergence in the mathematical expectation sense and/or mean square sense is obtained. No result on convergence in almost sure sense has been reported until now. In previous chapters, the convergence in almost sure sense of ILC input sequence is proved strictly for stochastic systems. However, data dropout in previous chapters is modeled by the random sequence model (RSM) with a finite length requirement. A finite length requirement means that the data dropout is not completely stochastic. Therefore, the Bernoulli random variable model is revisited in this chapter, and the convergence in almost sure sense is expected to be obtained.

If the measurement packet is successfully transmitted, then one can compare it with the desired reference value and compute the tracking error $e_k(t) \triangleq y_d(t) - y_k(t)$ for the updating. Otherwise, no output information is received, and no tracking error can be obtained for further updating.

The conventional control objective for a deterministic system is to build an ILC algorithm that generates the input sequence, so that the actual output of the system $y_k(t)$ can track some given trajectory $y_d(t)$ asymptotically. However, when considering stochastic systems, system noises and measurement noises, which cannot be predicted and eliminated by any algorithm, are observed. Thus, we cannot expect that $y_k(t) \to y_d(t)$, $\forall t$, for stochastic systems, as the iteration number k goes to infinity. Therefore, for stochastic systems, the best achievable tracking performance is that the tracking error only consists of the noise terms. To this end, the control objective of this chapter is to design an ILC algorithm such that $u_k(t) \to u_d(t)$, $\forall t$, as $k \to \infty$.

Remark 5.3 As stochastic noises cannot be predicted and eliminated, an intuitive idea is to minimize the following averaged tracking error index:

$$V_t = \limsup_{n \to \infty} \frac{1}{n} \sum_{k=1}^{n} \|y_d(t) - y_k(t)\|^2. \tag{5.5}$$

Through simple calculations, the above index is found to be minimized if $u_k(t) \to u_d(t)$, $\forall t$ as $k \to \infty$. Thus, in what follows, we show the direct convergence of the input sequence to the desired input.

5.2 Intermittent Update Scheme and Its Almost Sure Convergence

To achieve this control objective under stochastic noises, we first consider the following intermittent update scheme (IUS):

$$u_{k+1}(t) = u_k(t) + a_k \gamma_k(t+1) L_t e_k(t+1), \tag{5.6}$$

where a_k is the learning step size and L_t is the learning gain matrix.

The learning step size $\{a_k\}$ is a decreasing sequence, and it should satisfy the following:

$$a_k > 0, \; a_k \to 0, \; \sum_{k=1}^{\infty} a_k = \infty, \; \sum_{k=1}^{\infty} a_k^2 < \infty. \tag{5.7}$$

Remark 5.4 The decreasing sequence is added to the algorithms as the stochastic system is considered. Clearly, $a_k = \alpha/k$ meets all the requirements in (5.7), where the constant $\alpha > 0$ can be regarded as a tuning parameter. This learning step size a_k is introduced to suppress the effect of stochastic noises as the iteration number goes to infinity and to guarantee a zero-error convergence of the input sequence. Specifically, the tracking error contains stochastic noises as its part. After enough learning iteration, the stochastic noise dominates the tracking error. If no mechanism is available to suppress the effect of stochastic noises, the algorithm will fail to converge to a stable limitation. Therefore, we add the decreasing sequence to the conventional P-type learning algorithm.

Remark 5.5 The inherent mechanism of IUS is that the algorithm updates the input when the output is successfully received and stops updating otherwise. That is, the input can keep the latest one if no new output is received; this is why it is called IUS. In addition, the updating frequency is equal to the successful transmission rate. Thus, the learning step size a_k goes to 0 fast when the data dropout rate is large, and a slow learning speed is obtained. An alternative to (5.6) is given as follows to improve the performance:

$$u_{k+1}(t) = u_k(t) + a_{\kappa_k(t)} \gamma_k(t+1) L_t e_k(t+1),$$

$$\kappa_k(t) = \sum_{i=1}^{k} \gamma_i(t+1).$$

5.2 Intermittent Update Scheme and Its Almost Sure Convergence

Now, we present the technical convergence analysis of the IUS. Compared with that of the successive update scheme (SUS) given in the next section, the proof of the IUS is more intuitive as this scheme will keep the input signal invariant if the corresponding output of the last iteration is lost. Thus, one only needs to focus on the updating iterations.

In observing the probability of (5.4), clearly, $\mathbb{E}\gamma_k(t) = \rho$ and $\mathbb{E}\gamma_k^2(t) = \rho$. Denote $\delta x_k(t) = x_d(t) - x_k(t)$ and $\delta u_k(t) = u_d(t) - u_k(t)$. Then, the update law (5.6) is rewritten as

$$\begin{aligned}
u_{k+1}(t) &= u_k(t) + a_k\gamma_k(t+1)L_t e_k(t+1) \\
&= u_k(t) + a_k\gamma_k(t+1)L_t(y_d(t+1) - y_k(t+1)) \\
&= u_k(t) + a_k\gamma_k(t+1)L_t C(t+1)\delta x_k(t+1) \\
&\quad - a_k\gamma_k(t+1)L_t v_k(t+1) \\
&= u_k(t) + a_k\gamma_k(t+1)L_t C^+ B(t)\delta u_k(t) \\
&\quad + a_k\gamma_k(t+1)L_t C^+ A(t)\delta x_k(t) \\
&\quad - a_k\gamma_k(t+1)L_t C(t+1)w_k(t+1) \\
&\quad - a_k\gamma_k(t+1)L_t v_k(t+1) \\
&= u_k(t) + a_k\rho L_t C^+ B(t)\delta u_k(t) \\
&\quad + a_k(\gamma_k(t+1) - \rho)L_t C^+ B(t)\delta u_k(t) \\
&\quad + a_k\gamma_k(t+1)L_t C^+ A(t)\delta x_k(t) \\
&\quad - a_k\gamma_k(t+1)L_t C(t+1)w_k(t+1) \\
&\quad - a_k\gamma_k(t+1)L_t v_k(t+1),
\end{aligned}$$

where $C^+A(t) \triangleq C(t+1)A(t)$.

Subtracting both sides of the last equation from $u_d(t)$ leads to

$$\begin{aligned}
\delta u_{k+1}(t) &= \delta u_k(t) - a_k\rho L_t C^+ B(t)\delta u_k(t) \\
&\quad - a_k(\gamma_k(t+1) - \rho)L_t C^+ B(t)\delta u_k(t) \\
&\quad - a_k\gamma_k(t+1)L_t C^+ A(t)\delta x_k(t) \\
&\quad + a_k\gamma_k(t+1)L_t C(t+1)w_k(t+1) \\
&\quad + a_k\gamma_k(t+1)L_t v_k(t+1).
\end{aligned} \quad (5.8)$$

In the following, the argument t or its specific value may be omitted if no confusion exists to make the expressions concise.

Theorem 5.1 *Consider the stochastic system (5.1) and update law (5.6). Design $L_t \in \mathbf{R}^{p \times q}$ such that all eigenvalues of $L_t C^+ B(t)$ are with positive real parts. Then, the input $u_k(t)$ generated by (5.6) converges to $u_d(t)$ in almost sure sense as k goes to infinity, $\forall t$.*

Proof The proof is performed by mathematical induction along the time axis t. The steps for $t = 1, 2, \ldots, N - 1$ are identical to those in the case of $t = 0$, which will be expressed in detail in the following.

Initial Step. Consider the case $t = 0$.
For $t = 0$, (5.8) is

$$\begin{aligned}\delta u_{k+1}(0) = &\delta u_k(0) - a_k \rho L_0 C^+ B(0) \delta u_k(0) \\ &- a_k(\gamma_k(1) - \rho) L_0 C^+ B(0) \delta u_k(0) \\ &- a_k \gamma_k(1) L_0 C^+ A(0) \delta x_k(0) \\ &+ a_k \gamma_k(1) L_0 C(1) w_k(1) + a_k \gamma_k(1) L_0 v_k(1).\end{aligned} \quad (5.9)$$

Considering Assumption 5.3 and Remark 5.1, we find that $\delta x_k(0) = w_k(0)$. On the other side, $\gamma_k(1)$ is independent of $\delta x_k(0)$. Both $\{\gamma_k(1)\}$ and $\{\delta x_k(0)\}$ are i.i.d. sequences along the iteration axis with a finite second moment. In addition, $\mathbb{E}\delta x_k(0) = \mathbb{E} x_k(0) - x_d(0) = 0$. Therefore, if we denote $\varepsilon_k \triangleq \gamma_k(1) L_0 C^+ A(0) \delta x_k(0)$, then $\{\varepsilon_k\}$ is an i.i.d. sequence with zero mean and a finite second moment.

Through direct calculations, we have

$$\sum_{k=1}^{\infty} \mathbb{E}(a_k \varepsilon_k)^2 = \sum_{k=1}^{\infty} \mathbb{E}(a_k \gamma_k(1) L_0 C^+ A(0) \delta x_k(0))^2$$

$$= \sum_{k=1}^{\infty} a_k^2 \cdot \mathbb{E}\gamma_k^2(1) \cdot L_0 C^+ A(0) \cdot \mathbb{E}(\delta x_k(0))^2$$

$$\leq c_0 \sum_{k=1}^{\infty} a_k^2 < \infty,$$

where $c_0 > 0$ is a suitable constant. Then, we have $\sum_{k=1}^{\infty} a_k \gamma_k(1) L_0 C^+ A(0) \delta x_k(0) < \infty$ in almost sure sense by the Khintchine–Kolmogorov convergence theorem [6].

Similarly, both $\{w_k(1)\}$ and $\{v_k(1)\}$ are i.i.d. sequences with zero means and finite second moments. In addition, they are independent of $\gamma_k(1)$. Clearly,

$$\sum_{k=1}^{\infty} \mathbb{E}(a_k \gamma_k(1) L_0 C(1) w_k(1))^2$$

$$\leq \|L_0 C(1)\|^2 \sum_{k=1}^{\infty} a_k^2 \cdot \mathbb{E}\gamma_k^2(1) \cdot \mathbb{E} w_k^2(1) < \infty,$$

$$\sum_{k=1}^{\infty} \mathbb{E}(a_k \gamma_k(1) L_0 v_k(1))^2$$

$$\leq \|L_0\|^2 \sum_{k=1}^{\infty} a_k^2 \cdot \mathbb{E}\gamma_k^2(1) \cdot \mathbb{E} v_k^2(1) < \infty,$$

5.2 Intermittent Update Scheme and Its Almost Sure Convergence

which further leads to $\sum_{k=1}^{\infty} a_k \gamma_k(1) L_0 C(1) w_k(1) < \infty$, $\sum_{k=1}^{\infty} \mathbb{E} a_k \gamma_k(1) L_0 v_k(1) < \infty$ in almost sure sense.

In the third term on the right-hand side of (5.9), $a_k(\gamma_k(1) - \rho) L_0 C^+ B(0) \delta u_k(0)$. The sequence of this term is no longer mutually independent, unlike the last three terms of (5.9). To deal with this term, let $\mathscr{F}_k$ be the increasing σ-algebra generated by $y_j(t), x_j(t), w_j(t), v_j(t), 0 \leq t \leq N, 0 \leq j \leq k$, i.e., $\mathscr{F}_k \triangleq \sigma\{y_j(t), x_j(t), w_j(t), v_j(t), \gamma_j(t), 0 \leq j \leq k, t \in \{0, \ldots, N\}\}$. According to the update law (5.6), $u_k(t) \in \mathscr{F}_{k-1}$. Note that $\gamma_k(1)$ is independent of $\mathscr{F}_{k-1}$ and is thus independent of $\delta u_k(0)$. Therefore,

$$\mathbb{E}\{(\gamma_k(1) - \rho) L_0 C^+ B(0) \delta u_k(0) | \mathscr{F}_{k-1}\}$$
$$= L_0 C^+ B(0) \delta u_k(0) \mathbb{E}\{\gamma_k(1) - \rho | \mathscr{F}_{k-1}\}$$
$$= 0.$$

That is, $(a_k(\gamma_k(1) - \rho) L_0 C^+ B(0) \delta u_k(0), \mathscr{F}_k, k \geq 1)$ is a martingale difference sequence. In addition,

$$\sum_{k=1}^{\infty} \mathbb{E}\{\|a_k(\gamma_k(1) - \rho) L_0 C^+ B(0) \delta u_k(0)\|^2 | \mathscr{F}_{k-1}\}$$
$$\leq \sum_{k=1}^{\infty} a_k^2 \mathbb{E}\{(\gamma_k(1) - \rho)^2 | \mathscr{F}_{k-1}\}$$
$$\leq \sup_k \|L_0 C^+ B(0) \delta u_k(0)\|^2 \sum_{k=1}^{\infty} a_k^2 \mathbb{E}(\gamma_k(1) - \rho)^2$$
$$\leq c_1 \sum_{k=1}^{\infty} a_k^2 < \infty,$$

where $c_1 > 0$ is a suitable constant. Then, by the Chow convergence theorem of martingale [6], we have $\sum_{k=1}^{\infty} a_k(\gamma_k(1) - \rho) L_0 C^+ B(0) \delta u_k(0) < \infty$ in almost sure sense.

If the learning gain matrix is designed such that all eigenvalues of $L_0 C^+ B(0)$ are with positive real parts, then $-\rho L_0 C^+ B(0)$ is clearly stable. Thus, applying Lemma A.3 to the recursion (5.9), we have $\delta u_k(0) \to 0$ as $k \to \infty$ in almost sure sense.

Inductive Step. Assume that the convergence of $u_k(t)$ has been proved for $t = 0, 1, \ldots, s - 1$ and that the purpose in this step is to show the convergence for $t = s$. From (5.1) and (5.2), we have

$$\delta x_k(s)$$
$$= A(s-1)\delta x_k(s-1) + B(s-1)\delta u_k(s-1) - w_k(s)$$
$$= A(s-1)A(s-2)\delta x_k(s-2) + A(s-1)B(s-2)\delta u_k(s-2)$$
$$+ B(s-1)\delta u_k(s-1) - A(s-1)w_k(s-1) - w_k(s)$$
$$= \sum_{i=0}^{s-1}\left(\prod_{j=i+1}^{s-1} A(j)\right) B(i)\delta u_k(i) - \sum_{i=0}^{s}\left(\prod_{j=i}^{s-1} A(j)\right) w_k(i),$$

where $\prod_{k=i}^{j} A(k) = A(j)A(j-1)\cdots A(i), \forall j \geq i$ and $\prod_{k=i}^{i-1} A(k) = I$. Replacing all t in (5.9) with s and substituting the above equation, we have

$$\delta u_{k+1}(s)$$
$$= \delta u_k(s) - a_k\rho L_s C^+ B(s)\delta u_k(s)$$
$$\quad - a_k(\gamma_k(s+1) - \rho)L_s C^+ B(s)\delta u_k(s)$$
$$\quad - a_k\gamma_k(s+1)L_s C^+ A(s)\sum_{i=0}^{s-1}\left(\prod_{j=i+1}^{s-1} A(j)\right) B(i)\delta u_k(i)$$
$$\quad - a_k\gamma_k(s+1)L_s C^+ A(s)\sum_{i=0}^{s}\left(\prod_{j=i}^{s-1} A(j)\right) w_k(i)$$
$$\quad + a_k\gamma_k(s+1)L_s(C(s+1)w_k(s+1) + v_k(s+1)). \tag{5.10}$$

Through the induction assumption, we have that $\delta u_k(t) \to 0$ as $k \to \infty$ in almost sure sense for $t = 0, 1, \ldots, s-1$. Thus, $\gamma_k(s+1)L_s C^+ A(s)\sum_{i=0}^{s-1}\left(\prod_{j=i+1}^{s-1} A(j)\right)$ $B(i)\delta u_k(i) \to 0$ in almost sure sense.

Following the same step as the initial step, we can obtain the following results:

$$\sum_{k=1}^{\infty} a_k(\gamma_k(s+1) - \rho)L_s C^+ B(s)\delta u_k(s) < \infty,$$

$$\sum_{k=1}^{\infty} a_k\gamma_k(s+1)L_s C^+ A(s)\sum_{i=0}^{s}\left(\prod_{j=i}^{s-1} A(j)\right) w_k(i) < \infty,$$

$$\sum_{k=1}^{\infty} a_k\gamma_k(s+1)L_s C(s+1)w_k(s+1) < \infty,$$

$$\sum_{k=1}^{\infty} a_k\gamma_k(s+1)L_s v_k(s+1) < \infty \text{ in almost sure sense.}$$

Again, using Lemma A.3, we can easily conclude that $\delta u_k(s) \to 0$ as $k \to \infty$. The proof is completed by using the mathematical induction method.

5.2 Intermittent Update Scheme and Its Almost Sure Convergence

Remark 5.6 As is shown in Theorem 5.1, the condition on the design of the learning matrix L_t is relaxed, because it can be solved by finding a feasible solution for an LMI $L_t C^+ B(t) > 0$. If the system matrices $C(t)$ and $B(t)$ are known, then an intuitive selection of L_t is $L_t = (C^+ B(t))^T$, which leads $L_t C^+ B(t)$ to become a positive definite matrix. In addition, this condition can be assured under some uncertainties of the system model.

Remark 5.7 The proof of Theorem 5.1 indicates that the condition in the initial state Assumption 5.3 can be replaced by the following: the initial state $x_k(0)$ is independent of stochastic noises $w_k(t)$ and $v_k(t)$. Moreover, the deviation between $x_k(0)$ and $x_d(0)$ approaches zero, i.e., $\delta x_k(0) \to 0$ as $k \to \infty$. By incorporating some learning strategy of initial state such as [7], the applications can be further enlarged.

Remark 5.8 With slight modifications of the above proof, the convergence in almost sure sense of the alternative algorithm proposed in Remark 5.5 is easily shown. If the learning step size a_k is eliminated from (5.6), i.e.,

$$u_{k+1}(t) = u_k(t) + \gamma_k(t+1) L_t e_k(t+1), \tag{5.11}$$

then the convergence in almost sure sense can also be obtained as long as L_t satisfies that the spectral norm of $I - \rho L_t C^+ B(t)$ is less than 1. The selection of L_t is more restrictive than the one given in Theorem 5.1. The tracking performance comparisons of the proposed algorithm (5.6) and the conventional P-type update law (5.11) are simulated in Sect. 5.5.

5.3 Successive Update Scheme and Its Almost Sure Convergence

In this section, we proceed to consider the SUS (i.e., successive update scheme) as follows:

$$u_{k+1}(t) = u_k(t) + a_k L_t e_k^*(t+1), \tag{5.12}$$

where a_k and L_t mean the same as those in the IUS, and $e_k^*(t)$ is the latest available tracking error defined as

$$e_k^*(t) = \begin{cases} e_k(t), & \text{if } \gamma_k(t) = 1, \\ e_{k-1}^*(t), & \text{if } \gamma_k(t) = 0. \end{cases} \tag{5.13}$$

The conditions on the learning step size $\{a_k\}$ is the same to those given in the last section; that is,

$$a_k > 0, a_k \to 0, \sum_{k=1}^{\infty} a_k = \infty, \sum_{k=1}^{\infty} a_k^2 < \infty. \tag{5.14}$$

Remark 5.9 The inherent mechanism of SUS is that the algorithm keeps updating by using the latest available packet. In other words, if the output of the last iteration is received, then the algorithm will update its input using this information. If the output of the last iteration is lost, then the algorithm will update its input using the latest available output packet received previously. That is, algorithm (5.12) can be rewritten as

$$\begin{aligned}u_{k+1}(t) &= u_k(t) + a_k \gamma_k(t+1) L_t e_k(t+1) \\ &\quad + a_k(1 - \gamma_k(t+1)) L_t e_{k-1}^*(t+1).\end{aligned} \tag{5.15}$$

Note that no finite length condition in the successive data dropouts is required. In fact, the number of successive data dropout iteration can be arbitrarily large.

Remark 5.10 The differences between IUS and SUS are listed as follows. First, IUS is an event-triggered updating, whereas SUS is an iteration-triggered updating. That is, IUS only updates its signal when the measurement output of the last iteration is received, whereas SUS updates its signal in every iteration. Moreover, the updating frequency of IUS depends on the rate of successful transmission and is usually low. Generally, the larger the data dropout rate is, the lower the updating frequency is. Conversely, SUS keeps updating all the time. In sum, the respective convergence speeds of IUS and SUS are markedly different.

In the following, we present the convergence analysis of the SUS. The proof of this case is more technically complex than that of the IUS, as the update information in (5.12) or (5.15) is no longer relatively definitive. In other words, if the measurement output of the last iteration is lost during transmission, then the one used in (5.12) will be unknown because of the possibility of successive data dropouts. Thus, update information can come from any previous iteration.

To form this situation, stochastic stopping times $\{\tau_k^t, k = 1, 2, \ldots, 0 \le t \le N\}$ are introduced to denote the random iteration delays of the update caused by random data dropouts. Then, the updating scheme (5.12) is rewritten as

$$u_{k+1}(t) = u_k(t) + a_k L_t e_{k-\tau_k^{t+1}}(t+1), \tag{5.16}$$

where the stopping time $\tau_k^{t+1} \le k$. In other words, for the updating of input at t of $(k+1)$th iteration, no information of $e_m(t+1)$ with $m > k - \tau_k^{t+1}$ is received and only $e_{k-\tau_k^{t+1}}(t+1)$ is available. According to the SUS settings, for the iterations $k - \tau_k^{t+1} < m \le k$, the input $u_m(t)$ is successively updated with the same error $e_{k-\tau_k(t+1)}(t+1)$.

The coupling of stochastic stopping times and the successive updating mechanism make the convergence analysis much complex. Thus, the analysis will be proved by

5.3 Successive Update Scheme and Its Almost Sure Convergence

two steps: we first show the convergence of (5.16) with $\tau_k^t = 0$, $\forall k, t$, and then we consider the effect of stopping times τ_k^t.

We first show the convergence of the following updating scheme:

$$u_{k+1}(t) = u_k(t) + a_k L_t e_k(t+1). \tag{5.17}$$

This case actually is the conventional ILC for systems without random data dropouts. We have the following theorem.

Theorem 5.2 *Consider the stochastic system (5.1) and update law (5.17). Design $L_t \in \mathbf{R}^{p \times q}$ such that all eigenvalues of $L_t C^+ B(t)$ are with positive real parts. Then, the input $u_k(t)$ generated by (5.17) converges to $u_d(t)$ in almost sure sense as k goes to infinity, $\forall t$.*

Proof The proof is carried out by mathematical induction similar to the proof of Theorem 5.1.

Initial Step. Consider the case $t = 0$.
For $t = 0$, subtracting both sides of (5.17) from $u_d(0)$ leads to

$$\begin{aligned}
\delta u_{k+1}(0) &= \delta u_k(0) - a_k L_0 e_k(1) \\
&= \delta u_k(0) - a_k L_0 C^+ B(0) \delta u_k(0) \\
&\quad - a_k L_0 C^+ A(0) \delta x_k(0) \\
&\quad + a_k L_0 C(1) w_k(1) + a_k L_0 v_k(1).
\end{aligned} \tag{5.18}$$

Similar to Theorem 1, $\sum_{k=1}^{\infty} a_k L_0 C^+ A(0) \delta x_k(0) < \infty$, $\sum_{k=1}^{\infty} a_k L_0 C(1) w_k(1) < \infty$, and $\sum_{k=1}^{\infty} a_k L_0 v_k(1) < \infty$, in almost sure sense. Note that $-L_0 C^+ B(0)$ is stable through the suitable selection of L_0. Thus, applying Lemma A.3 to the recursion (5.18), we have that $\delta u_k(0) \to 0$ as $k \to \infty$ in almost sure sense.

Inductive Step. Assume that the convergence of $u_k(t)$ has been proved for $t = 0, 1, \ldots, s-1$ and that the purpose in this step is to show the convergence for $t = s$. The following recursion is easy to establish:

$$\begin{aligned}
\delta u_{k+1}(s) &= \delta u_k(s) - a_k L_s C^+ B(s) \delta u_k(s) \\
&\quad - a_k L_s C^+ A(s) \sum_{i=0}^{s-1} \left(\prod_{j=i+1}^{s-1} A(j) \right) B(i) \delta u_k(i) \\
&\quad - a_k L_s C^+ A(s) \sum_{i=0}^{s} \left(\prod_{j=i}^{s-1} A(j) \right) w_k(i) \\
&\quad + a_k L_s C(s+1) w_k(s+1) + a_k L_s v_k(s+1).
\end{aligned} \tag{5.19}$$

Through the induction assumption, we have

$$L_s C^+ A(s) \sum_{i=0}^{s-1} \left(\prod_{j=i+1}^{s-1} A(j) \right) B(i) \delta u_k(i) \to 0$$

in almost sure sense and the infinite summation of the last three terms on the right-hand side of (5.19) is finite. Thus, using Lemma A.3 again leads to the conclusion. The proof is completed.

Now it comes to the general case (5.16).

Theorem 5.3 *Consider the stochastic system* (5.1) *and update law* (5.16). *Design $L_t \in \mathbf{R}^{p \times q}$ such that all eigenvalues of $L_t C^+ B(t)$ are with positive real parts. Then, the input $u_k(t)$ generated by* (5.16) *converges to $u_d(t)$ in almost sure sense as k goes to infinity, $\forall t$.*

Proof Comparing (5.16) and (5.17), the effect of the random data dropout is an additional error:

$$a_k L_t (e_k(t+1) - e_{k-\tau_k^{t+1}}(t+1)). \tag{5.20}$$

Thus, the objective of the proof is to show that the above term satisfies the condition (A.8). Specifically, we have

$$\begin{aligned} &a_k L_t (e_k(t+1) - e_{k-\tau_k^{t+1}}(t+1)) \\ &= a_k L_t C^+ B(t)[\delta u_k(t) - \delta u_{k-\tau_k^{t+1}}(t)] \\ &\quad + a_k L_t (C^+ A(t)[\delta x_k(t) - \delta x_{k-\tau_k^{t+1}}(t)] \\ &\quad - a_k L_t C(t+1)[w_k(t) - w_{k-\tau_k^{t+1}}(t)] \\ &\quad - a_k L_t [v_k(t) - v_{k-\tau_k^{t+1}}(t)]. \end{aligned}$$

Undoubtedly, the last two terms satisfy condition (A.8). Similar to the proofs in Theorems 5.1 and 5.2, they can be proved by mathematical induction that the second term on the right-hand side of the last equation satisfies condition (A.8). Thus, only the first term, i.e., $a_k L_t C^+ B(t)[\delta u_k(t) - \delta u_{k-\tau_k^{t+1}}(t)]$, is left for further analysis.

Recalling update (5.16), the difference is expanded to

$$\begin{aligned} &\delta u_k(t) - \delta u_{k-\tau_k^{t+1}}(t) \\ &= \sum_{m=k-\tau_k^{t+1}}^{k-1} a_m L_t C^+ B(t) \delta u_{m-\tau_m^{t+1}}(t) \\ &\quad + \sum_{m=k-\tau_k^{t+1}}^{k-1} a_m L_t C^+ A(t) \delta x_{m-\tau_m^{t+1}}(t) \\ &\quad - \sum_{m=k-\tau_k^{t+1}}^{k-1} a_m L_t C(t+1) w_{m-\tau_m^{t+1}}(t+1) \end{aligned}$$

5.3 Successive Update Scheme and Its Almost Sure Convergence

$$+ \sum_{m=k-\tau_k^{t+1}}^{k-1} a_m L_t v_{m-\tau_m^{t+1}}(t+1). \tag{5.21}$$

To analyze the effect of (5.21), we need to estimate the number of successive data dropout iterations, i.e., τ_k^t. As data dropouts are modeled by a Bernoulli random variable, τ_k^t obeys the geometric distribution. To make the notations concise, we let τ denote a random variable satisfying the same geometric distribution, i.e., $\tau \sim G(\rho)$. Clearly, $\mathbb{E}\tau = \frac{1}{\rho}$ and $\text{Var}(\tau) = \frac{1-\rho}{\rho^2}$. As $\text{Var}(\tau) = \mathbb{E}(\tau - \mathbb{E}\tau)^2$, then $\mathbb{E}\tau^2 = \frac{2-\rho}{\rho^2}$. By direct calculations, we have

$$\sum_{n=1}^{\infty} \mathbb{P}\{\tau \geq n^{\frac{1}{2}}\} = \sum_{n=1}^{\infty} \mathbb{P}\{\tau^2 \geq n\}$$

$$= \sum_{n=1}^{\infty} \sum_{j=n}^{\infty} \mathbb{P}\{j \leq \tau^2 < j+1\}$$

$$= \sum_{j=1}^{\infty} j \mathbb{P}\{j \leq \tau^2 < j+1\}$$

$$\leq \mathbb{E}\tau^2 < \infty.$$

Using the Borel–Cantelli lemma, we have

$$\mathbb{P}\{\tau \geq n^{\frac{1}{2}} \text{ i.o.}\} = 0.$$

Consequently, we have

$$\frac{\tau_n^t}{n} \to 0 \text{ a.s. } \forall t. \tag{5.22}$$

Therefore,

$$\frac{n - \tau_n^t}{n} \to 1 \text{ a.s. } \forall t, \tag{5.23}$$

and

$$n - \tau_n^t \to \infty \text{ a.s.} \tag{5.24}$$

Now, let us prove that the three terms on the right-hand side of (5.21) satisfy the condition (A.8) of Lemma A.3.

Through the same steps of the proof of Theorem 5.1, $\sum_{m=0}^{k-1} a_m L_t (C(t+1) w_{m-\tau_m^{t+1}}(t+1) - v_{m-\tau_m^{t+1}}(t+1))$ converges to an unknown constant a.s., $\forall t$. Therefore, in view of (5.24), we have

$$\left\| \sum_{m=k-\tau_k^{t+1}}^{k-1} a_m L_t (C(t+1) w_{m-\tau_m^{t+1}}(t+1) - v_{m-\tau_m^{t+1}}(t+1)) \right\| = o(1).$$

Therefore, the last term of (5.21) satisfies condition (A.8) of Lemma A.3.

Again, as the same steps of the proof for Theorem 5.1, the second term on the right-hand side of (5.21) can be split into two parts: a finite summation of input error of past time instants and a finite summation of stochastic noises.

$$\sum_{m=k-\tau_k^{t+1}}^{k-1} a_m L_t C^+ A(t) \delta x_{m-\tau_m^{t+1}}(t)$$

$$= \sum_{m=k-\tau_k^{t+1}}^{k-1} a_m L_t C^+ A(t) \sum_{i=0}^{t-1} \left(\prod_{j=i+1}^{t-1} A(j) \right) B(i) \delta u_{m-\tau_m^{t+1}}(i)$$

$$- \sum_{m=k-\tau_k^{t+1}}^{k-1} a_m L_t C^+ A(t) \sum_{i=0}^{t} \left(\prod_{j=i}^{t-1} A(j) \right) w_{m-\tau_m^{t+1}}(i).$$

In consideration of (5.24), the former part can be proven convergent to zero following the typical mathematical induction steps, and the latter part can be shown to satisfy condition (A.8) similar to the last term of (5.21).

The first term on the right-hand side of (5.21) is the only one that remains. This term can be almost surely bounded from above by a sample path dependent on constant times $\sum_{m=k-\tau_k^{t+1}}^{k-1} a_m$. Noting the selection of a_k, this term is then bounded by $c_0 a_{k-\tau_k^{t+1}} \tau_k^{t+1}$, where c_0 is a suitable constant. Thus, this quantity is $o(1)$. For simplicity, we directly select the standard step size of $a_k = \frac{1}{k}$. The general case is similar but with a complicated explanation. In this case,

$$a_{k-\tau_k^{t+1}} \tau_k^{t+1} = \frac{1}{k - \tau_k^{t+1}} \tau_k^{t+1}$$

$$= \frac{k}{k - \tau_k^{t+1}} \times \frac{1}{k} \tau_k^{t+1}$$

$$= O(\frac{1}{k}) O(k^{\frac{1}{2}})$$

$$= O(\frac{1}{k^{\frac{1}{2}}}) = o(1).$$

In sum, we show that the effect of random data dropouts, i.e., (5.21), satisfies condition (A.8). The convergence proof of this theorem is also achieved using the same steps of Theorem 5.1 and using Lemma A.3.

5.3 Successive Update Scheme and Its Almost Sure Convergence

Remark 5.11 Unlike in the IUS, the update is conducted in each iteration using the latest available information in the SUS. As shown in (5.16), two random factors are involved: the random data dropout and the uncertain length of successive iterations that data dropouts occur. As data dropout is described by a Bernoulli random variable and is independent in different iterations, the length of successive dropout iterations is not a bounded variable. Therefore, we have to make an intensive estimation of the effect of this factor. This part is the kernel step of our proof, and it is not dealt with in previous ILC results.

5.4 Mean Square Convergence of Intermittent Update Scheme

In this section, similar to Sect. 2.3, we present the mean square convergence of the IUS for linear systems with arbitrary relative degree using the lifting form. To this end, we recall the formulation of linear systems with relative degree τ, which is given in Sect. 2.3, as follows:

$$\begin{aligned} x_k(t+1) &= A_t x_k(t) + B_t u_k(t) + w_k(t+1), \\ y_k(t) &= C_t x_k(t) + v_k(t). \end{aligned} \quad (5.25)$$

Again, we assume the system relative degree is τ, $\tau \geq 1$; that is, for any $t \geq \tau$,

$$C_t A_{t+1-i}^{t-1} B_{t-i} = 0, \quad 1 \leq i \leq \tau - 1, \quad (5.26)$$

$$C_t A_{t+1-\tau}^{t-1} B_{t-\tau} \neq 0, \quad (5.27)$$

where $A_i^j \triangleq A_j A_{j-1} \cdots A_i$, $j \geq i$, and $A_{i+1}^i \triangleq I_n$.

Given the desired reference as $y_d(t)$, $t \in \{0, 1, \ldots, N\}$, without loss of any generality, we assume the reference is achievable; that is, with suitable initial value of $x_d(0)$, there exists a unique input $u_d(t)$ such that

$$\begin{aligned} x_d(t+1) &= A_t x_d(t) + B_t u_d(t), \\ y_d(t) &= C_t x_d(t). \end{aligned} \quad (5.28)$$

Then, the input $u_d(t)$ can be recursively computed from the nominal model (5.28) for $t \geq \tau$ as follows:

$$\begin{aligned} u_d(t-\tau) &= \left[\left(C_t A_{t+1-\tau}^{t-1} B_{t-\tau}\right)^T \left(C_t A_{t+1-\tau}^{t-1} B_{t-\tau}\right) \right]^{-1} \left(C_t A_{t+1-\tau}^{t-1} B_{t-\tau}\right)^T \\ &\quad \times \left(y_d(t) - C_t A_{t-\tau}^{t-1} x_d(t-\tau)\right). \end{aligned} \quad (5.29)$$

For clear notations, we still apply the assumptions given in Sect. 2.3, i.e., Assumptions 2.5 and 2.6. However, the derivations are valid for random initial state case (an extension of Assumption 2.5). For smooth reading, we copy these two assumptions here.

Assumption 5.4 The system initial value satisfies that $x_k(0) = x_d(0)$, where $x_d(0)$ is consistent with the desired reference $y_d(0)$ in the sense that $y_d(0) = C_0 x_d(0)$.

Assumption 5.5 The stochastic noises $\{w_k(t)\}$ and $\{v_k(t)\}$ are martingale difference sequences along the iteration axis with finite conditional second moments. That is, for $t \in \{0, 1, \ldots, N\}$, $\mathbb{E}\{w_{k+1}(t) \mid \mathscr{F}_k\} = 0$, $\sup_k \mathbb{E}\{\|w_{k+1}(t)\|^2 \mid \mathscr{F}_k\} < \infty$, $\mathbb{E}\{v_{k+1}(t) \mid \mathscr{F}_k\} = 0$, $\sup_k \mathbb{E}\{\|v_{k+1}(t)\|^2 \mid \mathscr{F}_k\} < \infty$, where the σ-algebra is defined as $\mathscr{F}_k = \sigma\{x_i(t), u_i(t), y_i(t), w_i(t), v_i(t), 1 \leq i \leq k, 0 \leq t \leq N\}$ (i.e., the set of all events induced by these random variables) for $k \geq 1$.

For the system, we employ the same lifting forms given in Sect. 2.3; that is, we use

$$U_k = \left[u_k^T(0), u_k^T(1), \ldots, u_k^T(N-\tau)\right]^T, \tag{5.30}$$

$$Y_k = \left[y_k^T(\tau), y_k^T(\tau+1), \ldots, y_k^T(N)\right]^T, \tag{5.31}$$

and the associated transfer matrix **H**

$$\mathbf{H} = \begin{bmatrix} C_\tau A_1^{\tau-1} B_0 & \mathbf{0}_{q \times p} & \cdots & \mathbf{0}_{q \times p} \\ C_{\tau+1} A_1^\tau B_0 & C_{\tau+1} A_2^\tau B_1 & \cdots & \mathbf{0}_{q \times p} \\ \vdots & \vdots & \ddots & \vdots \\ C_N A_1^{N-1} B_0 & C_N A_2^{N-1} B_1 & \cdots & C_N A_{N-\tau+1}^{N-1} B_{N-\tau} \end{bmatrix}. \tag{5.32}$$

Similarly, U_d and Y_d can be defined by replacing the subscript k in the above equations with d. Then, we have

$$Y_k = \mathbf{H} U_k + M x_k(0) + \xi_k \tag{5.33}$$

and

$$Y_d = \mathbf{H} U_d + M x_d(0), \tag{5.34}$$

where $M = [(C_\tau A_0^{\tau-1})^T, \ldots, (C_N A_0^{N-1})^T]^T$ and

$$\xi_k = \left[\left(\sum_{i=1}^{\tau} C_\tau A_i^{\tau-1} w_k(i) + v_k(\tau)\right)^T, \left(\sum_{i=1}^{\tau+1} C_{\tau+1} A_i^\tau w_k(i) + v_k(\tau+1)\right)^T, \ldots, \left(\sum_{i=1}^{N} C_N A_i^{N-1} w_k(i) + v_k(N)\right)^T\right]^T. \tag{5.35}$$

5.4 Mean Square Convergence of Intermittent Update Scheme

Denote the lifted tracking error $E_k \triangleq Y_d - Y_k$. Then, it is evident that

$$E_k = Y_d - Y_k = \mathbf{H}(U_d - U_k) - \xi_k. \tag{5.36}$$

5.4.1 Noise-Free System Case

In this subsection, we consider the case that the stochastic noises are absent in (5.25); that is, we consider the noise-free system

$$\begin{aligned} x_k(t+1) &= A_t x_k(t) + B_t u_k(t), \\ y_k(t) &= C_t x_k(t). \end{aligned} \tag{5.37}$$

The P-type ILC update law is designed as follows:

$$u_{k+1}(t) = u_k(t) + \theta \gamma_k(t+\tau) L_t e_k(t+\tau), \tag{5.38}$$

for $t = 0, \ldots, N - \tau$, where θ is a positive constant to be specified later and $L_t \in \mathbf{R}^{p \times q}$ is the learning gain matrix for regulating the control direction. It should note that $\gamma_k(t)$ is modeled by a Bernoulli random variable, which is different from that in Sect. 2.3.

Similarly, we lift the input along the time axis as (5.30). The update law (5.38) can be rewritten as follows:

$$U_{k+1} = U_k + \theta \Gamma_k \mathbf{L} E_k, \tag{5.39}$$

where Γ_k and $\mathbf{L}$ are denoted as

$$\Gamma_k = \begin{bmatrix} \gamma_k(\tau) I_q & & & \\ & \gamma_k(\tau+1) I_q & & \\ & & \ddots & \\ & & & \gamma_k(N) I_q \end{bmatrix}, \tag{5.40}$$

$$\mathbf{L} = \begin{bmatrix} L_0 & & & \\ & L_1 & & \\ & & \ddots & \\ & & & L_{N-\tau} \end{bmatrix}. \tag{5.41}$$

Obviously, $\Gamma_k = \text{diag}\{\gamma_k(\tau), \gamma_k(\tau+1), \ldots, \gamma_k(N)\} \otimes I_q$.

Recalling that $E_k = \mathbf{H}(U_d - U_k)$ and substituting this equation into (5.39), we have

$$U_{k+1} = U_k + \theta \Gamma_k \mathbf{L} \mathbf{H}(U_d - U_k). \tag{5.42}$$

We define $\Lambda_k \triangleq \Gamma_k \mathbf{L} \mathbf{H}$ and

$$\mathbf{LH} = \begin{bmatrix} L_0 C_\tau A_1^{\tau-1} B_0 & \mathbf{0}_p & \cdots & \mathbf{0}_p \\ L_1 C_{\tau+1} A_1^\tau B_0 & L_1 C_{\tau+1} A_2^\tau B_1 & \cdots & \mathbf{0}_p \\ \vdots & \vdots & \ddots & \vdots \\ L_{N-\tau} C_N A_1^{N-1} B_0 & L_{N-\tau} C_N A_2^{N-1} B_1 & \cdots & L_{N-\tau} C_N A_{N-\tau+1}^{N-1} B_{N-\tau} \end{bmatrix}. \quad (5.43)$$

Denote $\Lambda_k = \Gamma_k \mathbf{LH}$, which has κ possible outcomes from $\mathfrak{S} = \{\Lambda^{(1)}, \ldots, \Lambda^{(\kappa)}\}$. Following completely the same discussions in Sect. 2.3.1 and the learning gain matrix condition, we obtain that there exists a positive matrix $\mathbf{Q}$ such that

$$(\mathbf{LH})^T \mathbf{Q} + \mathbf{Q}\mathbf{LH} = I, \quad (5.44)$$

and

$$\left(\Lambda^{(i)}\right)^T \mathbf{Q} + \mathbf{Q}\Lambda^{(i)} \geq 0, \quad (5.45)$$

for $i = 2, \ldots, \kappa - 1$.

When considering the Bernoulli variable model (BVM), the technical lemma for the RSM (i.e., random sequence model) case in Sect. 2.3.1, i.e., Lemma 2.2, is no longer valid due to the inherent randomness of data dropouts. However, in such case, the statistical property of the random variable $\gamma_k(t)$ is valuable for establishing the convergence results. Moreover, in BVM, the data dropout variable $\gamma_k(t)$ is independent along the iteration axis; that is, for different iteration numbers $k \neq l$, $\gamma_k(t)$ is independent of $\gamma_l(t), \forall t$. Such independence will be used in the convergence analysis as follows.

Theorem 5.4 *Consider the noise-free linear system* (5.37) *and the ILC update law* (5.38), *where the random data dropouts follow BVM. Assume that Assumption 5.4 holds. Then, the input sequence* $\{u_k(t)\}, t = 0, \ldots, N - \tau$, *achieves mean square convergence to the desired input* $u_d(t), t = 0, \ldots, N - \tau$, *if the learning gain matrix* L_t *satisfies that* $-L_t C_{t+\tau} A_{t+1}^{t+\tau-1} B_t$ *is a Hurwitz matrix and* θ *is small enough.*

Proof We still apply the weighted norm of δU_k, $V_k = \|\delta U_k\|_{\mathbf{Q}} = (\delta U_k)^T \mathbf{Q} \delta U_k$, where $\mathbf{Q}$ is a positive definite matrix given in (5.44). Then, we have

$$V_{k+1} = (\delta U_{k+1})^T \mathbf{Q} \delta U_{k+1}$$
$$= (\delta U_k)^T (I - \theta \Lambda_k)^T \mathbf{Q} (I - \theta \Lambda_k) \delta U_k. \quad (5.46)$$

In BVM, the data dropout is independent along the iteration axis, while δU_k is constructed based on the information of the $(k-1)$th iteration, thus $I - \theta \Lambda_k$ is independent of δU_k in (5.46). Consequently, taking the mathematical expectation to both sides of (5.46) leads to that

5.4 Mean Square Convergence of Intermittent Update Scheme

$$\mathbb{E}\|\delta U_{k+1}\|_{\mathbf{Q}} = \mathbb{E}\left[(\delta U_k)^T(I-\theta\Lambda_k)^T\mathbf{Q}(I-\theta\Lambda_k)\delta U_k\right]$$
$$= \mathbb{E}\left[(\delta U_k)^T \mathbb{E}\left((I-\theta\Lambda_k)^T\mathbf{Q}(I-\theta\Lambda_k)\right)\delta U_k\right]. \quad (5.47)$$

Notice that

$$\mathbb{E}\left((I-\theta\Lambda_k)^T\mathbf{Q}(I-\theta\Lambda_k)\right) = \mathbb{E}\left(\mathbf{Q}-\theta(\Lambda_k^T\mathbf{Q}+\mathbf{Q}\Lambda_k^T)+\theta^2\Lambda_k^T\mathbf{Q}\Lambda_k^T\right)$$
$$= \mathbf{Q}-\theta\mathbb{E}(\Lambda_k^T\mathbf{Q}+\mathbf{Q}\Lambda_k^T)+\theta^2\mathbb{E}\Lambda_k^T\mathbf{Q}\Lambda_k^T. \quad (5.48)$$

Recalling the definition of Λ_k^T, it is evident that $\mathbb{E}\Lambda_k = \rho\mathbf{LH}$. Incorporating with (5.44) leads to that

$$\mathbb{E}(\Lambda_k^T\mathbf{Q}+\mathbf{Q}\Lambda_k^T) = \rho I. \quad (5.49)$$

On the other hand, there exists a suitable constant $c_1 > 0$ such that

$$\mathbb{E}\Lambda_k^T\mathbf{Q}\Lambda_k^T = \sum_{i=1}^{\kappa}\mathbb{P}(\Lambda_k = \Lambda^{(i)})(\Lambda^{(i)})^T\mathbf{Q}\Lambda^{(i)} \leq c_1 I, \quad (5.50)$$

where $\mathbb{P}(\Lambda_k = \Lambda^{(i)})$ denotes the probability that Λ_k is valued to be $\Lambda^{(i)}$ and $\sum_{i=1}^{\kappa}\mathbb{P}(\Lambda_k = \Lambda^{(i)}) = 1$.

From (5.48), (5.49), and (5.50), it follows that

$$\mathbb{E}\left((I-\theta\Lambda_k)^T\mathbf{Q}(I-\theta\Lambda_k)\right) \leq \mathbf{Q}-\theta(\rho-\theta c_1)I. \quad (5.51)$$

Note that $\mathbf{Q}$ is a positive definite matrix, thus there exist $c_2 > 0$ such that $c_2\mathbf{Q} \leq I$. Substituting (5.51) into (5.47), we have

$$\mathbb{E}\|\delta U_{k+1}\|_{\mathbf{Q}} \leq \left(1-\theta(\rho-\theta c_1)c_2\right)\mathbb{E}\|\delta U_k\|_{\mathbf{Q}}, \quad (5.52)$$

and consequently, we have a contraction mapping of $\mathbb{E}\|\delta U_k\|_{\mathbf{Q}}$ as

$$\mathbb{E}\|\delta U_{k+1}\|_{\mathbf{Q}} \leq \eta_1 \mathbb{E}\|\delta U_k\|_{\mathbf{Q}}, \quad 0 < \eta_1 < 1,$$

where $\eta_1 \triangleq 1-\theta(\rho-\theta c_1)c_2$, as long as we select the parameter θ small enough such that $\rho - \theta c_1 > 0$ and $\theta\rho c_2 < 1$.

Consequently, we have that $\mathbb{E}\|\delta U_k\|_{\mathbf{Q}} \to 0$ as $k \to \infty$. Note that $\mathbf{Q}$ is a positive definite matrix, thus $\lim_{k\to\infty}\mathbb{E}\|\delta U_k\| = 0$. The mean square convergence is thus established. This completes the proof.

Remark 5.12 Note that there is a strict contraction of the weighted norm $\|\delta U_k\|_{\mathbf{Q}}$. Similar to Remark 2.8, we can derive that the infinite summation of $\mathbb{E}\|\delta U_k\|$ is bounded, and therefore, the almost sure convergence can also be proved by using Borel–Cantelli lemma. The details are similar to Remark 2.8.

Remark 5.13 The condition on parameter θ is given by two inequalities, $\rho - \theta c_1 > 0$ and $\theta \rho c_2 < 1$, which leads to $\theta < \rho c_1^{-1}$ and $\theta < \rho^{-1} c_2^{-1}$. Since $\rho < 1$, the second range can be reduced to $\theta < c_2^{-1}$. Thus, $\theta < \min\{\rho c_1^{-1}, c_2^{-1}\}$. From this formulation, we find that the successful transmission rate, i.e., the average level of data dropouts along the iteration axis, has an important influence on the selection of the parameter θ. Roughly speaking, the smaller the successful transmission rate ρ, the smaller the parameter θ. Meanwhile, as we have explained above, smaller selection of θ renders a slower convergence speed. This observation coincides with our intuitive recognition of the phenomenon that heavy data dropouts would lead to a slower convergence of the ILC algorithms.

5.4.2 Stochastic System Case

In the subsection, we come back to the stochastic system (5.25). In such case, the independence of data dropouts would help to establish the convergence similar to the last subsection. The update law is copied here,

$$u_{k+1}(t) = u_k(t) + a_k \gamma_k(t+\tau) L_t e_k(t+\tau), \tag{5.53}$$

where the learning step size $\{a_k\}$ is a decreasing sequence satisfying that

$$a_k \in (0,1), \quad a_k \to 0, \quad \sum_{k=1}^{\infty} a_k = \infty, \quad \sum_{k=1}^{\infty} a_k^2 < \infty, \quad \frac{1}{a_{k+1}} - \frac{1}{a_k} \to \chi > 0. \tag{5.54}$$

The lifted form is defined as

$$U_{k+1} = U_k + a_k \Gamma_k \mathbf{L} E_k. \tag{5.55}$$

Theorem 5.5 *Consider the stochastic linear system (5.25) and the ILC update law (5.53), where the random data dropouts follow BVM. Assume Assumptions 5.4 and 5.5 hold. Then, the input sequence $\{u_k(t)\}$, $t = 0, \ldots, N-\tau$, achieves mean square convergence to the desired input $u_d(t)$, $t = 0, \ldots, N-\tau$, if the learning gain matrix L_t satisfies that $-L_t C_{t+\tau} A_{t+1}^{t+\tau-1} B_t$ is a Hurwitz matrix.*

Proof Let us recall the update law (5.55) as follows:

$$U_{k+1} = U_k + a_k \Gamma_k \mathbf{L} E_k.$$

Subtracting both sides of the last equation from U_d, we have

5.4 Mean Square Convergence of Intermittent Update Scheme

$$\begin{aligned}\delta U_{k+1} &= \delta U_k - a_k \Gamma_k \mathbf{L} E_k \\ &= \delta U_k - a_k \Gamma_k \mathbf{LH} \delta U_k + a_k \Gamma_k \xi_k \\ &= \delta U_k - a_k \rho \mathbf{LH} \delta U_k + a_k (\rho I - \Gamma_k) \mathbf{LH} \delta U_k + a_k \Gamma_k \xi_k. \end{aligned} \quad (5.56)$$

Note that ρI is the mathematical expectation of Γ_k. Now let us apply the weighted norm of δU_k, $V_k = \|\delta U_k\|_{\mathbf{Q}}$,

$$\begin{aligned}V_{k+1} &= \delta U_{k+1}^T \mathbf{Q} \delta U_{k+1} \\ &= \delta U_k^T \mathbf{Q} \delta U_k + a_k^2 \rho^2 \delta U_k^T (\mathbf{LH})^T \mathbf{QLH} \delta U_k + a_k^2 \xi_k^T \Gamma_k^T \mathbf{Q} \Gamma_k \xi_k \\ &\quad + a_k^2 \delta U_k^T (\mathbf{LH})^T (\rho I - \Gamma_k) \mathbf{Q} (\rho I - \Gamma_k)(\mathbf{LH}) \delta U_k \\ &\quad - a_k \rho \delta U_k^T \left[(\mathbf{LH})^T \mathbf{Q} + \mathbf{QLH}\right] \delta U_k \\ &\quad + 2 a_k \left(\delta U_k - a_k \rho \mathbf{LH} \delta U_k\right)^T \mathbf{Q} (\rho I - \Gamma_k)(\mathbf{LH}) \delta U_k \\ &\quad + 2 a_k \left(\delta U_k - a_k \rho \mathbf{LH} \delta U_k\right)^T \mathbf{Q} \Gamma_k \xi_k \\ &\quad + 2 a_k^2 \delta U_k^T (\mathbf{LH})^T (\rho I - \Gamma_k) \mathbf{Q} \Gamma_k \xi_k. \end{aligned} \quad (5.57)$$

Note that U_k is constructed on the basis of the data from the $(k-1)$th iteration, thus it is independent of the data dropout variable at the kth iteration, i.e., Γ_k. This fact gives that

$$\mathbb{E}\left[\left(\delta U_k - a_k \rho \mathbf{LH} \delta U_k\right)^T \mathbf{Q} (\rho I - \Gamma_k)(\mathbf{LH}) \delta U_k\right] = 0. \quad (5.58)$$

Similarly, the independence of U_k, Γ_k and ξ_k yields

$$\mathbb{E}\left[\left(\delta U_k - a_k \rho \mathbf{LH} \delta U_k\right)^T \mathbf{Q} \Gamma_k \xi_k\right] = 0, \quad (5.59)$$

$$\mathbb{E}\left[\delta U_k^T (\mathbf{LH})^T (\rho I - \Gamma_k) \mathbf{Q} \Gamma_k \xi_k\right] = 0, \quad (5.60)$$

where Assumption 5.5 is applied.

Taking mathematical expectation to both sides of (5.57) and substituting (5.58)–(5.60) as well as the Lyapunov equation $(\mathbf{LH})^T \mathbf{Q} + \mathbf{QLH} = I$, we have

$$\begin{aligned}\mathbb{E} V_{k+1} &= \mathbb{E} V_k - a_k \rho \mathbb{E} \delta U_k^T \delta U_k + a_k^2 \rho^2 \mathbb{E}\left[\delta U_k^T (\mathbf{LH})^T \mathbf{QLH} \delta U_k\right] \\ &\quad + a_k^2 \mathbb{E}\left[\xi_k^T \Gamma_k^T \mathbf{Q} \Gamma_k \xi_k\right] + a_k^2 \mathbb{E}\left[\delta U_k^T (\mathbf{LH})^T (\rho I - \Gamma_k) \mathbf{Q}(\rho I - \Gamma_k)(\mathbf{LH}) \delta U_k\right]. \end{aligned} \quad (5.61)$$

According to Assumption 5.5, there exists a suitable constant $c_3 > 0$ such that

$$\mathbb{E}\left[\xi_k^T \Gamma_k^T \mathbf{Q} \Gamma_k \xi_k\right] < c_3. \quad (5.62)$$

Moreover, due to the positive definite property of **Q**, there are $c_4 > 0$ and $c_5 > 0$ such that

$$\mathbb{E}\left[\delta U_k^T (\mathbf{LH})^T \mathbf{Q} \mathbf{LH} \delta U_k\right] \le c_4 \mathbb{E} V_k, \tag{5.63}$$

$$\mathbb{E}\left[\delta U_k^T (\mathbf{LH})^T (\rho I - \Gamma_k) \mathbf{Q} (\rho I - \Gamma_k)(\mathbf{LH}) \delta U_k\right] \le c_5 \mathbb{E} V_k. \tag{5.64}$$

Substituting (5.62)–(5.64) and the inequality $c_2 \mathbf{Q} \le I$ into (5.61) leads to

$$\mathbb{E} V_{k+1} \le (1 - a_k \rho) \mathbb{E} V_k + a_k^2 \left(c_3 + (\rho^2 c_4 + c_5) \mathbb{E} V_k\right). \tag{5.65}$$

Then, it is evident that $\mathbb{E} V_k$ corresponds to ϑ_k in Lemma A.1. Applying Lemma A.1 it follows that $\lim_{k \to \infty} \mathbb{E} V_k = 0$, which further implies that $\mathbb{E} \|\delta U_k\|^2 = 0$ by the fact that **Q** is a positive definite matrix. The mean square convergence is thus obtained. This completes the proof.

Remark 5.14 Based on the above mean square convergence results, we can show the almost sure convergence with the help of Lemma A.2 as follows. Taking conditional expectation to both sides of (5.57) with respect to $\mathscr{F}'_{k-1}$, it follows

$$\mathbb{E}(V_{k+1} \mid \mathscr{F}'_{k-1}) \le V_k + a_k^2 (c_3 + (\rho^2 c_4 + c_5) V_k), \tag{5.66}$$

where $\mathscr{F}'_k = \sigma\{x_i(t), u_i(t), y_i(t), w_i(t), v_i(t), \gamma_i(t), 1 \le i \le k, 0 \le t \le N\}$. The condition (A.6) is easy to verify for the last term of the above inequality with the help of the mean square convergence. Therefore, by using Lemma A.2, it gives that δU_k converges almost surely. Then, the almost sure convergence of the input sequence $\{U_k\}$ is verified.

5.5 Illustrative Simulations

5.5.1 System Description

Let us consider a direct current motor control problem for velocity tracking. The dynamics of a permanent magnet linear motor (PMLM) is described as follows [8]:

$$\begin{cases} \dot{x}(t) = v(t), \\ u(t) = k_1 \psi_f \dot{x}(t) + Ri(t) + L\dot{i}(t), \\ f_l(t) = m\dot{v}(t) + f_{fri}(t) + f_{rip}(t) + f_{loa}(t) + f_w(t), \end{cases} \tag{5.67}$$

where $k_1 = \pi/\varsigma$. The definitions of the notations are listed in Table 5.1.

Following the simplification procedures in [8], the PMLM model can be transformed into the following case:

5.5 Illustrative Simulations

Table 5.1 List of notations

Notation	Meaning	Notation	Meaning
$x(t)$	Motor position	$f_l(t)$	Developed force
$v(t)$	Rotor velocity	$u(t)$	Stator voltage
$i(t)$	Current of stator	R	Resistance of stator
L	Inductance of stator	ς	Pole pitch
ψ_f	Flux linkage	m	Rotor mass
$f_{fri}(t)$	Frictional force	$f_{rip}(t)$	Ripple force
$f_{loa}(t)$	Applied load force	$f_w(t)$	Uncertainties/disturbances

$$\begin{cases} \dot{x}(t) = v(t), \\ \dot{v}(t) = -\frac{k_1 k_2 \psi_f^2}{Rm} v(t) + \frac{k_2 \psi_f}{Rm} u(t), \\ y(t) = v(t), \end{cases} \quad (5.68)$$

where $k_2 = 1.5\pi/\varsigma$.

According to our formulations, the discrete time interval is set to $\Delta = 10$ ms, and the whole iteration length is 1 s. That is, $N = 100$. The system is discretized using the Euler method. Taking the disturbances and noises into consideration, the following stochastic system is obtained:

$$\begin{cases} x(t+1) = x(t) + v(t)\Delta + \varepsilon_1(t+1), \\ v(t+1) = v(t) - \Delta \frac{k_1 k_2 \psi_f^2}{Rm} v(t) + \Delta \frac{k_2 \psi_f}{Rm} u(t) + \varepsilon_2(t+1), \\ y(t) = v(t) + \varepsilon(t), \end{cases} \quad (5.69)$$

where the parameters are given as follows: $\varsigma = 0.031$ m, $R = 8.6\,\Omega$, $m = 1.635$ kg, and $\psi_f = 0.35$ Wb. The product of the input/output coupling value is 0.0378.

The reference trajectory is $y_d(t) = 1/3(\sin(t/20) + 1 - \cos(3t/20))$, $0 \le t \le 100$. The initial control action is simply given as $u_0(t) = 0$, $\forall t$.

The stochastic noises $\varepsilon_1(t)$, $\varepsilon_2(t)$, and $\varepsilon(t)$ obey normal distribution $N(0, 0.02^2)$. Note that the upper bound of the tracking reference is small, and thus large noises dominate the output. To make the convergence clear, the stochastic noises here are set with a small derivation.

In the following, we present in detail the performance of the proposed algorithms and associated comparisons. Specifically, the tracking performances of IUS and SUS are illustrated in Sect. 5.5.2, which shows the advantage of SUS compared with IUS. The comparisons under different data dropout rates (DDRs) and different learning gains for both IUS and SUS are detailed in Sects. 5.5.3 and 5.5.4, respectively. The comparison between the proposed algorithms and the conventional P-type algorithm is discussed in Sect. 5.5.5, which reveals the effect of decreasing sequence $\{a_k\}$. These comparisons illustrate the application information of the proposed algorithms.

5.5.2 Tracking Performance of both Schemes

In this subsection, we first verify the convergence properties of the proposed updating schemes and then compare IUS and SUS. Any positive number L_t can satisfy the conditions in the convergence theorems.

We first set the probability as $\rho = 0.9$. That is, for any data packet, the probability of a successful transmission is 90%. To make expressions clear, let $\gamma = 1 - \rho$ denote the DDR, which is the average ratio of lost data to the whole data. In this case, the DDR is 10%, which is low. The decreasing sequence is set to $a_k = 1/(k+10)$, the learning gain is $L_t = 55$, and algorithms (5.6) and (5.12) are run for 300 iterations.

The final outputs of both algorithms are shown in Fig. 5.2, where the solid, dash, and dash-dotted lines denote the tracking reference, the final output of IUS, and the final output of SUS, respectively. As shown in the figure, three lines are almost coincident. Therefore, both schemes can converge to the desired target quickly under low DDR and stochastic noises. Moreover, the performances of both schemes are close to each other.

To understand the performances, we further plot the averaged absolute tracking error along the iteration axis. The average absolute tracking error (AATE) is defined as $\|e_k\| = (\sum_{t=1}^{N} \|e_k(t)\|)/N$ for the kth iteration. As shown in Fig. 5.3, little difference is observed between both schemes, thus confirms the above findings. Generally, the lower the DDR is, the closer the performances of IUS and SUS are. If no data dropout exists, then IUS and SUS will be the same.

Subsequently, we set $\rho = 0.3$, or equivalently $\gamma = 70\%$, to further compare the performance of both schemes. This percentage implies that the transmission

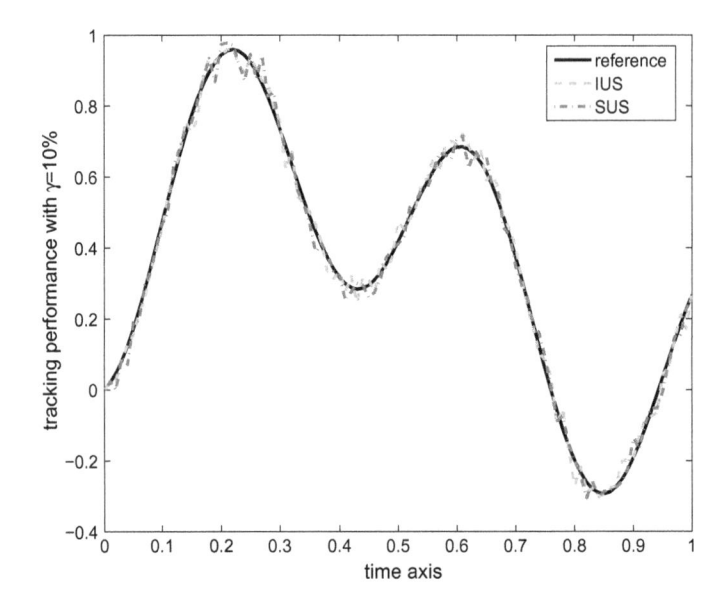

Fig. 5.2 Tracking performances of IUS and SUS at the final iteration: DDR $= 10\%$

5.5 Illustrative Simulations

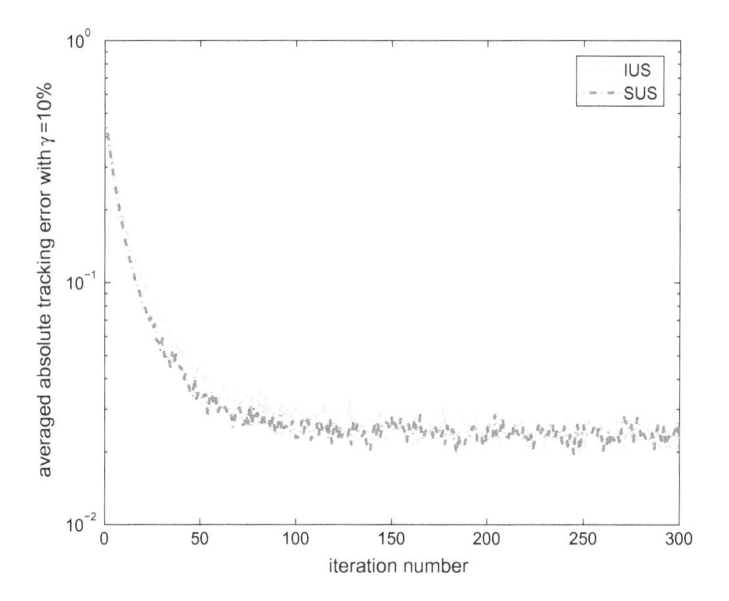

Fig. 5.3 AATE of IUS and SUS along the iteration axis: DDR = 10%

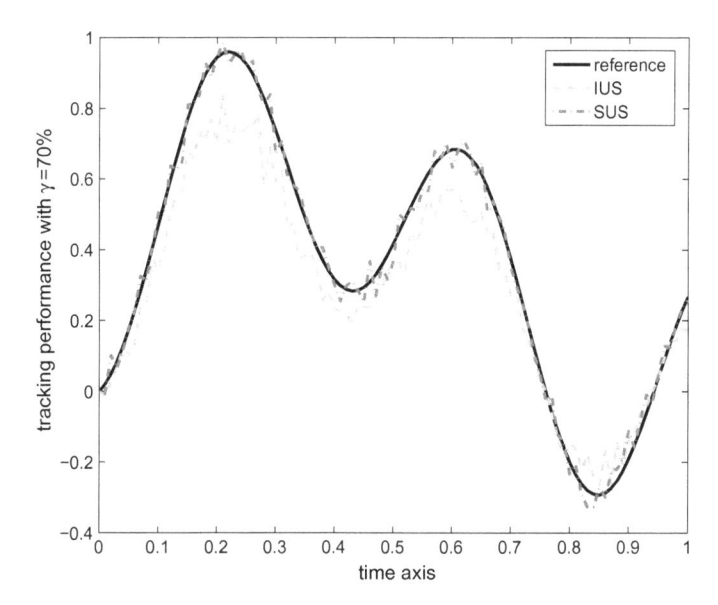

Fig. 5.4 Tracking performances of IUS and SUS at the final iteration: DDR = 70%

channel is poor, as about 70% of the data will be dropped. The parameters of both algorithms are the same as those in the low DDR case above. The final outputs of both algorithms are displayed in Fig. 5.4. Different from the low DDR case, SUS is more advantageous than IUS with respect to tracking accuracy for the same learning iterations under the high DDR condition.

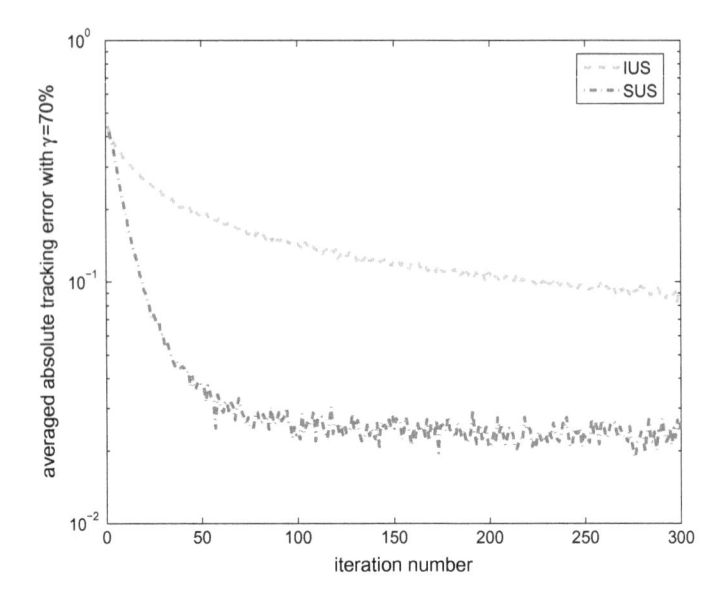

Fig. 5.5 AATE of IUS and SUS along the iteration axis: DDR = 70%

As presented in Fig. 5.5, where the AATE along the iteration axis is shown for both schemes, the tracking errors of SUS are much smaller than those of IUS. The inherent reason for this condition is that the IUS scheme will stop updating if the corresponding packet is lost, whereas the SUS keeps updating whether or not the corresponding packet is lost. In other words, SUS updates more iterations than IUS.

5.5.3 Comparison of Different Data Dropout Rates

To determine the influence of different DDRs, a comparison is made with respect to DDR between IUS and SUS. Here four DDRs are considered, namely, $\gamma = 10$, 30, 50, and 70%, respectively. Algorithms (5.6) and (5.12) are run for 300 iterations. The parameters are set to $a_k = 1/(k + 10)$ and $L_t = 55$.

The AATE of the IUS with respect to different DDRs is shown in Fig. 5.6, where the dash, dotted, solid, and dash-dotted lines denote the AATE from low DDR to high DDR, respectively. We find that tracking accuracy and convergence speed worsen distinctly as the DDR increases or the transmission condition worsens.

The SUS is illustrated in Fig. 5.7, where the lines have the same meaning as those in the IUS. The tracking accuracy of SUS has no visible changes for different DDRs. However, most data fails to be transmitted back when DDR increases. Thus, if the learning gain is quite large, there may be a large increase before the algorithm converges because of excessive updating of some available data.

5.5 Illustrative Simulations

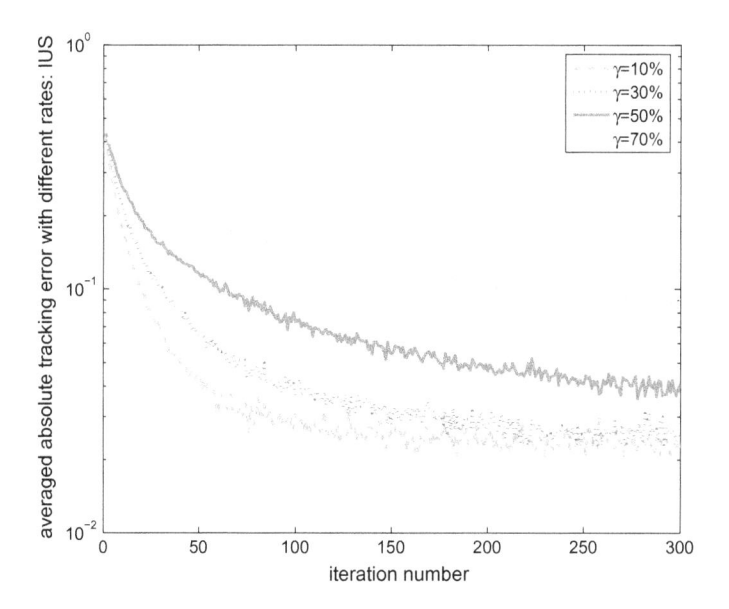

Fig. 5.6 AATE for IUS with respect to different DDRs: DDR = 10, 30, 50, and 70%

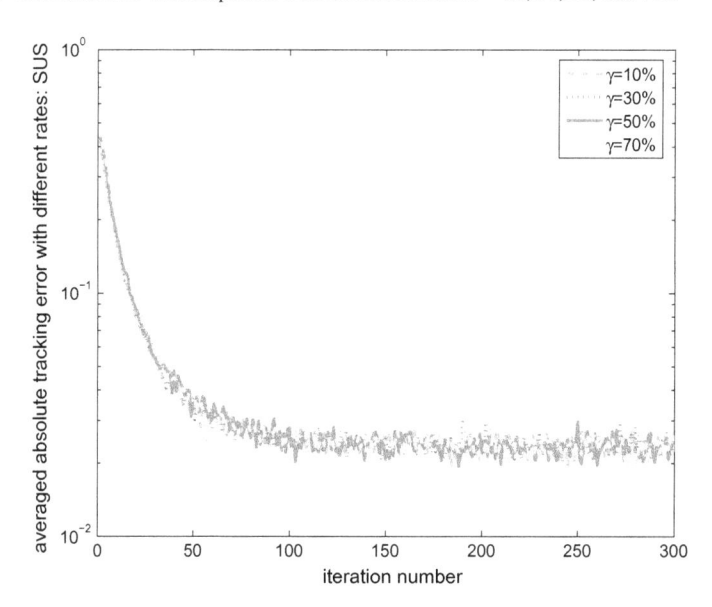

Fig. 5.7 AATE for SUS with respect to different DDRs: DDR = 10, 30, 50, and 70%

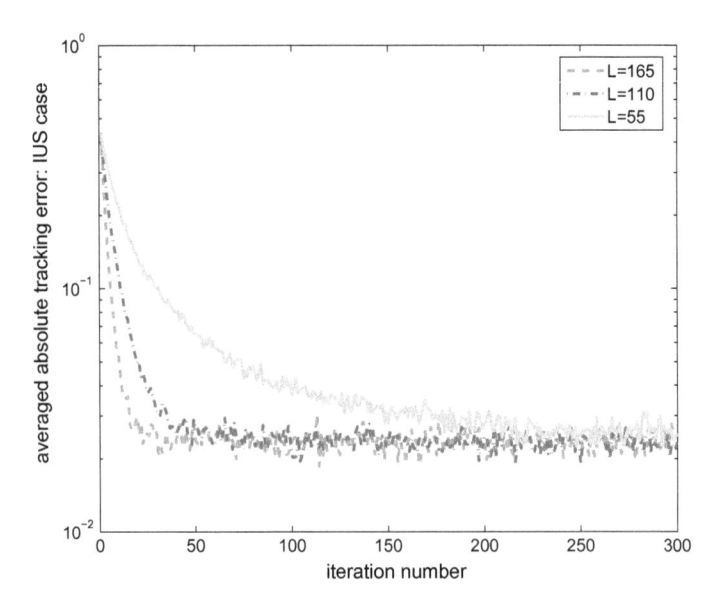

Fig. 5.8 AATE for IUS with respect to different gains: $L = 55, 110$, and 165 (DDR $= 30\%$)

5.5.4 Comparison of Different Learning Gains

To determine how the learning gain L_t is affected in practical applications, we compare the proposed algorithms with different learning gains. The parameters of the algorithms are given as $a_k = 1/(k+10)$ and $\gamma = 30\%$.

Three different learning gains are simulated: $L = 55, 110, 165$. The results for IUS and SUS are shown in Figs. 5.8 and 5.9, respectively, where the profiles of AATE are plotted. A large learning gain may lead to a fast convergence speed for both IUS and SUS. However, the final tracking accuracy is not significantly improved by increasing the learning gains.

5.5.5 Comparison with Conventional P-Type Algorithm

This subsection analyzes the effect of decreasing the sequence a_k and the comparison between the proposed algorithms and the conventional P-type algorithms. As has been analyzed in previous sections, a_k relaxes the selection of learning gains and guarantees a zero-convergence under stochastic noises. When the decreasing sequence a_k in (5.6) and (5.12) is taken out, algorithms (5.6) and (5.12) become conventional P-type algorithms.

We first verify the advantage of learning gain selection with the introduction of a_k. To do so, we select $L_t = 40$. As shown in Fig. 5.10, the conventional P-type

5.5 Illustrative Simulations 111

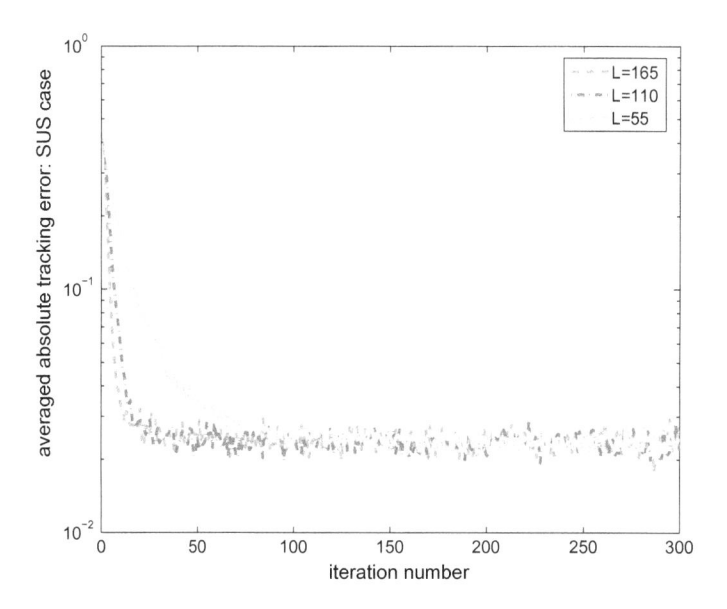

Fig. 5.9 AATE for SUS with respect to different gains: $L = 55$, 110 and 165 (DDR $= 30\%$)

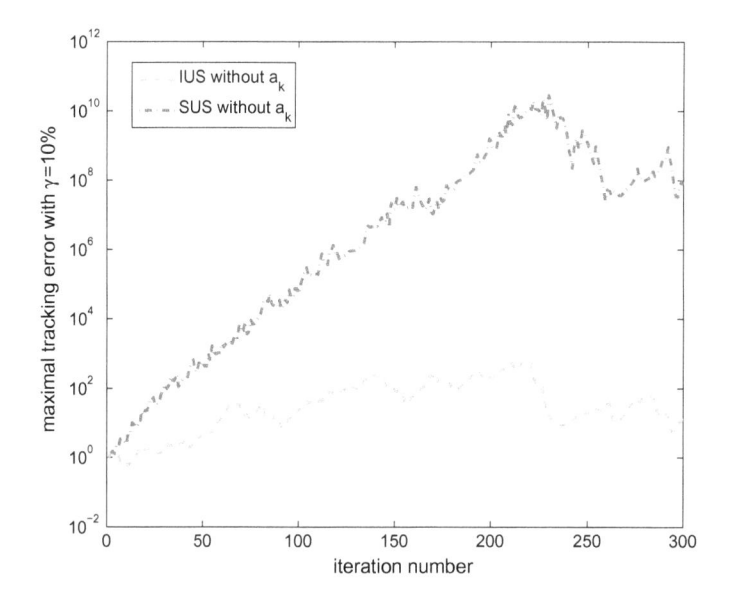

Fig. 5.10 Maximal error profiles for both IUS and SUS without a_k along the iteration axis: DDR $= 10\%$

algorithms fail to converge whether they are intermittent updating or successive updating. In this case, the DDR is set to a low value, $\gamma = 10\%$. The lines denote the profiles of maximal errors, defined as $\max_t \|e_k(t)\|$, along the iteration axis.

However, the proposed algorithms still guarantee convergence because of the introduction of a_k. In this example, we let the decreasing sequence be $a_k = 1/(k+10)$.

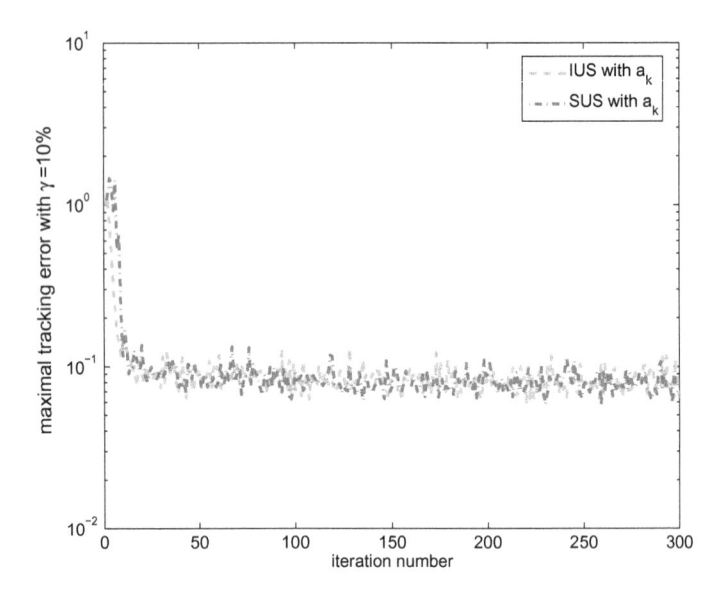

Fig. 5.11 Maximal error profiles for both IUS and SUS with a_k along the iteration axis: DDR = 10%

To make a fair comparison, the learning gain is set to $L_t = 440$, so that for the first iteration, the first coupling gain $a_1 L_t = 40$. The convergence is shown in Fig. 5.11.

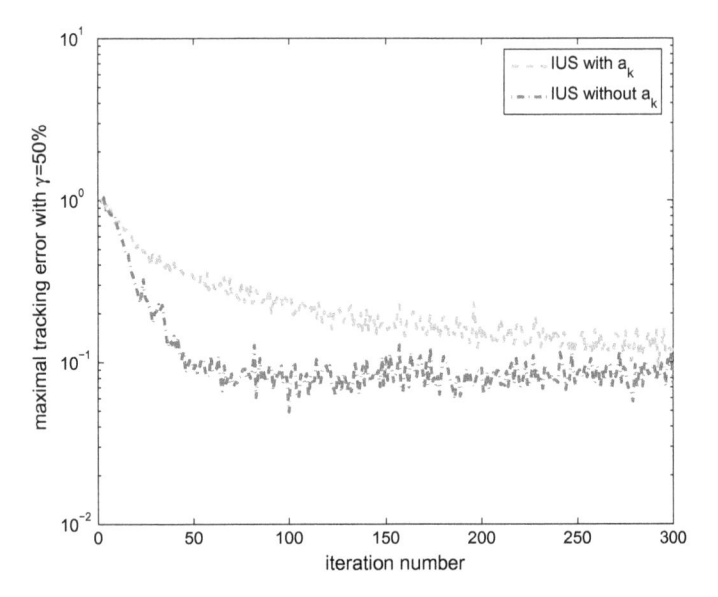

Fig. 5.12 Maximal error profiles for IUS along the iteration axis: DDR = 50%

5.5 Illustrative Simulations 113

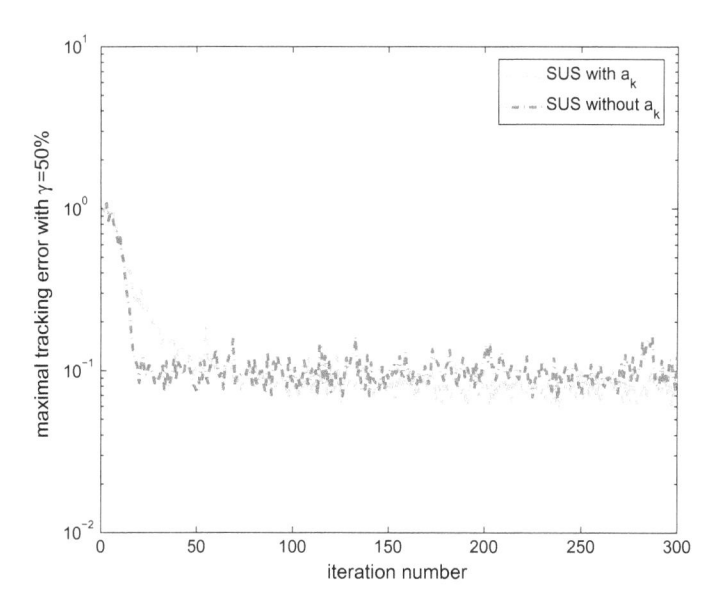

Fig. 5.13 Maximal error profiles for SUS along the iteration axis: DDR = 50%

Next, we compare the learning gain with the convergence of the proposed algorithms and the conventional P-type algorithms. For the conventional P-type algorithms, the learning gain is set to $L_t = 5$; for the proposed algorithms, the learning gain is set to $L_t = 55$ and $a_k = 1/(k+10)$. Therefore, the initial coupling learning gain is the same as that in the conventional P-type learning algorithms. The comparisons are illustrated in Figs. 5.12 and 5.13, in which the DDR is 50%.

In Fig. 5.12, the proposed IUS converges slower than the conventional P-type algorithm, as the decreasing sequence a_k weakens the effect of the learning process. However, the difference in the SUS is not obvious in Fig. 5.13.

5.6 Summary

In this chapter, the ILC is considered for stochastic linear systems with random data dropouts. The data dropout is modeled by a Bernoulli random variable, which is valued 1 or 0 denoting that the data are successfully transmitted or not, respectively. Two schemes are proposed to deal with the random data dropouts: the intermittent updating scheme, which updates its control signal if data is transmitted and does nothing otherwise, and the successive updating scheme, which always updates its control signal no matter whether the data is transmitted or not. That is, if the data is successfully transmitted, then the algorithm will use this data; if the data is lost, then the algorithm will use the latest available data that has been stored. The SUS has an advantage over IUS in tracking accuracy within the same learning iterations under a high data dropout rate, as it can successively improve its performance. However, the difference is rather subtle. In both schemes, the input sequences are proved to converge to the desired input in an almost sure sense under stochastic noises and random data dropouts. The mean square convergence is also established for the intermittent update scheme. The results in this chapter are mainly based on [9, 10].

References

1. Ahn, H.S., Chen, Y.Q., Moore, K.L.: Intermittent iterative learning control. In: Proceedings of the 2006 IEEE International Symposium on Intelligent Control, pp. 832–837 (2006)
2. Ahn, H.S., Moore, K.L., Chen, Y.Q.: Discrete-time intermittent iterative learning controller with independent data dropouts. In: Proceedings of the 2008 IFAC World Congress, pp. 12442–12447 (2008)
3. Ahn, H.S., Moore, K.L., Chen, Y.Q.: Stability of discrete-time iterative learning control with random data dropouts and delayed controlled signals in networked control systems. In: Proceedings the IEEE International Conference on Control Automation, Robotics, and Vision, pp. 757–762 (2008)
4. Bu, X., Hou, Z.-S., Yu, F.: Stability of first and high order iterative learning control with data dropouts. Int. J. Control Autom. Syst. **9**(5), 843–849 (2011)
5. Bu, X., Yu, F., Hou, Z.-S., Wang, F.: Iterative learning control for a class of nonlinear systems with random packet losses. Nonlinear Anal.: Real World Appl. **14**(1), 567–580 (2013)
6. Chow, Y.S., Teicher, H.: Probability Theory: Independence, Interchangeability, Martingales. Springer, New York (1978)
7. Chen, Y., Wen, C., Gong, Z., Sun, M.: An iterative learning controller with initial state learning. IEEE Trans. Autom. Control **44**(2), 371–376 (1999)
8. Zhou, W., Yu, M., Huang, D.Q.: A high-order internal model based iterative learning control scheme for discrete linear time-varying systems. Int. J. Autom. Comput. **12**(3), 330–336 (2015)
9. Shen, D., Zhang, C., Xu, Y.: Two compensation schemes of iterative learning control for networked control systems with random data dropouts. Inf. Sci. **381**, 352–370 (2017)
10. Shen, D., Xu, J.-X.: A framework of iterative learning control under random data dropouts: mean square and almost sure convergence. Int. J. Adaptive Control Signal Process. **31**(12), 1825–1852 (2017)

Chapter 6
Bernoulli Variable Model for Nonlinear Systems

6.1 Problem Formulation

Consider the following time-varying nonlinear system with measurement noise:

$$\begin{aligned} x_k(t+1) &= f(t, x_k(t)) + \mathbf{b}(t, x_k(t))u_k(t), \\ y_k(t) &= \mathbf{c}(t)x_k(t) + w_k(t), \end{aligned} \quad (6.1)$$

where $k = 1, 2, \ldots$ denote different iteration numbers, $t = 0, 1, \ldots, N$ denote different time instants in an iteration, and N is the length of each iteration. $x_k(t) \in \mathbf{R}^n$, $u_k(t) \in \mathbf{R}$, and $y_k(t) \in \mathbf{R}$ denote the state, the input, and the output of the system, respectively. $f(t, x_k(t))$, $\mathbf{b}(t, x_k(t))$, and $\mathbf{c}(t)$ denote unknown system information. The random variable $w_k(t)$ is the measurement noise. Many practical systems can be modeled by the affine nonlinear model, such as mass–spring system, single-link manipulator system, and two-link planar robot arm. However, as will be shown below, the proposed algorithms require little information on system model, which shows that ILC is a favorable data-driven approach to deal with nonlinear systems.

The setup of the control system is illustrated in Fig. 6.1, where the plant and learning controller locate separately and communicate via networks. Due to network congestion, linkage interrupt, and transmission error, the data may be dropped out through the networks. However, for concise expression without loss of any generality, the data dropouts are only considered for the side of output, i.e., the random data dropouts only happen on the network from the measurement output to the memory, whereas the network from learning controller to control plant is assumed to work well. This formulation is adopted to make our following expressions clear and the focal point highlighted. When considering the general data dropouts at both sides, the asynchronous update between the control signal generated by the learning controller and the one fed to the plant should be taken into account and more detailed analysis is required. Such case will be elaborated in Part II of this monograph.

Let the desired reference be $y_d(t)$, $t = 0, 1, \ldots, N$, with initial state $x_d(0)$, where $y_d(0) = \mathbf{c}(0)x_d(0)$.

Fig. 6.1 Block diagram of networked control system

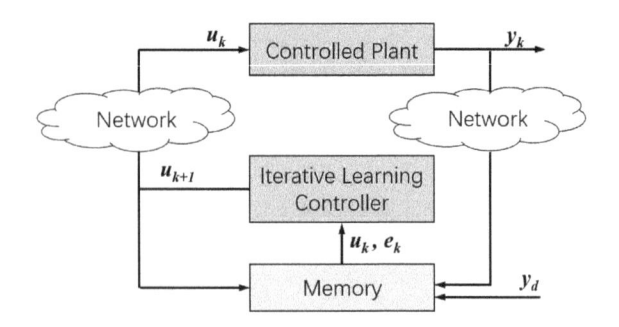

The following mild assumptions are given for system (6.1).

Assumption 6.1 For any $t = 0, 1, \ldots, N$, the functions $f(t, x)$ and $\mathbf{b}(t, x)$ are continuous with respect to the second argument x.

Remark 6.1 Assumption 6.1 could be relaxed to the case that the functions $f(t, x)$ and $\mathbf{b}(t, x)$ are allowed to have discontinuities with respect to x away from $x_d(t)$, where $x_d(t)$ is defined later in Remark 6.4. Since $x = x_d(t)$ is unknown prior, Assumption 6.1 is assumed for simplicity.

Assumption 6.2 The input/output coupling value $\mathbf{c}(t+1)\mathbf{b}(t, x)$ is unknown, but it is nonzero and does not change its sign during learning processes. Without loss of any generality, it is assumed known that $\mathbf{c}(t+1)\mathbf{b}(t, x) > 0$ for expression convenience in the rest of this chapter.

Remark 6.2 The sign of the input/output coupling value denotes the control direction. Control direction is a necessary information for the design of controller. This is why we assume that $\mathbf{c}(t+1)\mathbf{b}(t, x)$ does not change its sign. Otherwise, the controller would be very complex since we have to design a scheme to find the right control direction adaptively. The techniques in Chap. 4 can be used to handle this issue. Since it is out of the scope of this chapter, we simply give Assumption 6.2.

Remark 6.3 In Assumption 6.2, the assumption that $\mathbf{c}(t+1)\mathbf{b}(t, x)$ is nonzero implies that the relative degree of the system (6.1) is 1. However, this case can be extended to the high relative degree case with slight revisions to the learning algorithms. To be specific, assume the system is of high relative degree τ, that is, for any t, $\mathbf{c}\frac{\partial f^{\tau-1}(f+\mathbf{b}u)}{\partial u}$ is nonzero and $\mathbf{c}\frac{\partial f^i(f+\mathbf{b}u)}{\partial u} = 0$, $0 \leq i \leq \tau - 2$, where $f^i(x) = f^{i-1} \circ f(x)$ and $\circ$ denotes the composite operator of functions [1]. For this case, when updating the input at time instant t, the tracking error at time instant $t + \tau$ is used for the learning algorithms given in the next section instead of the one at time instant $t + 1$.

Remark 6.4 If the system (6.1) is noise free, then based on Assumption 6.2 we could recursively define the optimal/desired input $u_d(t)$ as follows, $t = 0, 1, \ldots, N - 1$:

6.1 Problem Formulation

$$u_d(t) = [\mathbf{c}(t+1)\mathbf{b}(t, x_d(t))]^{-1} \\ \times (y_d(t+1) - \mathbf{c}(t+1)f(t, x_d(t))), \\ x_d(t+1) = f(t, x_d(t)) + \mathbf{b}(t, x_d(t))u_d(t),$$

with the initial state $x_d(0)$. It is obvious that the following relationship holds for the desired reference $y_d(t)$:

$$\begin{aligned} x_d(t+1) &= f(t, x_d(t)) + \mathbf{b}(t, x_d(t))u_d(t), \\ y_d(t) &= \mathbf{c}(t)x_d(t). \end{aligned} \quad (6.2)$$

It is worth pointing out that (6.2) is the well-known realizable condition for ILC [2, 3]. Here, with the help of assumption on input/output coupling value, i.e., Assumption 6.2, we can establish this realizable condition directly. However, due to the fact that nonlinear functions $f(\cdot, \cdot)$, $\mathbf{b}(\cdot, \cdot)$ and output coefficient vector $\mathbf{c}(\cdot)$ are unknown, the recursive defined optimal input $u_d(t)$ cannot be actually used; thus, we have to design ILC update algorithms such that the generated input sequence converges to the optimal input.

Assumption 6.3 The initial values can be precisely reset asymptotically in the sense that $x_k(0) \to x_d(0)$ as $k \to \infty$.

Remark 6.5 In many papers, the initial state usually is required to be $x_d(0)$. In Assumption 6.3, it is required that the accurate initial state could be reset asymptotically. This is a technical condition, which aims to leave a space to design suitable initial value learning algorithms to realize this asymptotical re-initialization condition such as the one given in [4, 5]. It is obvious that the classic identical initialization condition (i.i.c.) is a special case of Assumption 6.3. For further discussions on initial condition, we refer to [6].

Assumption 6.4 For each time instant t, the measurement noise $\{w_k(t), k = 0, 1, \ldots\}$ is a sequence of independent and identically distributed (i.i.d.) random variables with $\mathbb{E}w_k(t) = 0$, $\sup_k \mathbb{E}\|w_k(t)\|^2 < \infty$, and $\lim_{n\to\infty} \frac{1}{n}\sum_{k=1}^n w_k(t)w_k^T(t) = R_w^t$ a.s., where R_w^t is an unknown matrix.

Remark 6.6 In Assumption 6.4, the condition on measurement noises is made according to the iteration axis, rather than the time axis. Thus, this requirement is not rigorous, as the process would be performed repeatedly and independently.

In this chapter, we adopt Bernoulli random variable to model the random data dropouts. To be specific, a random variable $\gamma_k(t)$ is introduced to indicate whether the measurement packet $y_k(t)$ is successfully transmitted or not. Denote $\gamma_k(t) = 1$ if $y_k(t)$ is successfully transmitted and $\gamma_k(t) = 0$ otherwise. Without loss of any generality,

$$\mathbb{P}(\gamma_k(t) = 1) = \rho, \quad \mathbb{P}(\gamma_k(t) = 0) = 1 - \rho, \quad (6.3)$$

where $0 < \rho < 1$. That is, the probability that the measurement $y_k(t)$ is successfully transmitted is ρ, $\forall k, t$.

Based on the above assumptions, the control objective of this chapter is to design ILC schemes to generate the input sequence such that the following averaged tracking index is minimized, $\forall t = 0, 1, \ldots, N$, under random data dropouts:

$$V_t = \limsup_{n \to \infty} \frac{1}{n} \sum_{k=1}^{n} \|y_d(t) - y_k(t)\|^2, \tag{6.4}$$

where $y_d(t)$ is the desired reference. If we define the control output as $z_k(t) = \mathbf{c}(t)x_k(t)$, then it is easy to see that $z_k(t) \to y_d(t)$ as $k \to \infty$ whenever the tracking index (6.4) is minimized and vice versa. That is, the index (6.4) implies that the precise tracking performance could be achieved if measurements noises are eliminated.

6.2 Intermittent Update Scheme and Its Almost Sure Convergence

In this section, we apply the following intermittent update scheme (IUS):

$$u_{k+1}(t) = u_k(t) + a_k \gamma_k(t+1) e_k(t+1), \tag{6.5}$$

where $e_k(t) \triangleq y_d(t) - y_k(t)$ is the tracking error and a_k is the learning step size satisfying

$$a_k > 0, a_k \to 0, \sum_{k=1}^{\infty} a_k = \infty, \sum_{k=1}^{\infty} a_k^2 < \infty. \tag{6.6}$$

Remark 6.7 The reason why the first algorithm (6.5) is called IUS is that the algorithm only updates its signal when the output is successfully received. In other words, the input signal would stop updating if the corresponding output is lost. As a result, the algorithm (6.5) would update in some iterations and keep the latest values in other iterations. In addition, it is noticed that the updating frequency is equal to the successful transmission rate due to the inherent mechanism of (6.5). Therefore, roughly speaking, the larger the data dropout rate (DDR) is, the slower the algorithm converges. This motivates us to find whether a faster convergence speed could be achieved under large DDR.

Remark 6.8 In this chapter, to make our idea for the convergence analysis clear, we adopt the SISO formulation to reduce the expression complexity. However, the results can be easily extended to the MIMO case with slight modifications to the algorithms following similar steps given below. The major modification to the proposed algorithms is to multiply the tracking error from the left by a learning gain matrix L_t such that all eigenvalues of $L_t C(t+1) B(t, x)$ are with positive real parts,

6.2 Intermittent Update Scheme and Its Almost Sure Convergence

where $C(t+1)B(t,x)$ denotes the multidimensional input/output coupling matrix, i.e., the counterpart of $\mathbf{c}(t+1)\mathbf{b}(t,x)$.

For simplicity of writing, let us set $f_k(t) = f(t, x_k(t))$, $f_d(t) = f(t, x_d(t))$, $\mathbf{b}_k(t) = \mathbf{b}(t, x_k(t))$, $\mathbf{b}_d(t) = \mathbf{b}(t, x_d(t))$, $\delta u_k(t) = u_d(t) - u_k(t)$, $\delta f_k(t) = f_d(t) - f_k(t)$, $\delta \mathbf{b}_k(t) = \mathbf{b}_d(t) - \mathbf{b}_k(t)$, and $\mathbf{c}^+\mathbf{b}_k(t) = \mathbf{c}(t+1)\mathbf{b}_k(t)$.

For further analysis, the following lemmas are needed while the proofs have been given in previous chapters.

Lemma 6.1 *Assume Assumptions 6.1–6.3 hold for system (6.1). If $\lim_{k\to\infty} \delta u_k(s) = 0$, $s = 0, 1, \ldots, t$, then at time instant $t+1$, $\|\delta x_k(t+1)\| \xrightarrow[k\to\infty]{} 0$, $\|\delta f_k(t+1)\| \xrightarrow[k\to\infty]{} 0$, $\|\delta \mathbf{b}_k(t+1)\| \xrightarrow[k\to\infty]{} 0$.*

Lemma 6.2 *Assume Assumptions 6.1–6.4 hold for system (6.1) and tracking reference $y_d(t)$, then the index (6.4) will be minimized for arbitrary time instant $t+1$ if the control sequence $\{u_k(t)\}$ is admissible and satisfies $u_k(i) \xrightarrow[k\to\infty]{} u_d(i)$, $i = 0, 1, \ldots, t$. In this case, $\{u_k(t)\}$ is called the optimal control sequence.*

Recalling (6.3), we have that $\mathbb{E}\gamma_k(t) = \rho$ and $\mathbb{E}\gamma_k^2(t) = \rho$. Denote $\delta x_k(t) = x_d(t) - x_k(t)$ and $\delta u_k(t) = u_d(t) - u_k(t)$. Subtracting both side of (6.5), we have

$$\delta u_{k+1}(t) = \delta u_k(t) - a_k \gamma_k(t+1) e_k(t+1).$$

Notice that $e_k(t) = y_d(t) - y_k(t)$, then

$$\begin{aligned}
\delta u_{k+1}(t) &= \delta u_k(t) - a_k \gamma_k(t+1) e_k(t+1) \\
&= \delta u_k(t) - a_k \gamma_k(t+1)(y_d(t+1) - y_k(t+1)) \\
&= \delta u_k(t) - a_k \gamma_k(t+1) \mathbf{c}(t+1) \delta x_k(t+1) \\
&\quad + a_k \gamma_k(t+1) w_k(t+1) \\
&= \delta u_k(t) - a_k \gamma_k(t+1) \mathbf{c}^+ \mathbf{b}_k(t) \delta u_k(t) \\
&\quad - a_k \gamma_k(t+1)[\mathbf{c}^+ \delta f_k(t) + \mathbf{c}^+ \delta \mathbf{b}_k(t) u_d(t)] \\
&\quad + a_k \gamma_k(t+1) w_k(t+1).
\end{aligned}$$

Then, we have the following convergence theorem.

Theorem 6.1 *Consider the stochastic system (6.1), index (6.4), and update law (6.5), and assume Assumptions 6.1–6.4 hold, then the input $u_k(t)$ generated by (6.5) with learning gain sequence $\{a_k\}$ satisfying (6.6) converges to $u_d(t)$ almost surely as $k \to \infty$, $\forall t$.*

To prove the above theorem, we need the following technical lemma, which is a simplified version of Lemma A.3 (given in the Appendix).

Lemma 6.3 *Let $\{h_k\}$ be a sequence with $h_k \to h$ where h is a negative constant. Let a_k satisfy the conditions in (6.6) and both $\{\mu_k\}$ and $\{\nu_k\}$ satisfy the following conditions:*

$$\sum_{k=1}^{\infty} a_k \mu_k < \infty, \quad v_k \xrightarrow[k \to \infty]{} 0, \tag{6.7}$$

then $\{\alpha_k\}$ generated by the following recursion with arbitrary initial value α_0 converges to zero a.s.:

$$\alpha_{k+1} = \alpha_k + a_k h_k \alpha_k + a_k (\mu_k + v_k). \tag{6.8}$$

Now we present the proof of Theorem 6.1.

Proof The proof is carried out by mathematical induction along the time axis t. It should be indicated that the steps for $t = 1, 2, \ldots, N - 1$ are identical to the case of $t = 0$, which will be expressed in detail in the following.

Initial Step. Consider the case $t = 0$.

For $t = 0$, the input error recursion could be rewritten as

$$\begin{aligned}
\delta u_{k+1}(0) = {} & (1 - a_k \rho \mathbf{c}^+ \mathbf{b}_k(0)) \delta u_k(0) \\
& - a_k (\gamma_k(1) - \rho) \mathbf{c}^+ \mathbf{b}_k(0) \delta u_k(0) \\
& - a_k \gamma_k(1) [\mathbf{c}^+ \delta f_k(0) + \mathbf{c}^+ \delta \mathbf{b}_k(0) u_d(0)] \\
& + a_k \gamma_k(1) w_k(1).
\end{aligned} \tag{6.9}$$

Note that $\mathbf{b}_k(0)$ is continuous with respect to the initial state by Assumption 6.1, we have that $\mathbf{b}_k(0) \to \mathbf{b}_d(0)$ as $k \to \infty$ by Assumption 6.3. In addition, the coupling value $\mathbf{c}^+ \mathbf{b}_k(0)$ would converge to $\mathbf{c}^+ \mathbf{b}_d(0)$ by Assumption 6.2. Thus, it follows that $\rho \mathbf{c}^+ \mathbf{b}_k(0) > \varepsilon$ for sufficiently large k, say $k \geq k_0$, where $\varepsilon > 0$ is a suitable constant.

Note that the first term on the right-hand side of (6.9) is the main recursion term, while the others are structural and measurement noises. According to Lemma 6.3, it is sufficient to show that these noises satisfy the condition (6.7).

By Assumptions 6.1 and 6.3, it is easy to derive that $\delta f_k(0) \xrightarrow[k \to \infty]{} 0$ and $\delta \mathbf{b}_k(0) \xrightarrow[k \to \infty]{} 0$. Notice that both $\gamma_k(1)$ and $u_d(0)$ are bounded. Therefore, the third term on the right-hand side of (6.9) converges to 0 as $k \to \infty$.

Further, the sequence $\{w_k(1)\}$ is an i.i.d. sequence with zero mean and finite second moments. In addition, $w_k(1)$ is independent of $\gamma_k(1)$. Thus, it is obvious that $\sum_{k=1}^{\infty} \mathbb{E}[a_k \gamma_k(1) w_k(1)]^2 \leq \sup_k \mathbb{E} w_k^2(1) \cdot \mathbb{E} \gamma_k^2(1) \sum_{k=1}^{\infty} a_k^2 \leq \rho \|R_w^1\| \sum_{k=1}^{\infty} a_k^2 < \infty$ where $\|\cdot\|$ is a suitable matrix norm. This further leads to that $\sum_{k=1}^{\infty} a_k \gamma_k(1) w_k(1) < \infty$, a.s. by Khintchine–Kolmogorov convergence theorem [7]. In other words, the last term of (6.9) satisfies (6.7).

Now it comes to the second term on the right-hand side of (6.9), $a_k(\gamma_k(1) - \rho) \mathbf{c}^+ \mathbf{b}_k(0) \delta u_k(0)$. The sequence of this term is no longer mutual independent. To deal with this term, let $\mathscr{F}_k$ be the increasing σ-algebra generated by $y_j(t)$, $w_j(t)$, $\gamma_j(t)$, $x_j(0)$, $0 \leq j \leq k$, $\forall t$. That is, $\mathscr{F}_k \triangleq \sigma\{y_j(t), w_j(t), \gamma_j(t), x_j(0), 0 \leq j \leq k, \forall t\}$. Then according to the learning law (6.5), it is easy to find that $u_k(t) \in \mathscr{F}_{k-1}$ and $\mathbf{b}_k(0) \in \mathscr{F}_{k-1}$. In addition, $\gamma_k(1)$ is independent of $\mathscr{F}_{k-1}$ and thus is independent of $\delta u_k(0)$ and $\mathbf{b}_k(0)$. Therefore,

6.2 Intermittent Update Scheme and Its Almost Sure Convergence

$$\mathbb{E}\{(\gamma_k(1) - \rho)\mathbf{c}^+\mathbf{b}_k(0)\delta u_k(0)|\mathscr{F}_{k-1}\}$$
$$= \mathbf{c}^+\mathbf{b}_k(0)\delta u_k(0)\mathbb{E}\{\gamma_k(1) - \rho|\mathscr{F}_{k-1}\} = 0. \tag{6.10}$$

This means that $((\gamma_k(1) - \rho)\mathbf{c}^+\mathbf{b}_k(0)\delta u_k(0), \mathscr{F}_k, k \geq 1)$ is a martingale difference sequence [7]. In addition,

$$\sum_{k=1}^{\infty}\mathbb{E}\{[a_k(\gamma_k(1) - \rho)\mathbf{c}^+\mathbf{b}_k(0)\delta u_k(0)]^2|\mathscr{F}_{k-1}\}$$

$$\leq \sup_k[\mathbf{c}^+\mathbf{b}_k(0)\delta u_k(0)]^2 \sum_{k=1}^{\infty} a_k^2 \mathbb{E}\{(\gamma_k(1) - \rho)^2|\mathscr{F}_{k-1}\}$$

$$\leq c_1 \sum_{k=1}^{\infty} a_k^2 < \infty,$$

where $c_1 > 0$ is a suitable constant. Then by Chow convergence theorem of martingale [7], we have $\sum_{k=1}^{\infty} a_k(\gamma_k(1) - \rho)\mathbf{c}^+\mathbf{b}_k(0)\delta u_k(0) < \infty$. In other words, the second term on the right-hand side of (6.9) satisfies (6.7).

Then applying Lemma 6.3 to (6.9), we are now able to have that $\delta u_k(0) \to 0$ as $k \to \infty$ a.s.

Inductive Step. Assume that the convergence of $u_k(t)$ has been proved for $t = 0, 1, \ldots, s-1$ and the target is to show the convergence for $t = s$. From the inductive assumptions and Lemma 6.1, we have $\delta x_k(s) \xrightarrow[k\to\infty]{} 0$ and therefore $\delta f_k(s) \xrightarrow[k\to\infty]{} 0$ and $\delta \mathbf{b}_k(s) \xrightarrow[k\to\infty]{} 0$. On the other hand, the recursion for $t = s$ is as follows:

$$\delta u_{k+1}(s) = \delta u_k(s) - a_k \gamma_k(s+1)\mathbf{c}^+\mathbf{b}_k(s)\delta u_k(s)$$
$$- a_k \gamma_k(s+1)[\mathbf{c}^+\delta f_k(s) + \mathbf{c}^+\delta \mathbf{b}_k(s)u_d(s)]$$
$$+ a_k \gamma_k(s+1)w_k(s+1).$$

Then following similar steps of the case $t = 0$, we are with no further efforts to conclude that $\delta u_k(s) \to 0$ as $k \to \infty$ a.s. This completes the proof.

Remark 6.9 As has been pointed out in Remark 6.7, the algorithm (6.5) only updates itself when the corresponding output package is well received. Thus, if the DDR is large, then the learning step size a_k during the updating iterations will decrease to zero fast, which will further lead to a slow convergence speed. To overcome this disadvantage, one could change the learning step size only when the output is well received. In other words, the following algorithm is an alternative to (6.5):

$$u_{k+1}(t) = u_k(t) + a_{\mu_k(t)}\gamma_k(t+1)e_k(t+1),$$

$$\mu_k(t) = \sum_{i=1}^{k} \gamma_i(t+1).$$

6.3 Successive Update Scheme and Its Almost Sure Convergence

In this section, we consider the following successive update scheme (SUS):

$$u_{k+1}(t) = u_k(t) + a_k e_k^*(t+1), \quad (6.11)$$

where $e_k^*(t)$ is the latest available tracking error, defined as

$$e_k^*(t) = \begin{cases} e_k(t), & \text{if } \gamma_k(t) = 1 \\ e_{k-1}^*(t), & \text{if } \gamma_k(t) = 0 \end{cases} \quad (6.12)$$

The learning step size $\{a_k\}$ is a decreasing sequence and it should satisfy

$$a_k > 0, \ a_k \to 0, \ \sum_{k=1}^{\infty} a_k = \infty, \ \sum_{k=1}^{\infty} a_k^2 < \infty. \quad (6.13)$$

Remark 6.10 Different from (6.5), the new scheme (6.11) always keeps updating no matter whether the corresponding output is lost or not. If the output of the last iteration is received, then the algorithm would update its input by using this output; while if the output is lost, then the algorithm would update its input by using the latest available output information of certain previous iteration. As a matter of fact, the algorithm (6.11) is

$$\begin{aligned} u_{k+1}(t) = u_k(t) &+ a_k \gamma_k(t+1) e_k(t+1) \\ &+ a_k (1 - \gamma_k(t+1)) e_{k-1}^*(t+1). \end{aligned} \quad (6.14)$$

Comparing with (6.5), the updating of (6.11) is deterministic in the sense that the algorithm updates itself every iteration. However, the technical proof of the convergence is more complex than that of the IUS case because the error information in (6.11) is no longer straightforward. As one could see, if the output of the last iteration is lost during transmission, then the error used in (6.11) is unknown for analysis because of successive data dropouts. That is, the error information could come from any previous iteration with different probabilities.

To form this situation, the stochastic stopping time sequence $\{\tau_k^t, k = 1, 2, \ldots, 0 \le t \le N\}$ is introduced to denote the random iteration delay of the update due to random data dropouts. The algorithm (6.11) is reformulated as follows:

$$u_{k+1}(t) = u_k(t) + a_k e_{k-\tau_k^{t+1}}(t+1), \quad (6.15)$$

where the stopping time $\tau_k^{t+1} \le k$. In other words, for the updating of input at t of the $(k+1)$th iteration, no information of $e_m(t+1)$ with $m > k - \tau_k^{t+1}$ is received and only $e_{k-\tau_k^{t+1}}(t+1)$ is available. In addition, according to the SUS settings, for

6.3 Successive Update Scheme and Its Almost Sure Convergence

the mth iteration with $k - \tau_k^{t+1} < m \le k$, the input $u_m(t)$ is successively updated with the same error $e_{k-\tau_k^{t+1}}(t+1)$.

For the convergence analysis, the major difficulty lies in the technical analysis of the influences caused by random iteration delays or stochastic stopping time τ_k^t. Therefore, the analysis is completed in two steps. The first step is to show the convergence of (6.15) without any iteration delay, i.e., $\tau_k^t = 0, \forall k, t$. The second step devotes to the effect of stopping time τ_k^t.

When there is no iteration delay, i.e., $\tau_k^t = 0$, the algorithm (6.15) turns into

$$u_{k+1}(t) = u_k(t) + a_k e_k(t+1). \tag{6.16}$$

This actually is the conventional ILC for systems without any data dropout. The convergence analysis of this algorithm could be derived directly following the similar steps of Theorem 6.1 by letting $\gamma_k(t) \equiv 1, \forall t, k$. Thus, we could give the following theorem without proof.

Theorem 6.2 *Consider the stochastic system* (6.1) *without any data dropout, index* (6.4), *and update law* (6.16), *assume Assumptions 6.1–6.4 hold, then the input $u_k(t)$ generated by* (6.16) *with learning gain sequence $\{a_k\}$ satisfying* (6.13) *converges to $u_d(t)$ almost surely as $k \to \infty, \forall t$.*

Now, we are able to give the following convergence theorem for the SUS.

Theorem 6.3 *Consider the stochastic system* (6.1), *index* (6.4), *and update law* (6.15), *assume Assumptions 6.1–6.4 hold, then the input $u_k(t)$ generated by* (6.15) *with learning gain sequence $\{a_k\}$ satisfying* (6.13) *converges to $u_d(t)$ almost surely as $k \to \infty, \forall t$.*

Proof Comparing (6.15) and (6.16), we find that the effect of the random data dropouts acts as an additional error $e_{k-\tau_k^{t+1}}(t+1) - e_k(t+1)$. Taking the main idea of the proof for convergence into account and recalling the preliminary result of Theorem 6.2, it is sufficient to show that this error satisfies the condition (6.7). Specifically, we have

$$\begin{aligned}
& e_{k-\tau_k^{t+1}}(t+1) - e_k(t+1) \\
&= y_k(t+1) - y_{k-\tau_k^{t+1}}(t+1) + w_k(t+1) - w_{k-\tau_k^{t+1}}(t+1) \\
&= \mathbf{c}^+ \mathbf{b}_k(t)[u_k(t) - u_{k-\tau_k^{t+1}}(t)] + [\mathbf{c}^+ f_k(t) - \mathbf{c}^+ f_{k-\tau_k^{t+1}}(t)] \\
&\quad + [\mathbf{c}^+ \mathbf{b}_k(t) - \mathbf{c}^+ \mathbf{b}_{k-\tau_k^{t+1}}(t)] u_{k-\tau_k^{t+1}}(t) \\
&\quad + w_k(t+1) - w_{k-\tau_k^{t+1}}(t+1).
\end{aligned} \tag{6.17}$$

It is no doubt that the last term satisfies the condition (6.7). In addition, it could be proved by mathematical induction, similar to the proof of Theorem 6.1, that the second and the third terms on the right-hand side of (6.17) satisfy the condition (6.7) with the help of Lemma 6.1. Thus, only the first term, i.e., $\mathbf{c}^+ \mathbf{b}_k(t)[u_k(t) - u_{k-\tau_k^{t+1}}(t)]$, is left for further analysis. It is also easy to prove boundedness and convergence of $c^+ b_k(t)$ by the mathematical induction principle.

Recalling the learning algorithm (6.15), we find that the difference is expanded as

$$u_k(t) - u_{k-\tau_k^{t+1}}(t) = \sum_{m=k-\tau_k^{t+1}}^{k-1} a_m c^+ b_m(t) \delta u_{m-\tau_m^{t+1}}(t)$$
$$+ \sum_{m=k-\tau_k^{t+1}}^{k-1} a_m c^+ \delta f_{m-\tau_m^{t+1}}(t)$$
$$- \sum_{m=k-\tau_k^{t+1}}^{k-1} a_m w_{m-\tau_m^{t+1}}(t+1). \quad (6.18)$$

In order to analyze the effect of (6.18), we need to give an estimation on the number of successive data dropout iterations, i.e., τ_k^t. Noticing that the data dropouts are modeled by a Bernoulli random variable, we find that τ_k^t obeys the geometric distribution. Here, for concise notations, we let τ denote a random variable satisfying the same distribution, i.e., $\tau \sim G(\rho)$. Then it is obvious that $\mathbb{E}\tau = 1/\rho$ and $\text{Var}(\tau) = (1-\rho)/\rho^2$. Then, we further have that $\mathbb{E}\tau^2 = 1/\rho$. By direct calculations, we have that

$$\sum_{n=1}^{\infty} \mathbb{P}\{\tau \geq n^{\frac{1}{2}}\} = \sum_{n=1}^{\infty} \mathbb{P}\{\tau^2 \geq n\}$$
$$= \sum_{n=1}^{\infty} \sum_{j=n}^{\infty} \mathbb{P}\{j \leq \tau^2 < j+1\}$$
$$= \sum_{j=1}^{\infty} j \mathbb{P}\{j \leq \tau^2 < j+1\} \leq \mathbb{E}\tau^2 < \infty.$$

By the Borel–Cantelli lemma, it further leads to that $\mathbb{P}\{\tau \geq n^{\frac{1}{2}} \text{ i.o.}\} = 0$. Consequently, we have $\tau_n^t/n \xrightarrow[n\to\infty]{} 0$, a.s., $\forall t$. That is, $(n - \tau_n^t)/n \xrightarrow[n\to\infty]{} 1$ and $n - \tau_n^t \to \infty$, a.s., $\forall t$.

Based on this observation, now we can prove that the terms on the right-hand side of (6.18) satisfy the condition (6.7).

Similar to the proof of Theorem 6.1, it is concluded that $\sum_{m=0}^{k} a_m w_{m-\tau_m^{t+1}}(t+1)$ converges to an unknown constant, a.s., $\forall t$. Therefore, noticing that $n - \tau_n^t \to \infty$, we have that $\|\sum_{m=k-\tau_k^{t+1}}^{k-1} a_m w_{m-\tau_m^{t+1}}(t+1)\| = o(1)$. This further yields that the last term of (6.18) satisfies the condition (6.7). On the other hand, by mathematical induction principle, it can be proved that the state function error, $\delta f_{m-\tau_m^{t+1}}(t)$, in the

6.3 Successive Update Scheme and Its Almost Sure Convergence

second term on the right-hand side of (6.18), converges to zero as iteration number goes to infinity. That is, the condition (6.7) is also satisfied for the second term.

Therefore, only the first term on the right-hand side of (6.18) is left to discuss. As a matter of fact, this term can be almost surely bounded by a term, which is the product of $\sum_{m=k-\tau_k^{t+1}}^{k-1} a_m$ and a sample path-dependent constant, according to Assumption 6.1. Further, the selection of a_k leads to that this term is bounded by $c_0 a_{k-\tau_k^{t+1}} \tau_k^{t+1}$ where c_0 is a suitable constant. Thus, it is sufficient to show that $c_0 a_{k-\tau_k^{t+1}} \tau_k^{t+1} = o(1)$. For expressions easy to understand, here we select $a_k = 1/k$. The general case is similar but with more complicated derivations on the quantity estimation. For this case, we can directly calculate the term as follows:

$$\begin{aligned}
a_{k-\tau_k^{t+1}} \tau_k^{t+1} &= 1/(k - \tau_k^{t+1}) \cdot \tau_k^{t+1} \\
&= k/(k - \tau_k^{t+1}) \times \tau_k^{t+1}/k \\
&= O(1/k) O(k^{\frac{1}{2}}) = O(1/k^{\frac{1}{2}}) \to 0
\end{aligned}$$

as $k \to \infty$. Therefore, the first term is also verified.

In sum, we have proved that the effect of random data dropouts, i.e., (6.17), satisfies the condition (6.7). Then, the convergence proof of this theorem could be completed by similar steps of Theorem 6.1.

Remark 6.11 The key step in the proof of the above theorem is to show that the effect of random data dropouts is asymptotically negligible. In other words, as the iteration number goes to infinity, the random iteration delay is not with the same magnitude of iteration number. That is, the random iteration delay is negligible comparing with the large enough iteration number. Consequently, the behaviors of SUS are close to those of the conventional learning algorithm (6.16) as iteration number increases.

6.4 Illustrative Simulations

In order to show the effectiveness of the proposed ILC algorithm and verify the convergence analysis, a DC-motor driving a single rigid link through a gear is taken as an example [8]. The single-link mechanism is shown in Fig. 6.2, while the dynamics is expressed as the following second-order differential equation:

$$\left(J_m + \frac{J_l}{n^2}\right)\ddot{\theta}_m + \left(B_m + \frac{B_l}{n^2}\right)\dot{\theta}_m + \frac{Mgl}{n}\sin\left(\frac{\theta_m}{n}\right) = u, \quad (6.19)$$

where the notations are described in Table. 6.1.

By Euler's approximation, we have the discrete-time state-space expression with state and output being $x = (x_1, x_2)^T = (\theta_m, \dot{\theta}_m)^T$ and $y = \dot{\theta}_l$, respectively, and the system function and matrices are

Fig. 6.2 Single-link mechanism

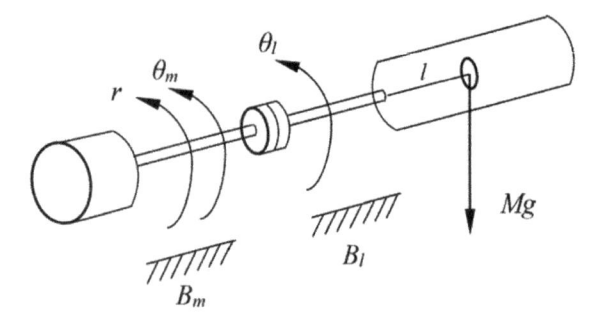

Table 6.1 Notations meaning of (6.19)

Notation	Meaning
J_m	Motor inertia
B_m	Motor damping coefficient
θ_m	Motor angle
J_l	Link inertia
B_l	Link damping coefficient
θ_l	Link angle, $\theta_l = \theta_m/n$
n	Gear ratio
u	Motor torque
M	Lumped mass
g	Gravitational acceleration
l	The center of mass from the axis of motion

$$f(x,t) = \begin{bmatrix} x_1(t) + \Delta x_2(t) \\ x_2(t) + \frac{\Delta}{J_m + J_l/n^2}\left[-\left(B_m + \frac{B_l}{n^2}\right)x_2(t) - \frac{Mgl}{n}\sin\left(\frac{x_1(t)}{n}\right) \right] \end{bmatrix}$$

$$B = \begin{bmatrix} 0 \\ \frac{\Delta}{J_m + J_l/n^2} \end{bmatrix}, \quad C = [0, \frac{1}{n}],$$

where Δ is the discrete time interval. In this simulation, let $\Delta = 50ms$ and let the operation period be $3s$, thus iteration length is $N = 60$. Other parameters are given as follows: $J_m = 0.3$, $J_l = 0.44$, $B_m = 0.3$, $B_l = 0.25$, $M = 0.5$, $g = 9.8$, $n = 1.6$, and $l = 0.15$.

The desired trajectory is $y_d(t) = \frac{1}{3}\sin(t/20) + 1 - \cos(3t/20)$, $0 \leq t \leq 60$. The initial input, i.e., input for the first iteration, is simply assumed to be $u_0(t) = 0$. The initial state is first fixed at $x_k(0) = [0, 0]^T$. The output is involved with a stochastic noise $w_k(t) \sim N(0, 0.1^2)$. The learning gain is set as $a_k = 5/k$. The algorithms have been run for 500 iterations.

6.4 Illustrative Simulations

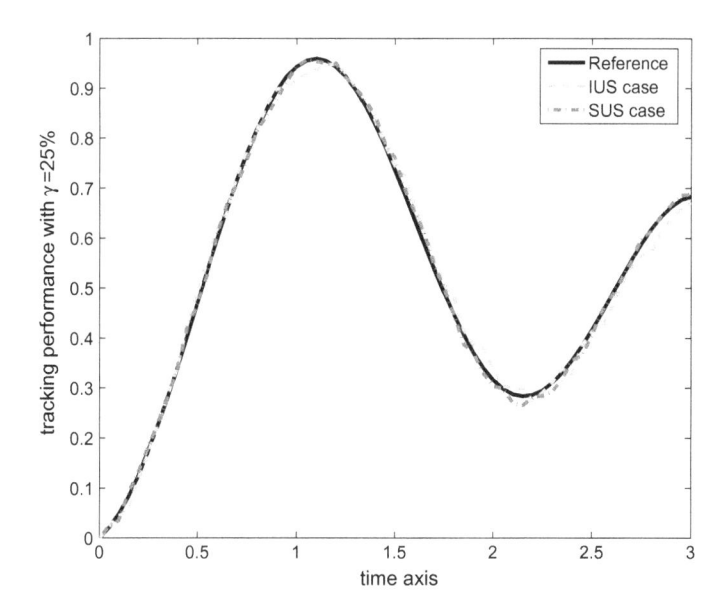

Fig. 6.3 Tracking performance of IUS and SUS with $\gamma = 0.25$

We first set the probability as $\rho = 0.75$. In other words, for any given time instance, the data of about 25% iterations might be lost during transmission. To make expression simple, let $\gamma = 1 - \rho$ denote the DDR. The tracking performance for the last iteration is shown in Fig. 6.3 with $\gamma = 0.25$, where the solid line, dashed line, and dash-dotted line denote the desired reference, final outputs of IUS and SUS, respectively. It is seen that well tracking is achieved for both schemes. Moreover, the maximal tracking errors, $\max_t |e_k(t)|$, are shown in Fig. 6.4 for both schemes. It should be pointed out that due to the existence of stochastic noises, the maximal errors generally do not converge to zero. All results reveal that the performances of IUS and SUS are similar under low DDR.

Next, we set $\rho = 0.25$ or equivalently $\gamma = 0.75$ to further compare the performance of both schemes. It means the transmission function is rather bad. The final outputs of both schemes are displayed in Fig. 6.5, while the maximal errors along iteration axis are shown in Fig. 6.6. It is noticed that under high DDR, SUS is superior to IUS. The inherent reason is that IUS would stop updating if the corresponding data is dropped while SUS keeps updating no matter whether the corresponding data is dropped or not. Thus, SUS updates more times than IUS within the same iteration amount. Moreover, one may find from Fig. 6.6 that there is a trade-off between IUS and SUS. However, this trade-off does not exist in Fig. 6.4. In conclusion, the trade-off is due to high DDR. To be specific, when the DDR is high and the learning gain is also large, SUS may lead to slightly excessive updating, which further generates a large crest of its maximal error profile along iteration axis.

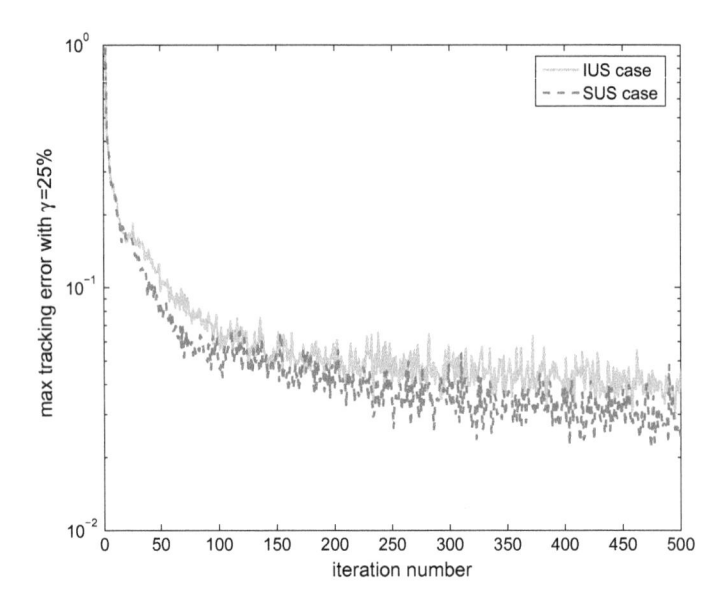

128　　　　　　　　　　　　　　　　　　6　Bernoulli Variable Model for Nonlinear Systems

Fig. 6.4 Maximal errors of IUS and SUS along iteration axis with $\gamma = 0.25$

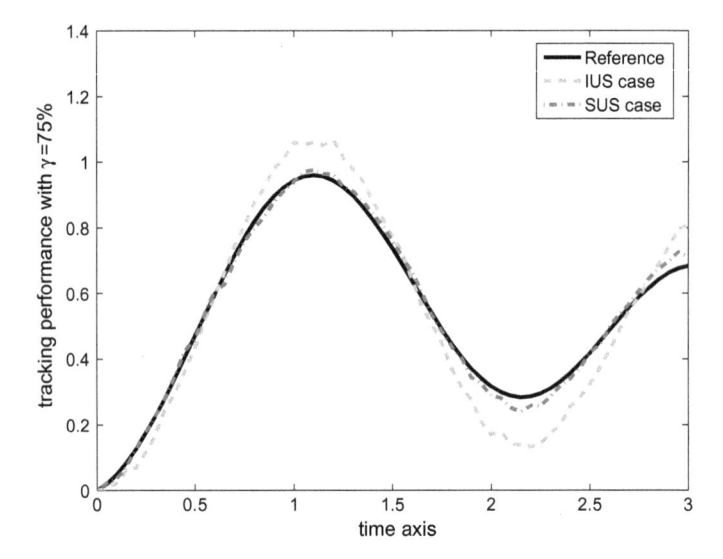

Fig. 6.5 Tracking performance of IUS and SUS with $\gamma = 0.75$

6.4 Illustrative Simulations

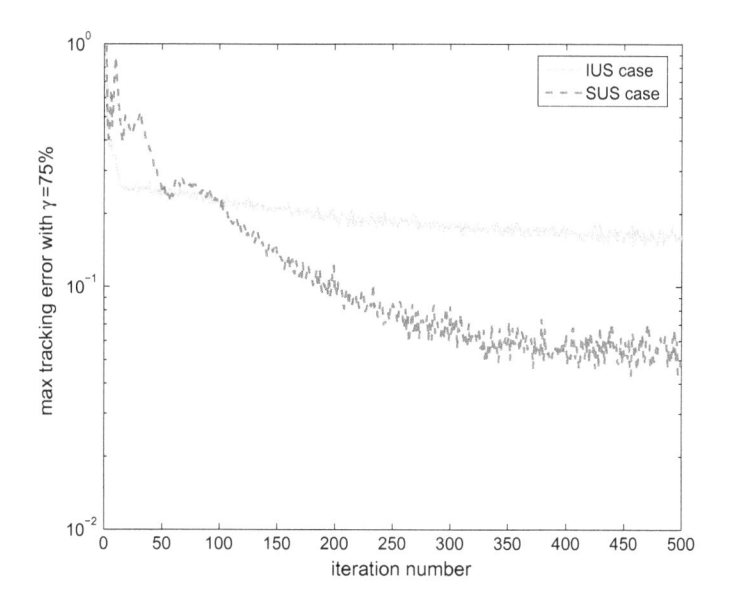

Fig. 6.6 Maximal errors of IUS and SUS along iteration axis with $\gamma = 0.75$

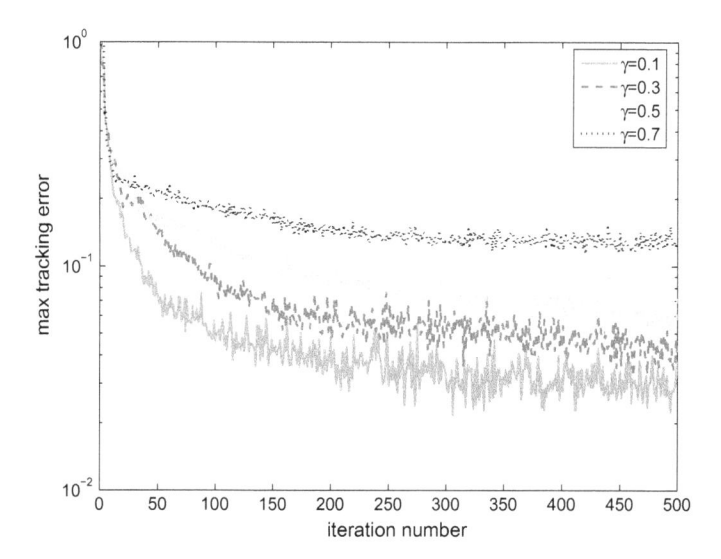

Fig. 6.7 Tracking performance comparison of IUS for different DDRs

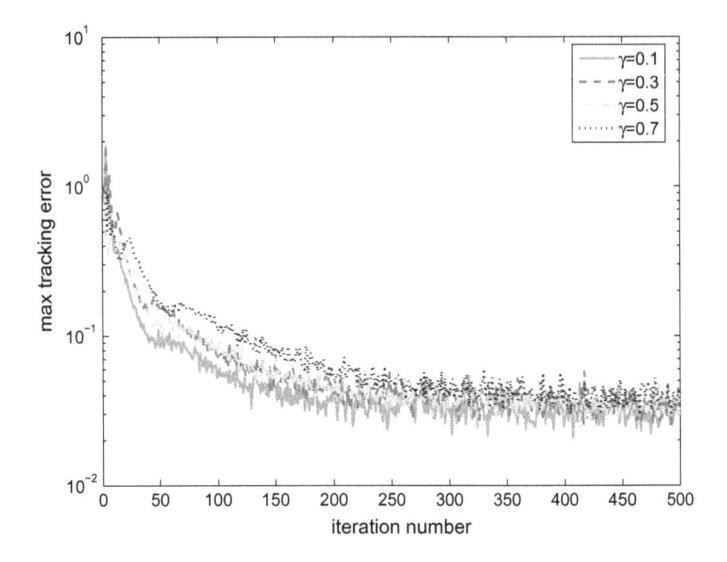

Fig. 6.8 Tracking performance comparison of SUS for different DDRs

To see the influence of DDR, we make comparisons for different DDRs. Here, we consider four cases of different DDRs, 0.1, 0.3, 0.5, and 0.7. The performance of IUS and SUS is shown in Figs. 6.7 and 6.8, respectively. It is seen from Fig. 6.7 that the tracking performance worsens as the DDR increases for IUS at the same iteration. In contrast, SUS can maintain similar performance after several iterations even though the rate increases, as shown in Fig. 6.8.

6.5 Summary

In this chapter, two data-driven schemes, i.e., the intermittent and successive update schemes, are addressed for networked nonlinear systems with random data dropouts. Here, the intermittent type only updates the input when the new packet is successfully received, whereas the successive type would keep updating with latest available data no matter whether the data is dropped or not. The Bernoulli random variable is taken to describe the data dropouts. Stochastic measurement noises are also considered. The almost sure convergence of the proposed schemes for any time instant is strictly proved based on mathematical induction principle. The results in this chapter are mainly based on [9].

References

1. Sun, M., Wang, D.: Analysis of nonlinear discrete-time systems with higher-order iterative learning control. Dyn. Control **11**, 81–96 (2001)
2. Bu, X., Yu, F., Hou, Z.-S., Wang, F.: Iterative learning control for a class of nonlinear systems with random packet losses. Nonlinear Anal.: Real World Appl. **14**(1), 567–580 (2013)
3. Tan, Y., Dai, H.-H., Huang, D., Xu, J.-X.: Unified iterative learning control schemes for nonlinear dynamic systems with nonlinear input uncertainties. Automatica **48**(12), 3173–3182 (2012)
4. Chen, Y., Wen, C., Gong, Z., Sun, M.: An iterative learning controller with initial state learning. IEEE Trans. Autom. Control **44**(2), 371–376 (1999)
5. Yang S., Xu J.-X., Huang D.: Iterative learning control for multi-agent systems consensus tracking. In: Proceedings of the 51st IEEE Conference on Decision and Control, pp. 4672–4677 (2012)
6. Xu, J.-X., Yan, R.: On initial conditions in iterative learning control. IEEE Trans. Autom. Control **50**(9), 1349–1354 (2005)
7. Chow, Y.S., Teicher, H.: Probability Theory: Independence, Interchangeability, Martingales. Springer, New York (1978)
8. Wang, D.: Convergence and robustness of discrete time nonlinear systems with iterative learning control. Automatica **34**(11), 1445–1448 (1998)
9. Shen, D., Zhang, C., Xu, Y.: Intermittent and successive ILC for stochastic nonlinear systems with random data dropouts. Asian J. Control (2018). https://doi.org/10.1002/asjc.1480

Chapter 7
Markov Chain Model for Linear Systems

7.1 Problem Formulation

This section first formulates the system configuration, and then proceeds to the data dropout models and the ILC algorithm.

Consider the following linear time-varying system:

$$\begin{aligned} x_k(t+1) &= A_t x_k(t) + B_t u_k(t), \\ y_k(t) &= C_t x_k(t) + w_k(t), \end{aligned} \quad (7.1)$$

where k is the iteration number, $k = 1, 2, \ldots$, and t is the time instant, $t = 0, 1, \ldots, N$, with N being the iteration length. The variables $x_k(t) \in \mathbf{R}^n$, $u_k(t) \in \mathbf{R}^p$, and $y_k(t) \in \mathbf{R}^q$ denote the system state, input, and output, respectively. Moreover, A_t, B_t, and C_t are system matrices with appropriate dimensions. The random variable $w_k(t) \in \mathbf{R}^q$ denotes the measurement noise or system disturbance. To economize on the notations, we omit random disturbance from the state equation; instead, we incorporate it into the above formulation as part of $w_k(t)$.

Without loss of generality, we assume the relative degree of the system as $\tau = 1$; that is, for any $t \geq 1$, $C_t B_{t-1} \neq 0$. In this case, the input at time instant t will affect the output at time instant $t+1$ but not the output at time instant t. In other words, the relative degree defines the smallest delay after which the input affects the corresponding output.

Remark 7.1 The relative degree τ is an inherent property of the system, so is usually time invariant. Generally, if we assume $\tau > 1$, then the following relationships hold for any $t \geq \tau$: $C_t A_{t+1-i}^{t-1} B_{t-i} = 0$, $1 \leq i \leq \tau - 1$, and $C_t A_{t+1-\tau}^{t-1} B_{t-\tau} \neq 0$, where $A_i^j \triangleq A_j A_{j-1} \cdots A_i$, $j \geq i$, and $A_{i+1}^i \triangleq I_n$. In addition, in the multi-input multi-output system formulation, the relative degree may depend on the dimension of the output vector; that is, different dimensions of the output vector may have different relative degree values. The results derived below are easily extended to the general relative degree case.

Denote the desired reference as $y_d(t)$, $t \in \{0, 1, \ldots, N\}$. Without loss of generality, we assume an achievable reference; that is, for a suitable initial value of $x_d(0)$, there exists a unique input $u_d(t)$ such that

$$x_d(t+1) = A_t x_d(t) + B_t u_d(t), \qquad (7.2)$$
$$y_d(t) = C_t x_d(t).$$

Remark 7.2 The uniqueness of the desired input $u_d(t)$ is guaranteed if the matrix $C_{t+1} B_t$ is of full-column rank for $t \geq 0$. That is, the input $u_d(t)$ can be recursively computed from the nominal model (7.2) as follows:

$$u_d(t) = \left[(C_{t+1} B_t)^T (C_{t+1} B_t) \right]^{-1} (C_{t+1} B_t)^T$$
$$\times \left(y_d(t+1) - C_{t+1} A_t x_d(t) \right). \qquad (7.3)$$

This conclusion has been stated in existing papers, such as [1, 2]. Note that the full-column rank requirement is not strict, because it is necessary for perfect tracking (as proved in [3, 4]). Consequently, (7.2) is a mild assumption for the system, as applied in many ILC papers. In addition, the desired input (7.3) is not accessible because the system information is unknown. Therefore, we must design ILC algorithms that approximate the desired input $u_d(t)$.

We make the following assumptions for the system (7.1).

Assumption 7.1 The initial state satisfies $x_k(0) = x_d(0)$, where $x_d(0)$ is consistent with the desired reference $y_d(0)$; that is, $y_d(0) = C_0 x_d(0)$.

Assumption 7.2 At an arbitrary time instant t, the random measurement noise $\{w_k(t)\}$ is a martingale difference sequence along the iteration axis with bounded second moments with respect to the σ-algebra $\mathscr{F}_k \triangleq \sigma\{x_i(t), u_i(t), y_i(t), w_i(t), 1 \leq i \leq k, 0 \leq t \leq N\}$ (i.e., the set of all events induced by these random variables up to the kth iteration). That is, $\mathbb{E}\{w_{k+1}(t) \mid \mathscr{F}_k\} = 0$, $\sigma_w^2 \triangleq \sup_k \mathbb{E}\{\|w_k(t)\|^2\} < \infty$.

Remark 7.3 Assumption 7.1 is the well-known identical initialization condition, which is a basic requirement for both time and space resetting of the system operation. This condition can be relaxed to the case $\mathbb{E}\{x_k(0)\} = x_d(0)$ because the initialization error $x_k(0) - x_d(0)$ can be reformulated as part of $w_k(0)$ satisfying Assumption 7.2. Moreover, since the system is considered to be repetitive, the stochastic noises are usually independent along the iteration axis. Therefore, Assumption 7.2 is mild and widely fulfilled in applications.

The system configuration is illustrated in Fig. 7.1. The data is transmitted through unreliable communication channels at both the measurement and actuator sides. To clarify our main idea, we consider that the communication network is unreliable only at the measurement side in this chapter, and operates well at the actuator side.

7.1 Problem Formulation

Fig. 7.1 System configuration of a networked ILC

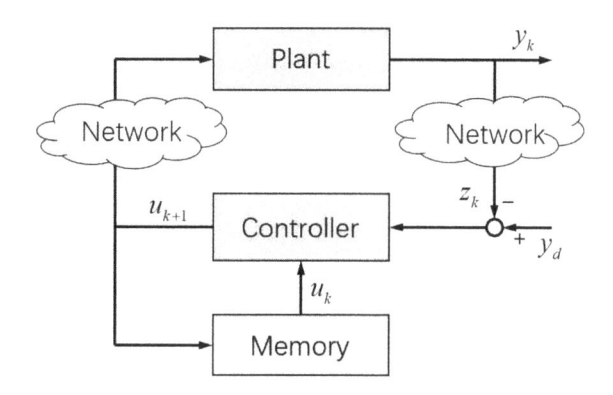

In this configuration, the output $y_k(t)$ of the system is transmitted back to the learning controller through an unreliable communication channel. A successfully transmitted and received signal is denoted by $z_k(t)$

$$z_k(t) = \alpha_k(t)y_k(t) + v_k(t), \qquad (7.4)$$

where $\alpha_k(t)$ represents the channel fading and $v_k(t)$ is the stochastic additive communication noise. In other words, the unreliable communication channel appends both multiplicative and additive randomness to the original measured output. Consequently, the actual signal for the learning update is $z_k(t)$ rather than $y_k(t)$. Similarly to [5, 6], we impose further assumptions on the fading channel as follows.

Assumption 7.3 Assume that $\alpha_k(t)$ is an independent scalar-valued white noise process with

$$\bar{\alpha}_t \triangleq \mathbb{E}\{\alpha_k(t)\}, \quad \sigma_t^2 \triangleq \mathbb{E}\{(\alpha_k(t) - \bar{\alpha}_t)(\alpha_k(t) - \bar{\alpha}_t)\} \qquad (7.5)$$

satisfying $\bar{\alpha}_t > 0$, $\sigma_t^2 < \infty$ for $t \geq 0$.

Assumption 7.4 The channel noise $v_k(t)$ is independent of $w_k(t)$ and $\alpha_k(t)$. Moreover, at an arbitrary time instant, $v_k(t)$ has zero-mean and bounded second moment, i.e., $\mathbb{E}\{v_k(t)\} = 0$, $\sigma_v^2 = \sup_k \mathbb{E}\{\|v_k(t)\|^2\} < \infty$.

In practice, transmission of a data packet may fail randomly under various communication conditions such as limited bandwidth and data congestion. In other words, $z_k(t)$ is not always available during the whole iteration interval and over all iterations. The data may be dropped while transmitting through an unreliable communication channel. To model this phenomenon, we introduce a binary random variable $\gamma_k(t)$ with state space $\{0, 1\}$ to describe its evolution, where $\gamma_k(t) = 1$ implies that the output $y_k(t)$ has been successfully received by the controller as $z_k(t)$, and $\gamma_k(t) = 0$ denotes that no information of $y_k(t)$ has been received by the controller. In this chapter, the randomness of $\gamma_k(t)$ is modeled by a Markov chain model.

Markov Chain Model (MCM). The random variable $\gamma_k(t)$ is independent for different time instants t. Moreover, for each time instant t, the evolution of $\gamma_k(t)$ along the iteration axis follows a two-state Markov chain with the probability transition matrix given by

$$P = \begin{bmatrix} p_{11} & p_{10} \\ p_{01} & p_{00} \end{bmatrix} = \begin{bmatrix} \mu & 1-\mu \\ 1-\nu & \nu \end{bmatrix} \tag{7.6}$$

with $0 < \mu, \nu < 1$, where $p_{ij} = \mathbb{P}(\gamma_{k+1}(t) = j \mid \gamma_k(t) = i)$, $i, j = 1, 0$.

We note that in the classical MCM, the independence of the data dropout variable $\gamma_k(t)$ is required with respect to time instants. That is, the random variable $\gamma_k(t)$ is independently given at different time instants. Such independence allows a positive probability that the outputs from the same iteration, i.e., $\{y_k(t), t = 1, \ldots, N\}$, are all successfully received by the controller (in the form of $z_k(t)$), and are used in the subsequent updating. To clarify this point, we make a specific description of the MCM for the entire iteration.

In an arbitrary iteration k, the random variable $\gamma_k(t)$ is either 1 or 0 at time instant t. We collect the time instants in which $\gamma_k(t) = 1$, i.e., in which the output is successfully received by the controller, and assemble them into a set ζ_k. That is, $\zeta_k \triangleq \{t \mid \gamma_k(t) = 1, 1 \le t \le N\}$. The set ζ_k contains the indices of all available outputs to the controller at the kth iteration. Hereafter, we call ζ_k a collection of data index at the kth iteration, or simply a *collection*. Because $\gamma_k(t)$ is independently given at different time instants, the collection ζ_k has $\iota \triangleq 2^N$ possible outcomes. We denote the set of all possible outcomes of ζ_k as $\varXi$, i.e., $\varXi \triangleq \{\zeta^{(1)}, \zeta^{(2)}, \ldots, \zeta^{(\iota)}\}$. Without loss of generality, we denote $\zeta^{(1)} = \mathbf{Z}_1^N$ as the collection of all time instants, and $\zeta^{(\iota)} = \emptyset$ excluding all time instants. In other words, $\varXi$ is a set of sets $\zeta^{(i)}$. Moreover, it is easily verified that the evolution of ζ_k (taking values from $\varXi$) is also a homogeneous ergodic Markov chain with positive transition probability matrix $[p_{ij}]_{1 \le i,j \le \iota}$, where p_{ij} denotes the transition probability from state $\zeta^{(i)}$ to state $\zeta^{(j)}$. In particular, we have $\min_{1 \le i \le \iota} p_{i1} > 0$. This property is extremely important in the convergence proof (see Theorems 7.1 and 7.2 in Sect. 7.3).

However, some practical scenarios cannot be described by the above models. First, the limited transmission bandwidth may preclude the transmission of all output packets. In other words, $\zeta^{(1)}$ cannot be constructed. Moreover, the data dropouts may occur time-dependently rather than time-independently during each iteration. In other words, early data dropouts may influence the occurrence of latter dropouts. Such cases invalidate the classical MCM, but can be accommodated in the general model proposed here.

General Markov Data Dropout Model (GMDDM). All possible collections of time instants at which the data is successfully transmitted comprise a subset $\varXi_G \subseteq \varXi$. That is, we have only κ possible data dropout collections from $\varXi$, i.e., $\zeta_G^{(i)} \in \varXi$, $1 \le i \le \kappa$ with $\kappa \le \iota$. Moreover, the union of all $\zeta_G^{(i)} \in \varXi_G$ is $\zeta^{(1)}$, i.e., $\bigcup_{i=1}^{\kappa} \zeta_G^{(i)} = \zeta^{(1)}$.

To extend the analysis in the GMDDM, we assume a Markov property. The Markov state at the kth iteration, denoted by m_k, has a finite state space. The states

7.1 Problem Formulation

of this space are sequenced in a set $\{1,\ldots,\kappa\}$ corresponding to the collection set $\Xi = \{\zeta_G^{(1)},\ldots,\zeta_G^{(\kappa)}\}$. That is, $m_k = i$ if and only if $\zeta_k = \zeta_G^{(i)}, i = 1,\ldots,\kappa$, where ζ_k is the Markov random collection at the kth iteration. Thus, the random process m_k completely describes the dynamic switching among different data collections.

Assumption 7.5 $\{m_k, k \geq 1\}$ is a homogeneous ergodic Markov chain with transition probability matrix $[p_{ij}]_{1 \leq i,j \leq \kappa}$, where $p_{ij} = \mathbb{P}\{m_{k+1} = j \mid m_k = i, m_0, \ldots, m_{k-1}\}$.

Remark 7.4 The newly defined GMDDM with Assumption 7.5 generalizes the classical MCM. To verify this point, let $\kappa = \iota$ or $\Xi_G = \Xi$. Clearly, GMDDM reduces to MCM. Indeed, MCM and GMDDM differ in that GMDDM handles incomplete data transmission over the whole iteration, and the occurrence of $\zeta^{(1)}$ is no longer necessary in GMDDM. In other words, GMDDM allows definitely incomplete data in all iterations, which cannot be covered by the classical models. Moreover, the time-dependence of the data dropouts is indicated by the selection of possible collections $\zeta_G^{(i)}$. A certain group of these collections implies a certain time-dependence among the data dropouts at different time instants.

Remark 7.5 GMDDM imposes a requirement $\bigcup_{i=1}^{\kappa} \zeta_G^{(i)} = \zeta^{(1)}$ on the information transmission. Intuitively, this condition implies that if the data dropout collection contains all possible alternatives in Ξ_G during successive multiple iterations, then the output can be transmitted back to the controller during these iterations. We emphasize that this condition is necessary to guarantee asymptotical convergence. Otherwise, the output will never be successfully received by the controller in at least one time instant, providing no information for improving the corresponding input. In addition, in a special case of GMDDM, the set Ξ_G includes $\zeta^{(1)}$. This case naturally satisfies the requirement $\bigcup_{i=1}^{\kappa} \zeta_G^{(i)} = \zeta^{(1)}$ and converges as specified in Corollary 7.2 in Sect. 7.4.

7.2 ILC Algorithms

The actual tracking error is denoted as $e_k(t) \triangleq y_d(t) - y_k(t)$, $t \in \{0, 1, \ldots, N\}$. Owing to the unreliable communication channel, this signal is not available for updating the control signal. Indeed, we can only access the received signal $z_k(t)$. We update the input $u_k(t)$ as follows:

$$u_{k+1}(t) = u_k(t) + a_k \gamma_k(t^+) L_t[y_d(t^+) - \bar{\alpha}_{t^+}^{-1} z_k(t^+)], \quad (7.7)$$

where t^+ is shorthand for $t + 1$, $\{a_k\}$ is a sequence of positive numbers denoting the step size, L_t is the learning gain matrix, and $\bar{\alpha}_t$ is the mean of the fading factor $\alpha_k(t)$ defined in Assumption 7.3. Clearly, the proposed algorithm is P-type. The step size $\{a_k\}$ requires the following assumption.

Assumption 7.6 The step size sequence $\{a_k\}$ satisfies: $a_k > 0$, $\sum_{k=1}^{\infty} a_k = \infty$, $\sum_{k=1}^{\infty} a_k^2 < \infty$, $a_k \to 0$ as $k \to \infty$.

To facilitate the convergence of (7.7), we rewrite it in a compact form by lifting all quantities from the entire iteration as follows:

$$U_{k+1} = U_k + a_k(\Gamma_k \otimes I_p)\mathscr{L}[Y_d - (\Upsilon^{-1} \otimes I_q)Z_k], \quad (7.8)$$

where $U_k = [u_k^T(0), \ldots, u_k^T(N-1)]^T \in \mathbf{R}^{pN}$, $\Gamma_k = \text{diag}\{\gamma_k(1), \ldots, \gamma_k(N)\} \in \mathbf{R}^{N \times N}$, $\mathscr{L} = \text{diag}\{L_0, \ldots, L_{N-1}\} \in \mathbf{R}^{pN \times qN}$, $Y_d = [y_d^T(1), \ldots, y_d^T(N)]^T \in \mathbf{R}^{qN}$, $\Upsilon = \text{diag}\{\bar{\alpha}_1, \ldots, \bar{\alpha}_N\} \in \mathbf{R}^{N \times N}$, and $Z_k = [z_k^T(1), \ldots, z_k^T(N)]^T \in \mathbf{R}^{qN}$. Similarly, we can define $Y_k = [y_k^T(1), \ldots, y_k^T(N)]^T \in \mathbf{R}^{qN}$, $W_k = [w_k^T(1), \ldots, w_k^T(N)]^T \in \mathbf{R}^{qN}$, and then we have

$$Y_k = \mathscr{H}U_k + Mx_k(0) + W_k, \quad (7.9)$$

where

$$\mathscr{H} = \begin{bmatrix} C_1 B_0 & 0 & \cdots & 0 \\ C_2 A_1 B_0 & C_2 B_1 & \cdots & 0 \\ \vdots & \vdots & \ddots & \vdots \\ C_N A_1^{N-1} B_0 & C_N A_2^{N-1} B_1 & \cdots & C_N B_{N-1} \end{bmatrix} \quad (7.10)$$

and $M = [(C_1 A_0)^T, \ldots, (C_N A_0^N)^T]^T \in \mathbf{R}^{qN \times n}$. Applying these notations and the desired reference dynamics (7.2), we have

$$Y_d = \mathscr{H}U_d + Mx_d(0), \quad (7.11)$$

where $U_d = [u_d^T(0), \ldots, u_d^T(N-1)]^T \in \mathbf{R}^{pN}$. Similarly, the transmission equation (7.4) is compactly expressed as

$$Z_k = (\Upsilon_k \otimes I_q)Y_k + V_k, \quad (7.12)$$

where $\Upsilon_k = \text{diag}\{\alpha_k(1), \ldots, \alpha_k(N)\} \in \mathbf{R}^{N \times N}$ and $V_k = [v_k^T(1), \ldots, v_k^T(N)]^T \in \mathbf{R}^{qN}$. Substituting (7.9), (7.11), and (7.12) into (7.8) and using Assumption 7.1, we have

$$\begin{aligned} U_{k+1} = & U_k + a_k(\Gamma_k \otimes I_p)\mathscr{L}\mathscr{H}(U_d - U_k) \\ & - a_k(\Gamma_k \otimes I_p)\mathscr{L}((\Upsilon^{-1}\widetilde{\Upsilon}_k) \otimes I_q)\mathscr{H}U_k \\ & - a_k(\Gamma_k \otimes I_p)\mathscr{L}((\Upsilon^{-1}\widetilde{\Upsilon}_k) \otimes I_q)Mx_k(0) \\ & - a_k(\Gamma_k \otimes I_p)\mathscr{L}(\Upsilon^{-1}\Upsilon_k \otimes I_q)W_k \\ & - a_k(\Gamma_k \otimes I_p)\mathscr{L}(\Upsilon^{-1} \otimes I_q)V_k, \end{aligned} \quad (7.13)$$

where $\widetilde{\Upsilon}_k \triangleq \Upsilon_k - \Upsilon$.

We emphasize that the super-vector formulation (7.8) only facilitates the analysis. The input is actually updated using the original formulation (7.7). Therefore, the learning update avoids the dimensional difficulty of handling the long operation length N. Note that the learning update law for the control input combines the input of the previous iteration (the first term on the right-hand side (RHS) of (7.13)) and an innovation correction term (the second term on the RHS of (7.13)). The remaining terms on the RHS of (7.13) describe the various stochastic noises/uncertainties in the system. We emphasize that the possible outcomes of Γ_k comprise a Markov chain corresponding to the data collection ζ_k, which introduces more uncertainties. These uncertainties will be considered later.

Our objective is to derive a strong convergence of the system output $y_k(t)$ to the desired reference $y_d(t)$ for $1 \leq t \leq N$. However, the measurement noise $w_k(t)$ prevents a precise tracking performance. Instead, we expect that the noise-free part $C_t x_k(t)$ can perfectly track the reference $y_d(t)$. Therefore, it is sufficient to show that the input sequence U_k strongly converges to the desired input U_d, as elaborated in the following sections.

7.3 ILC for Classical Markov Chain Model Case

In this section, we first derive the mean square and almost sure convergence of the input sequence to the desired input in the classical MCM case. The convergence proof in this case will assist our convergence analysis in the GMDDM. The convergence speed and extensions to time-varying step sizes are also discussed.

Under the assumption of the classical MCM, the data dropout variable $\gamma_k(t)$ evolves as a Markov chain. Therefore, at each iteration, the output of the entire iteration will be successfully transmitted with a positive probability. Convergence can then be proved by the Lyapunov method, as the energy function decay is guaranteed for each iteration. The mean square convergence is shown by the following theorem.

Theorem 7.1 *Consider system (7.1) and communication channel (7.4) under Assumptions 7.1–7.4 and 7.6. The data dropout is described by the classical Markov chain model. Then, the input sequence generated by Algorithm (7.7) converges to the desired input in the mean square sense,*

$$\lim_{k \to \infty} \mathbb{E}\{\|u_d(t) - u_k(t)\|^2\} = 0, \quad \forall t \in \mathbf{Z}_0^{N-1}, \tag{7.14}$$

if L_t is designed such that $-L_t C_{t+1} B_t$ is stable for all $t = 0, \ldots, N-1$.

Before conducting the convergence analysis, we need the following technical lemma.

Lemma 7.1 ([7]) *Let $\{r_k, k = 0, 1, \ldots\}$, $\{s_k, k = 0, 1, \ldots\}$, and $\{q_k, k = 0, 1, \ldots\}$ be real sequences, satisfying $0 < q_k \leq 1$, $s_k \geq 0$, $k = 0, 1, \ldots$, $\sum_{k=1}^{\infty} q_k = \infty$,*

$(s_k/q_k) \to 0$ as $k \to \infty$, and $r_{k+1} \leq (1-q_k)r_k + s_k$. Then, $\limsup_{k\to\infty} r_k \leq 0$. Particularly, if $r_k \geq 0$, $k = 0, 1, \ldots$, then $\lim_{k\to\infty} r_k = 0$.

We can now prove Theorem 7.1.

Proof Denote $\delta U_k = U_d - U_k$. Subtracting both side of (7.13) from U_d, we obtain

$$\begin{aligned}\delta U_{k+1} = & \left[I - a_k(\Gamma_k \otimes I_p)\mathscr{L}\mathscr{H}\right]\delta U_k \\ & + a_k(\Gamma_k \otimes I_p)\mathscr{L}((\Upsilon^{-1}\tilde{\Upsilon}_k) \otimes I_q)\mathscr{H}U_k \\ & + a_k(\Gamma_k \otimes I_p)\mathscr{L}((\Upsilon^{-1}\tilde{\Upsilon}_k) \otimes I_q)Mx_k(0) \\ & + a_k(\Gamma_k \otimes I_p)\mathscr{L}(\Upsilon^{-1}\Upsilon_k \otimes I_q)W_k \\ & + a_k(\Gamma_k \otimes I_p)\mathscr{L}(\Upsilon^{-1} \otimes I_q)V_k.\end{aligned} \quad (7.15)$$

Note that because $\mathscr{L}$ and $\mathscr{H}$ are block matrices, $\mathscr{L}\mathscr{H}$ is a lower triangular block matrix. The diagonal blocks of $\mathscr{L}\mathscr{H}$ are $L_t C_{t+1} B_t$, $t = 0, \ldots, N-1$. The designed learning gain matrix L_t must ensure the stability of $-L_t C_{t+1} B_t$. That is, all eigenvalues of $-L_t C_{t+1} B_t$ are with negative real parts, and then the lifted matrix $-\mathscr{L}\mathscr{H}$ is also stable. Consequently, there exists a positive definite matrix $\mathscr{Q}$ satisfying the Lyapunov equation

$$(\mathscr{L}\mathscr{H})^T \mathscr{Q} + \mathscr{Q}\mathscr{L}\mathscr{H} = I_{pN}. \quad (7.16)$$

Define a Lyapunov function $J_k = (\delta U_k)^T \mathscr{Q} \delta U_k$. Then, by (7.15), we obtain

$$\begin{aligned}J_{k+1} = & (\delta U_k)^T \left[I - a_k \Lambda_k\right]^T \mathscr{Q}\left[I - a_k \Lambda_k\right]\delta U_k \\ & + 2a_k(\delta U_k)^T \left[I - a_k \Lambda_k\right]^T \mathscr{Q}\Phi_k + a_k^2 \Phi_k^T \mathscr{Q}\Phi_k,\end{aligned} \quad (7.17)$$

where $\Lambda_k = (\Gamma_k \otimes I_p)\mathscr{L}\mathscr{H}$ and $\Phi_k = (\Gamma_k \otimes I_p)\mathscr{L}[((\Upsilon^{-1}\tilde{\Upsilon}_k) \otimes I_q)(\mathscr{H}U_k + Mx_k(0)) + (\Upsilon^{-1}\Upsilon_k \otimes I_q)W_k + (\Upsilon^{-1} \otimes I_q)V_k]$.

We further define the σ-algebra $\mathscr{G}_k = \sigma(x_i(0), u_i(t), y_i(t), w_i(t), \alpha_i(t), v_i(t), \gamma_i(t), 1 \leq i \leq k-1, 0 \leq t \leq N)$. Clearly, U_k is adapted to $\mathscr{G}_k$ (that is, $U_k \in \mathscr{G}_k$) by the updating mechanism (7.7). Below, we present a detailed estimation of $\mathbb{E}\{J_{k+1} \mid \mathscr{G}_k\}$.

We first note that

$$\begin{aligned}\mathbb{E}\left\{(\delta U_k)^T\right. & \left.\left[I - a_k \Lambda_k\right]^T \mathscr{Q}\left[I - a_k \Lambda_k\right]\delta U_k \mid \mathscr{G}_k\right\} \\ = & J_k - a_k(\delta U_k)^T \mathbb{E}\{\Lambda_k^T \mathscr{Q} + \mathscr{Q}\Lambda_k \mid \mathscr{G}_k\}\delta U_k \\ & + a_k^2 (\delta U_k)^T \mathbb{E}\{\Lambda_k^T \mathscr{Q}\Lambda_k \mid \mathscr{G}_k\}\delta U_k.\end{aligned} \quad (7.18)$$

Note that Γ_k is a diagonal matrix, whose possible outcomes have a one-to-one mapping to the set $\zeta^{(i)} \in \Xi$. In other words, the time-independence in the classical MCM admits ι possible cases of Γ_k. Moreover, as each diagonal entry $\gamma_k(t)$ of Γ_k is

7.3 ILC for Classical Markov Chain Model Case

a Markov random variable, Γ_k evolves as an irreducible and ergodic Markov chain. Thus, Γ_k can be the identity matrix I_N with a positive probability. In addition, all eigenvalues of the remaining $\iota - 1$ possible diagonal matrices of Γ_k are either 1 or 0. Consequently, there exists a constant $c_1 > 0$ such that

$$\mathbb{E}\{\Lambda_k^T \mathcal{Q} + \mathcal{Q}\Lambda_k \mid \mathcal{G}_k\} \geq c_1 I_{pN}. \tag{7.19}$$

Moreover, noting that $\|\Lambda_k\| = \|(\Gamma_k \otimes I_p)\mathcal{L}\mathcal{H}\| \leq \|\mathcal{L}\mathcal{H}\|$, we can find another constant $c_2 = \max\{\|\mathcal{L}\mathcal{H}\|, 1\}$ for which $(\delta U_k)^T \mathbb{E}\{\Lambda_k^T \mathcal{Q}\Lambda_k \mid \mathcal{G}_k\}\delta U_k \leq c_2 J_k$. Substituting these derivations into (7.18) yields

$$\mathbb{E}\left\{(\delta U_k)^T [I - a_k \Lambda_k]^T \mathcal{Q}[I - a_k \Lambda_k]\delta U_k \mid \mathcal{G}_k\right\}$$
$$\leq (1 - c_1 c_3 a_k + c_2 a_k^2) J_k, \tag{7.20}$$

where $c_3 > 0$ satisfies $I_{pN} \geq c_3 \mathcal{Q}$.

By Assumptions 7.2–7.4 we have $\mathbb{E}\{\widetilde{\Upsilon}_k \mid \mathcal{G}_k\} = \mathbb{E}\{W_k \mid \mathcal{G}_k\} = \mathbb{E}\{V_k \mid \mathcal{G}_k\} = 0$, which further leads to

$$\mathbb{E}\{\Phi_k \mid \mathcal{G}_k\}$$
$$= \mathbb{E}\{(\Gamma_k \otimes I_p)\mathcal{L}[((\Upsilon^{-1}\widetilde{\Upsilon}_k) \otimes I_q)(\mathcal{H}U_k + Mx_k(0))] \mid \mathcal{G}_k\}$$
$$+ \mathbb{E}\{(\Gamma_k \otimes I_p)\mathcal{L}[(\Upsilon^{-1}\Upsilon_k \otimes I_q)W_k + (\Upsilon^{-1} \otimes I_q)V_k] \mid \mathcal{G}_k\}$$
$$= \mathbb{E}\{(\Gamma_k \otimes I_p) \mid \mathcal{G}_k\}\mathcal{L}\left[\mathbb{E}\{\Upsilon^{-1}\widetilde{\Upsilon}_k \mid \mathcal{G}_k\} \otimes I_q\right](\mathcal{H}U_k + Mx_d(0))$$
$$+ \mathbb{E}\{(\Gamma_k \otimes I_p) \mid \mathcal{G}_k\}\mathcal{L}\mathbb{E}\{(\Upsilon^{-1}\Upsilon_k \otimes I_q) \mid \mathcal{G}_k\}\mathbb{E}\{W_k \mid \mathcal{G}_k\}$$
$$+ \mathbb{E}\{(\Gamma_k \otimes I_p) \mid \mathcal{G}_k\}\mathcal{L}(\Upsilon^{-1} \otimes I_q)\mathbb{E}\{V_k \mid \mathcal{G}_k\} = 0, \tag{7.21}$$

where we have used the signal independence among $\alpha_k(t)$, $w_k(t)$, $v_k(t)$, and $\gamma_k(t)$. We then deduce that

$$\mathbb{E}\{(\delta U_k)^T [I - a_k \Lambda_k]^T \mathcal{Q}\Phi_k \mid \mathcal{G}_k\} = 0. \tag{7.22}$$

By Assumption 7.3, Υ_k is independent along the iteration axis. Therefore, we have

$$\mathbb{E}\left\{\|(\Gamma_k \otimes I_p)\mathcal{L}((\Upsilon^{-1}\widetilde{\Upsilon}_k) \otimes I_q)\mathcal{H}U_k\|_{\mathcal{Q}}^2 \mid \mathcal{G}_k\right\}$$
$$= (\mathcal{H}U_k)^T \mathbb{E}'\left\{((\Upsilon^{-1}\widetilde{\Upsilon}_k) \otimes I_q)^T \mathcal{L}^T \mathbb{E}\{(\Gamma_k \otimes I_p)^T \mathcal{Q}(\Gamma_k \otimes I_p) \mid \mathcal{G}_k\}\right.$$
$$\times \mathcal{L}((\Upsilon^{-1}\widetilde{\Upsilon}_k) \otimes I_q)\right\}\mathcal{H}U_k, \tag{7.23}$$

where the expectation $\mathbb{E}'$ is computed with respect to $\widetilde{\Upsilon}_k$. Under the classical MCM and the second-order boundedness in Assumption 7.3, there exists a constant $c_4 > 0$ such that

$$\mathbb{E}\left\{\left\|(\varGamma_k \otimes I_p)\mathscr{L}((\varUpsilon^{-1}\widetilde{\varUpsilon}_k) \otimes I_q)\mathscr{H} U_k\right\|_{\mathscr{Q}}^2 \mid \mathscr{G}_k\right\} \leq c_4 \|\mathscr{H} U_k\|_{\mathscr{Q}}^2. \tag{7.24}$$

In addition, noting that $U_k = U_d - \delta U_k$ and applying Cauchy inequality, we obtain

$$c_4 \|\mathscr{H} U_k\|_{\mathscr{Q}}^2 \leq 2c_4(\|U_d\|_{\mathscr{Q}}^2 + J_k). \tag{7.25}$$

Similarly, we obtain

$$\mathbb{E}\left\{\left\|(\varGamma_k \otimes I_p)\mathscr{L}((\varUpsilon^{-1}\widetilde{\varUpsilon}_k) \otimes I_q)M x_k(0)\right\|_{\mathscr{Q}}^2 \mid \mathscr{G}_k\right\}$$
$$\leq c_4 \|M x_k(0)\|_{\mathscr{Q}}^2 = c_4 \|M x_d(0)\|_{\mathscr{Q}}^2, \tag{7.26}$$

$$\mathbb{E}\left\{\left\|(\varGamma_k \otimes I_p)\mathscr{L}(\varUpsilon^{-1}\varUpsilon_k \otimes I_q)W_k\right\|_{\mathscr{Q}}^2 \mid \mathscr{G}_k\right\} \leq c_5 \mathbb{E}\{\|W_k\|^2 \mid \mathscr{G}_k\}, \tag{7.27}$$

$$\mathbb{E}\left\{\left\|(\varGamma_k \otimes I_p)\mathscr{L}(\varUpsilon^{-1} \otimes I_q)V_k\right\|_{\mathscr{Q}}^2 \mid \mathscr{G}_k\right\} \leq c_6 \mathbb{E}\{\|V_k\|^2 \mid \mathscr{G}_k\}, \tag{7.28}$$

by Assumptions 7.1–7.4, where $c_5 > 0$ and $c_6 > 0$ are suitable constants. From (7.24)–(7.28) and applying Cauchy inequality, it follows that

$$\mathbb{E}\{\Phi_k^T \mathscr{Q} \Phi_k \mid \mathscr{G}_k\}$$
$$\leq 8c_4 J_k + 8c_4 \|U_d\|_{\mathscr{Q}}^2 + 4c_4 \|M x_d(0)\|_{\mathscr{Q}}^2$$
$$+ 4c_5 \mathbb{E}\{\|W_k\|^2 \mid \mathscr{G}_k\} + 4c_6 \mathbb{E}\{\|V_k\|^2 \mid \mathscr{G}_k\}. \tag{7.29}$$

Summarizing the above derivations, we obtain

$$\mathbb{E}\{J_{k+1} \mid \mathscr{G}_k\} \leq [1 - c_1 c_3 a_k + (c_2 + 8c_4)a_k^2] J_k$$
$$+ a_k^2 \left[8c_4 \|U_d\|_{\mathscr{Q}}^2 + 4c_4 \|M x_d(0)\|_{\mathscr{Q}}^2\right.$$
$$\left. + 4c_5 \mathbb{E}\{\|W_k\|^2 \mid \mathscr{G}_k\} + 4c_6 \mathbb{E}\{\|V_k\|^2 \mid \mathscr{G}_k\}\right], \tag{7.30}$$

and then take mathematical expectation to both sides of the above inequality,

$$\mathbb{E}\{J_{k+1}\} \leq [1 - c_1 c_3 a_k + (c_2 + 8c_4)a_k^2]\mathbb{E}\{J_k\} + c_7 a_k^2, \tag{7.31}$$

where $c_7 = 8c_4 \|U_d\|_{\mathscr{Q}}^2 + 4c_4 \|M x_d(0)\|_{\mathscr{Q}}^2 + 4c_5 N\sigma_w^2 + 4c_6 N\sigma_v^2$.

Next, we apply Lemma 7.1 to show the mean square convergence of the classical MCM. To this end, we note that $\mathbb{E}\{J_k\}$, $c_1 c_3 a_k - (c_2 + 8c_4)a_k^2$, and $c_7 a_k^2$ correspond to r_k, q_k, and s_k in Lemma 7.1, respectively. Noting also that $c_1 c_3 > 0$, $a_k \to 0$, there exists a sufficiently large number k_0 such that $(c_2 + 8c_4)a_k \leq c_1 c_3/2$ and $c_1 c_3 a_k < 1$, $\forall k \geq k_0$. Therefore, $0 \leq 1 - c_1 c_3 a_k + (c_2 + 8c_4)a_k^2 < 1$, $\forall k \geq k_0$. In addition,

7.3 ILC for Classical Markov Chain Model Case

$$\sum_{k=k_0}^{\infty} \left[c_1 c_3 a_k - (c_2 + 8c_4) a_k^2 \right] \geq \frac{1}{2} c_1 c_3 \sum_{k=k_0}^{\infty} a_k = \infty, \tag{7.32}$$

$$\frac{c_7 a_k^2}{c_1 c_3 a_k - (c_2 + 8c_4) a_k^2} \to 0, \quad \text{as } k \to \infty. \tag{7.33}$$

As the conditions of Lemma 7.1 are fulfilled, we obtain $\lim_{k \to \infty} \mathbb{E}\{J_k\} = 0$. Again, noting that $\mathscr{Q}$ is positive definite and $J_k = \|\delta U_k\|_{\mathscr{Q}}^2 \geq \sigma_{\min}(\mathscr{Q}) \|\delta U_k\|^2$, where $\sigma_{\min}(\cdot)$ denotes the smallest singular value, we obtain $\lim_{k \to \infty} \mathbb{E}\{\|\delta U_k\|^2\} = 0$, or equivalently, $\lim_{k \to \infty} \mathbb{E}\{\|u_d(t) - u_k(t)\|^2\} = 0$, $\forall t \in \mathbf{Z}_0^{N-1}$. Thus, the proof is completed.

Theorem 7.1 establishes the mean square convergence of the proposed algorithm (7.7) when the data dropouts evolve as a conventional Markov chain. Below we further give the almost sure analysis by showing more properties of the involved random variables including uniform integrability and certain moment conditions.

Theorem 7.2 *Consider system (7.1) and communication channel (7.4) under Assumptions 7.1–7.4 and 7.6. The data dropout is described by the classical Markov chain model. Then, the input sequence generated by Algorithm (7.7) converges to the desired input in the almost sure sense,*

$$\lim_{k \to \infty} u_k(t) = u_d(t) \ a.s., \quad \forall t \in \mathbf{Z}_0^{N-1}, \tag{7.34}$$

if L_t is designed such that $-L_t C_{t+1} B_t$ is stable for all $t = 0, \ldots, N-1$.

To prove the theorem, we need the following lemma.

Lemma 7.2 ([8]) *Let $\{\beta_k, \mathscr{F}_k\}$, $\{b_k, \mathscr{F}_k\}$, and $\{d_k, \mathscr{F}_k\}$ be nonnegative adapted sequences satisfying*

$$\mathbb{E}\{\beta_{k+1} \mid \mathscr{F}_k\} \leq \beta_k - b_k + d_k, \tag{7.35}$$

and $\mathbb{E}\{|\beta_1|\} < \infty$, $\sum_{k=1}^{\infty} d_k < \infty$, a.s. Then, β_k converges to a finite random variable in the almost sure sense, and $\sum_{k=1}^{\infty} b_k < \infty$, a.s.

We now provide the proof of Theorem 7.2.

Proof Recalling the recursive relation (7.30), we show the convergence by verifying the conditions of Lemma 7.2. To this end, by Assumptions 7.2 and 7.4, we have

$$\mathbb{E}\left\{\sum_{k=1}^{\infty} a_k^2 \mathbb{E}\{\|W_k\|^2 \mid \mathscr{G}_k\}\right\} = \sum_{k=1}^{\infty} a_k^2 \mathbb{E}\{\|W_k\|^2\} < \infty,$$

$$\mathbb{E}\left\{\sum_{k=1}^{\infty} a_k^2 \mathbb{E}\{\|V_k\|^2 \mid \mathscr{G}_k\}\right\} = \sum_{k=1}^{\infty} a_k^2 \mathbb{E}\{\|V_k\|^2\} < \infty.$$

Additionally, we have shown that $\lim_{k \to \infty} \mathbb{E}\{J_k\} = 0$, implying that $\mathbb{E}\{J_k\}$ is bounded. Thus,

$$\mathbb{E}\left\{\sum_{k=1}^{\infty} a_k^2 J_k\right\} = \sum_{k=1}^{\infty} a_k^2 \mathbb{E}\{J_k\} < \infty.$$

The above estimations indicate that $\sum_{k=1}^{\infty} a_k^2 \mathbb{E}\{\|W_k\|^2 \mid \mathscr{G}_k\} < \infty$, a.s., $\sum_{k=1}^{\infty} a_k^2 \mathbb{E}\{\|V_k\|^2 \mid \mathscr{G}_k\} < \infty$, a.s., $\sum_{k=1}^{\infty} a_k^2 J_k < \infty$, a.s. Thus, by Lemma 7.2 we confirm that J_k converges almost surely as $k \to \infty$. In other words, δU_k converges in the almost sure sense. On the other hand, δU_k converges to zero in the mean square sense (see proof of Theorem 7.1). Then, the almost surely convergent limit of δU_k should also be zero. That is, $\lim_{k\to\infty} u_k(t) = u_d(t)$, a.s., $\forall t$. The proof is completed.

Remark 7.6 Comparing (7.30) and (7.35), we observe that $c_1 c_3 a_k J_k$ corresponds to b_k. As all conditions of Lemma 7.2 are satisfied, it follows that

$$\sum_{k=1}^{\infty} a_k J_k < \infty, \quad a.s. \tag{7.36}$$

Furthermore, if the learning step size satisfies $a_k \downarrow 0$, by the Kronecker lemma we have

$$\lim_{k\to\infty} a_k \sum_{i=1}^{k} J_i = 0, \quad a.s. \tag{7.37}$$

Then, applying the Cauchy inequality, we obtain

$$\frac{1}{k}\sum_{i=1}^{k}\sqrt{J_i} \leq \frac{1}{\sqrt{ka_k}}\left(a_k \sum_{i=1}^{k} J_i\right)^{1/2} = o\left(\frac{1}{\sqrt{ka_k}}\right), \quad a.s. \tag{7.38}$$

The above estimation provides an insight into the convergence speed of the proposed algorithm. In particular, we know that the iteration average of the input error ($\|\delta U_k\|$) converges to zero no more slowly than the inverse of $\sqrt{ka_k}$. From this estimation, we can select the suitable learning step size in practical applications.

We now revisit the update algorithm (7.7). It is seen that the learning step size a_k is specified in advance, imitating the stochastic approximation method for stochastic systems. However, in practical applications with data dropout, the step size a_k decreases with increasing iteration number, regardless of whether the data is successfully transmitted or not. Therefore, a prespecified step size may result in premature convergence. In an unreliable communication channel with small successful transmission rate (representing the average number ratio of the update iterations to the total iterations, i.e., $\lim_{n\to\infty} 1/n \times \sum_{k=1}^{n} \gamma_k(t)$), the step size used in the update rapidly decreases to zero, and premature convergence is certain. To solve this problem, a possible way is to make the step size change only when the updating truly happens. For this purpose, we modify algorithm (7.7) as

7.3 ILC for Classical Markov Chain Model Case

$$u_{k+1}(t) = u_k(t) + a_{\tau(k,t)}\gamma_k(t^+)L_t[y_d(t^+) - \bar{\alpha}_{t^+}^{-1}z_k(t^+)], \quad (7.39)$$

where $\tau(k, t)$ depends on both the iteration and time indices, $\tau(k, t) = \sum_{i=1}^{k-1} \gamma_i(t^+)$. In other words, $\tau(k, t)$ denotes the number of true updates up to but excluding the kth iteration. Moreover, when $\gamma_k(t^+) = 0$ (implying data dropout), then $\tau(k+1, t) = \tau(k, t)$, indicating that the step size of the next iteration is unchanged. Indeed, $\tau(k, t)$ is a stopping time with respect to iteration index k, $\forall t$. Noting the newly introduced randomness of $\tau(k, t)$ and the time-dependent update of $a_{\tau(k,t)}$, it is important to analyze convergence properties of the modified algorithm (7.39), in which the step size may alert at different time instants.

In this case, algorithm (7.39) can be compacted as follows:

$$U_{k+1} = U_k + (\mathscr{A}_k \otimes I_p)(\Gamma_k \otimes I_p)\mathscr{L}\mathscr{H}\delta U_k - (\mathscr{A}_k \otimes I_p)\Phi_k, \quad (7.40)$$

where $\mathscr{A}_k = \mathrm{diag}\{a_{\tau(k,0)}, \ldots, a_{\tau(k,N-1)}\}$.

The convergence properties of Algorithm (7.40) are described below.

Theorem 7.3 *Consider system (7.1) and communication channel (7.4) under Assumptions 7.1–7.4 and 7.6. The data dropout is described by the classical Markov chain model. If L_t is designed such that $-L_t C_{t+1} B_t$ is stable for all $t = 0, \ldots, N-1$, then the input sequence generated by Algorithm (7.39) converges to the desired input in both the mean square and almost sure senses.*

Proof The proof of this theorem is similar to those of Theorems 7.1 and 7.2, so we outline the necessary steps only.

From (7.40), it follows that

$$\delta U_{k+1} = [I - (\mathscr{A}_k \otimes I_p)\Lambda_k]\delta U_k + (\mathscr{A}_k \otimes I_p)\Phi_k,$$

and therefore

$$\begin{aligned}J_{k+1} = &(\delta U_k)^T[I - (\mathscr{A}_k \otimes I_p)\Lambda_k]^T \mathscr{Q}[I - (\mathscr{A}_k \otimes I_p)\Lambda_k]\delta U_k \\&+ 2(\delta U_k)^T[I - (\mathscr{A}_k \otimes I_p)\Lambda_k]^T \mathscr{Q}(\mathscr{A}_k \otimes I_p)\Phi_k \\&+ \Phi_k^T(\mathscr{A}_k \otimes I_p)^T \mathscr{Q}(\mathscr{A}_k \otimes I_p)\Phi_k.\end{aligned} \quad (7.41)$$

We now estimate $\mathbb{E}\{J_{k+1} \mid \mathscr{G}_k\}$. Noting that $a_{\tau(k,t)} \in \mathscr{G}_k$, $\forall t$, we have $\mathscr{A}_k \in \mathscr{G}_k$, implying that

$$\mathbb{E}\left\{(\delta U_k)^T[I - (\mathscr{A}_k \otimes I_p)\Lambda_k]^T \mathscr{Q}(\mathscr{A}_k \otimes I_p)\Phi_k \mid \mathscr{G}_k\right\} = 0. \quad (7.42)$$

Now denote $\bar{a}_{\tau(k)} = \mathbb{E}\{a_{\tau(k,t)}\}$, $\forall t$ and $\tilde{a}_{\tau(k,t)} = a_{\tau(k,t)} - \bar{a}_{\tau(k)}$. The data dropout variable is assumed to be time independent with the same statistics, meaning that $\mathbb{E}\{a_{\tau(k,t)}\}$ is invariant at different time instants. Then, $\widetilde{\mathscr{A}}_k = \mathrm{diag}\{\tilde{a}_{\tau(k,0)}, \ldots, \tilde{a}_{\tau(k,N-1)}\} = \mathscr{A}_k - \bar{a}_k I_N$, and we obtain

$$\mathbb{E}\{J_{k+1} \mid \mathscr{G}_k\} = \mathbb{E}\{\xi_{1,k} \mid \mathscr{G}_k\} + \mathbb{E}\{\xi_{2,k} \mid \mathscr{G}_k\}, \tag{7.43}$$

where $\xi_{1,k} = (\delta U_k)^T [I - \bar{a}_{\tau(k)} \Lambda_k]^T \mathscr{Q} [I - \bar{a}_{\tau(k)} \Lambda_k] \delta U_k + \bar{a}_{\tau(k)}^2 \Phi_k^T \mathscr{Q} \Phi_k$, $\xi_{2,k} = -2(\delta U_k)^T [I - \bar{a}_{\tau(k)} \Lambda_k]^T \mathscr{Q} (\widetilde{\mathscr{A}_k} \otimes I_p) \Lambda_k \delta U_k + (\delta U_k)^T \Lambda_k^T (\widetilde{\mathscr{A}_k} \otimes I_p)^T \mathscr{Q} (\widetilde{\mathscr{A}_k} \otimes I_p) \Lambda_k \delta U_k + \bar{a}_{\tau(k)} \Phi_k^T [(\widetilde{\mathscr{A}_k} \otimes I_p)^T \mathscr{Q} + \mathscr{Q} (\widetilde{\mathscr{A}_k} \otimes I_p)] \Phi_k + \Phi_k^T (\widetilde{\mathscr{A}_k} \otimes I_p)^T \mathscr{Q} (\widetilde{\mathscr{A}_k} \otimes I_p) \Phi_k$.

Similar to (7.30) we have

$$\mathbb{E}\{\xi_{1,k} \mid \mathscr{G}_k\} \leq [1 - c_1 c_3 \bar{a}_{\tau(k)} + (c_2 + 8c_4) \bar{a}_{\tau(k)}^2] J_k + c_7 \bar{a}_{\tau(k)}^2. \tag{7.44}$$

Using similar derivations of (7.30) we can obtain

$$\mathbb{E}\{\xi_{2,k} \mid \mathscr{G}_k\} \leq -c_8 \min_{0 \leq t \leq N-1} \{\widetilde{a}_{\tau(k,t)}\} J_k + c_9 \bar{a}_{\tau(k)} \|\widetilde{\mathscr{A}_k}\| J_k + c_{10} \|\widetilde{\mathscr{A}_k}\|^2 J_k + (\bar{a}_{\tau(k)} \|\widetilde{\mathscr{A}_k}\| + \|\widetilde{\mathscr{A}_k}\|^2) c_7. \tag{7.45}$$

From the above two inequalities, it follows

$$\mathbb{E}\{J_{k+1} \mid \mathscr{G}_k\} \leq (1 - q_k') J_k + (\bar{a}_{\tau(k)}^2 + \bar{a}_{\tau(k)} \|\widetilde{\mathscr{A}_k}\| + \|\widetilde{\mathscr{A}_k}\|^2) c_7, \tag{7.46}$$

where $q_k' = c_1 c_3 \bar{a}_{\tau(k)} - (c_2 + 8c_4) \bar{a}_{\tau(k)}^2 - c_8 \min_{0 \leq t \leq N-1} \{\widetilde{a}_{\tau(k,t)}\} + c_9 \bar{a}_{\tau(k)} \|\widetilde{\mathscr{A}_k}\| + c_{10} \|\widetilde{\mathscr{A}_k}\|^2$. Taking mathematical expectation on both sides of the last inequality we have $\mathbb{E}\{J_{k+1}\} \leq (1 - q_k') \mathbb{E}\{J_k\} + (\bar{a}_{\tau(k)}^2 + \bar{a}_{\tau(k)} \mathbb{E}\{\|\widetilde{\mathscr{A}_k}\|\} + \mathbb{E}\{\|\widetilde{\mathscr{A}_k}\|^2\}) c_7$.

Note that $\bar{a}_{\tau(k)} > a_k$, thus $\sum_{k=1}^{\infty} \bar{a}_{\tau(k)} = \infty$. Meanwhile, in the classical MCM, the Markov chain is stationary along the iteration axis, so $\mathbb{E}\{\tau(k,t)\}$ linearly depends on the iteration number k. This implies that $\bar{a}_{\tau(k)} \to 0$ as $k \to \infty$. Moreover, the difference between $\bar{a}_{\tau(k)}$ and $a_{\tau(k,t)}$ is infinitesimal in higher order with respect to $\bar{a}_{\tau(k)}$, that is, $\|\widetilde{\mathscr{A}_k}\| = o(\bar{a}_{\tau(k)})$. Similar to the proof of Theorem 7.1, we thus verify that all conditions of Lemma 7.1 are fulfilled after sufficiently many iterations. Therefore, the convergence of $\mathbb{E}\{J_k\}$ is guaranteed. Moreover, as $\mathscr{Q}$ is positive definite, $u_k(t)$ converges in the mean square sense.

To show the almost sure convergence, we apply the property $\sum_{k=1}^{\infty} \bar{a}_{\tau(k)}^2 < \infty$. In the classical MCM, we find that $\lim_{k \to \infty} \tau(k,t)/k > 0$, $\forall t$. Therefore, there exists a suitable $\varepsilon > 0$ such that $\tau(k,t) > \varepsilon k$ for all $k \geq k_0$ with k_0 being a sufficiently large number. Consequently, we have $\sum_{k=k_0}^{\infty} \bar{a}_{\tau(k)}^2 \leq \sum_{k=k_0}^{\infty} a_{\varepsilon k}^2 < \infty$. Similar to the proof of Theorem 7.2, we can verify the conditions of Lemma 7.2, and hence confirm that J_k converges almost surely as $k \to \infty$. The proof is completed.

In this theorem, the data dropout distribution is assumed to be time invariant. In general, the dropout frequency or successful transmission rate may vary from one time instant to another. With the limited knowledge on the dropout frequency, we can vary the step size sequence at different time instants. When the successful transmission rate is low, we require a slowly decreasing step size sequence; when the successful transmission rate is high, the step size sequence can decrease more

7.3 ILC for Classical Markov Chain Model Case

rapidly. At each time instant t, we assign the sequence $\{a_k(t)\}$, and the ILC algorithm (in the following compact form) becomes

$$U_{k+1} = U_k + (\mathscr{A}_k^* \otimes I_p)(\Gamma_k \otimes I_p)\mathscr{L}\mathscr{H}\delta U_k - (\mathscr{A}_k^* \otimes I_p)\Phi_k, \qquad (7.47)$$

where $\mathscr{A}_k^* = \text{diag}\{a_k(0), \ldots, a_k(N-1)\}$. Similar to Theorem 7.3, we have the following corollary.

Corollary 7.1 *Consider system (7.1) and communication channel (7.4) under Assumptions 7.1–7.4. The data dropout is described by the classical Markov chain model. Then, the input sequence generated by Algorithm (7.47) converges to the desired input in both the mean square and almost sure senses, if L_t is designed such that $-L_t C_{t+1} B_t$ is stable for all $t = 0, \ldots, N-1$ and the step size sequences $\{a_k(t)\}$ satisfy Assumption 7.6 and*

$$\max_{0 \leq t, s \leq N-1} |a_k(t) - a_k(s)| = o\left(\sum_{t=0}^{N-1} a_k(t)\right), \quad as\ k \to \infty. \qquad (7.48)$$

This corollary can be proved similarly to Theorem 7.3, but the expectation of the step size in the proof of Theorem 7.3 is replaced by the average value $\bar{a}_k = (1/N)\sum_{t=0}^{N-1} a_k(t)$ to facilitate the decay of J_k. The details are omitted to conserve space.

To conclude this section, we note that the convergence analysis primarily depends on the positive probability of the event that all outputs of the entire iteration are successfully transmitted in the classical MCM. In certain applications, these requirements are precluded by limited transmission bandwidth and severe transmission circumstances. Therefore, we propose the GMDDM to describe this case. In such case, the Lyapunov function J_k is unlikely to decrease in a step-by-step manner. Instead, the convergence of J_k is ensured by the assumption that the union of all possible data collections is $\zeta^{(1)}$. The details will be elaborated in the next section.

7.4 ILC for General Markov Data Dropout Model Case

In this section, we are in the position to prove the corresponding results of Theorems 7.1 and 7.2 under GMDDM. To this end, we need to strengthen the assumption of the step size sequence $\{a_k\}$ as follows.

Assumption 7.7 In addition to Assumption 7.6, there exist positive constants $\rho_1 > 0$ and $\rho_2 > 0$ such that $\rho_1 a_k \leq a_{k+1} \leq \rho_2 a_k, \forall k \geq 1$.

The mean square convergence is demonstrated in the following theorem.

Theorem 7.4 *Consider system (7.1) and communication channel (7.4) with the data dropout being described by the general Markov data dropout model, and assume that*

Assumptions 7.1–7.7 hold. Then, the input sequence generated by Algorithm (7.7) converges to the desired input in the mean square sense,

$$\lim_{k\to\infty} \mathbb{E}\{\|u_d(t) - u_k(t)\|^2\} = 0, \quad \forall t \in \mathbf{Z}_0^{N-1}, \tag{7.49}$$

if L_t is designed such that $-L_t C_{t+1} B_t$ is stable for all $t = 0, \ldots, N-1$.

Proof From (7.15), it follows that

$$\delta U_{(n+1)h} = \Psi((n+1)h, nh)\delta U_{nh} + \Omega_{nh}, \tag{7.50}$$

where h is a positive integer describing the length of successive iterations, $\Psi(i+1, j)$ is defined as

$$\Psi(i+1, j) = (I - a_i \Lambda_i) \cdots (I - a_j \Lambda_j), \quad \forall i \geq j,$$
$$\Psi(i, i) = I_{pN}, \tag{7.51}$$

and $\Omega_k = \sum_{j=k}^{k+h-1} \Psi(k+h-1, j) a_j (\Gamma_j \otimes I_p) \mathscr{L}\big[((\Upsilon^{-1}\widetilde{\Upsilon}_j) \otimes I_q)(\mathscr{H} U_j + M x_j(0)) + (\Upsilon^{-1}\Upsilon_j \otimes I_q)W_j + (\Upsilon^{-1} \otimes I_q)V_j\big] = \sum_{j=k}^{k+h-1} \Psi(k+h-1, j) a_j \Phi_j$.

By the definition of $\mathscr{Q}$, we have that $\Lambda_k^T \mathscr{Q} + \mathscr{Q}\Lambda_k$ is semi-positive definite. Under Assumption 7.7, we know that a constant c_{11} exists such that

$$\Psi^T((n+1)h, nh)\mathscr{Q}\Psi((n+1)h, nh)$$
$$\leq \mathscr{Q} - \sum_{k=nh}^{(n+1)h-1} a_k \left(\Lambda_k^T \mathscr{Q} + \mathscr{Q}\Lambda_k\right) + c_{11} a_{nh}^2 \mathscr{Q}. \tag{7.52}$$

Therefore, we can obtain

$$J_{(n+1)h} = \delta U_{(n+1)h}^T \mathscr{Q} \delta U_{(n+1)h}$$
$$= \delta U_{nh}^T \Psi^T((n+1)h, nh)\mathscr{Q}\Psi((n+1)h, nh)\delta U_{(n+1)h}$$
$$+ 2\delta U_{nh}^T \Psi^T((n+1)h, nh)\mathscr{Q}\Omega_{nh} + \Omega_{nh}^T \mathscr{Q}\Omega_{nh}$$
$$\leq J_{nh} - \delta U_{nh}^T \left[\sum_{k=nh}^{(n+1)h-1} a_k \left(\Lambda_k^T \mathscr{Q} + \mathscr{Q}\Lambda_k\right)\right] \delta U_{nh}$$
$$+ c_{11} a_{nh}^2 J_{nh} + \Omega_{nh}^T \mathscr{Q}\Omega_{nh}$$
$$+ 2\delta U_{nh}^T \Psi^T((n+1)h, nh)\mathscr{Q}\Omega_{nh}$$
$$\leq J_{nh} - c_{12} a_{nh} \delta U_{nh}^T \left[\sum_{k=nh}^{(n+1)h-1} \left(\Lambda_k^T \mathscr{Q} + \mathscr{Q}\Lambda_k\right)\right] \delta U_{nh}$$
$$+ c_{11} a_{nh}^2 J_{nh} + \Omega_{nh}^T \mathscr{Q}\Omega_{nh}$$
$$+ 2\delta U_{nh}^T \Psi^T((n+1)h, nh)\mathscr{Q}\Omega_{nh}, \tag{7.53}$$

7.4 ILC for General Markov Data Dropout Model Case

where $c_{12} > 0$ is a constant that only depends on h.

Now we first prove that

$$\mathbb{E}\left\{2\delta U_{nh}^T \Psi^T((n+1)h, nh)\mathscr{Q}\Omega_{nh} \mid \mathscr{G}_{nh}\right\} = 0. \tag{7.54}$$

To this end, define the σ-algebras $\mathscr{G}'_{(n+1)h} = \sigma\{x_k(0), u_k(t), y_k(t), w_k(t), v_k(t), \alpha_k(t), m_l, k = 1, \ldots, nh - 1, l = 1, \ldots, (n+1)h - 1, 0 \leq t \leq N\}$, $\mathscr{G}''_{nh+j} = \sigma\{x_k(0), u_k(t), y_k(t), w_k(t), v_k(t), \alpha_k(t), m_l, k = 1, \ldots, nh + j - 1, l = 1, \ldots, (n+1)h - 1, 0 \leq t \leq N\}$. Then, we have

$$\mathbb{E}\left\{2\delta U_{nh}^T \Psi^T((n+1)h, nh)\mathscr{Q}\Omega_{nh} \mid \mathscr{G}_{nh}\right\}$$
$$= 2\delta U_{nh}^T \mathbb{E}\bigg\{\Psi^T((n+1)h, nh)\mathscr{Q}\sum_{j=nh}^{(n+1)h-1}\Psi((n+1)h-1, j)a_j$$
$$\times (\Gamma_j \otimes I_p)\mathscr{L} \cdot \mathbb{E}\left\{((\Upsilon^{-1}\widetilde{\Upsilon}_j) \otimes I_q)(\mathscr{H}U_j + Mx_j(0))\right.$$
$$+ (\Upsilon^{-1}\Upsilon_j \otimes I_q)W_j + (\Upsilon^{-1} \otimes I_q)V_j \mid \mathscr{G}'_{(n+1)h}\bigg\}\bigg|\mathscr{G}_{nh}\bigg\}. \tag{7.55}$$

Noticing

$$\mathbb{E}\left\{((\Upsilon^{-1}\widetilde{\Upsilon}_j) \otimes I_q)\mathscr{H}U_j \mid \mathscr{G}'_{(n+1)h}\right\}$$
$$= \mathbb{E}\left\{\mathbb{E}\left\{((\Upsilon^{-1}\widetilde{\Upsilon}_j) \otimes I_q) \mid \mathscr{G}''_{nh+j}\right\}\mathscr{H}U_j \mid \mathscr{G}'_{(n+1)h}\right\} = 0,$$
$$\mathbb{E}\left\{((\Upsilon^{-1}\widetilde{\Upsilon}_j) \otimes I_q)Mx_j(0) \mid \mathscr{G}'_{(n+1)h}\right\}$$
$$= \mathbb{E}\left\{((\Upsilon^{-1}\widetilde{\Upsilon}_j) \otimes I_q) \mid \mathscr{G}'_{(n+1)h}\right\}Mx_d(0) = 0,$$
$$\mathbb{E}\left\{(\Upsilon^{-1}\Upsilon_j \otimes I_q)W_j \mid \mathscr{G}'_{(n+1)h}\right\} = 0,$$
$$\mathbb{E}\left\{(\Upsilon^{-1} \otimes I_q)V_j \mid \mathscr{G}'_{(n+1)h}\right\} = 0,$$

we can obtain (7.54) from (7.55).

Next, we provide an estimation of $\Omega_{nh}^T \mathscr{Q}\Omega_{nh}$. Similarly to the steps in the proof of Theorem 7.1, by using the Cauchy inequality and Minkowski inequality, we have

$$\mathbb{E}\left\{\Omega_{nh}^T \mathscr{Q}\Omega_{nh} \mid \mathscr{G}_{nh}\right\}$$
$$\leq 4\mathbb{E}\bigg\{\sum_{j=nh}^{(n+1)h-1} a_j^2 \|\Psi((n+1)h-1, j)\|^2$$
$$\times \bigg(\|(\Gamma_j \otimes I_p)\mathscr{L}((\Upsilon^{-1}\widetilde{\Upsilon}_j) \otimes I_q)\mathscr{H}U_j\|_{\mathscr{Q}}$$
$$+ \|(\Gamma_j \otimes I_p)\mathscr{L}((\Upsilon^{-1}\widetilde{\Upsilon}_j) \otimes I_q)Mx_k(0)\|_{\mathscr{Q}}$$
$$+ \|(\Gamma_j \otimes I_p)\mathscr{L}(\Upsilon^{-1}\Upsilon_j \otimes I_q)W_j\|_{\mathscr{Q}}$$

$$+ \|(\Gamma_j \otimes I_p)\mathscr{L}(\Upsilon^{-1} \otimes I_q)V_j\|_{\mathscr{Q}}\Big)\Big|\mathscr{G}_{nh}\Big\}$$

$$\leq c_{13} \sum_{j=nh}^{(n+1)h-1} a_j^2 \bigg(c_{14} + \mathbb{E}\{J_j \mid \mathscr{G}_{nh}\}$$

$$+ \mathbb{E}\{\|W_j\|^2 \mid \mathscr{G}_{nh}\} + \mathbb{E}\{\|V_j\|^2 \mid \mathscr{G}_{nh}\}\bigg), \tag{7.56}$$

where $c_{13} > 0$ and $c_{14} > 0$ are suitable constants.

Under the Assumption 7.5, there exists $h > 0$ such that for any iteration number k and initial state, the Markov chain m_k will visit all states in the successive iterations $[k, k+h]$ with a positive probability $\varepsilon_0 > 0$, where ε_0 is independent of the iteration number k. Thus, by the definition of $\mathscr{Q}$ and the assumptions in general Markov data dropout model, we can conclude that

$$c_{12}\mathbb{E}\bigg\{\bigg[\sum_{k=nh}^{(n+1)h-1}(\Lambda_k^T\mathscr{Q} + \mathscr{Q}\Lambda_k)\bigg] \mid \mathscr{G}_{nh}\bigg\}$$

$$= c_{12}\mathbb{E}\bigg\{\bigg(\sum_{k=nh}^{(n+1)h-1}\Lambda_k^T\bigg)\mathscr{Q} + \mathscr{Q}\bigg(\sum_{k=nh}^{(n+1)h-1}\Lambda_k\bigg) \mid \mathscr{G}_{nh}\bigg\}$$

$$\geq c_{15}\mathscr{Q}, \tag{7.57}$$

for some suitable positive constant $c_{15} > 0$.

Combining (7.53), (7.54), (7.56), and (7.57), we obtain

$$\mathbb{E}\{J_{(n+1)h} \mid \mathscr{G}_{nh}\}$$

$$\leq (1 - c_{15}a_{nh} + c_{11}a_{nh}^2)J_{nh}$$

$$+ c_{13}\sum_{j=nh}^{(n+1)h-1} a_j^2 \bigg(c_{14} + \mathbb{E}\{J_j \mid \mathscr{G}_{nh}\}$$

$$+ \mathbb{E}\{\|W_j\|^2 \mid \mathscr{G}_{nh}\} + \mathbb{E}\{\|V_j\|^2 \mid \mathscr{G}_{nh}\}\bigg). \tag{7.58}$$

To further derive the convergence property, we need to give an estimation to $\mathbb{E}\{J_k\}$ for $k = nh+1, \cdots, (n+1)h - 1$. Similarly to (7.31) in Theorem 7.1, we conclude the existence of positive constants $c_{16} > 0$ and $c_{17} > 0$ such that

$$\mathbb{E}\{J_{k+1}\} \leq \big(1 + c_{16}a_k^2\big)\mathbb{E}\{J_k\} + c_{17}a_k^2$$

$$= \prod_{i=nh}^{k}\big(1 + c_{16}a_i^2\big)\mathbb{E}\{J_{nh}\}$$

7.4 ILC for General Markov Data Dropout Model Case

$$+ \sum_{i=nh}^{k} \prod_{j=i+1}^{k} \left(1 + c_{16} a_j^2\right) c_{17} a_i^2$$

$$\leq c_{18} \mathbb{E}\{J_{nh}\} + c_{19}, \quad \forall k = nh, \ldots, (n+1)h - 2, \tag{7.59}$$

where $c_{18} > 0$ and $c_{19} > 0$ are sufficiently large numbers under Assumption 7.6.

Now, taking the expectation to both sides of (7.58), substituting (7.59), and using Assumptions 7.2, 7.4, 7.6, and 7.7, we observe that for some $c_{20} > 0$ and $c_{21} > 0$,

$$\mathbb{E}\{J_{(n+1)h}\} \leq (1 - c_{15} a_{nh} + c_{20} a_{nh}^2) \mathbb{E}\{J_{nh}\} + c_{21} a_{nh}^2. \tag{7.60}$$

From Assumption 7.7, it is valid that $\sum_{n=0}^{\infty} a_{nh} \geq \bar{\rho}_1 \sum_{n=0}^{\infty} \sum_{i=nh+1}^{(n+1)h} a_i = \bar{\rho}_1 \sum_{k=1}^{\infty} a_k = \infty$ and $\sum_{n=0}^{\infty} a_{nh}^2 \leq \sum_{k=1}^{\infty} a_k^2 < \infty$. Therefore, applying Lemma 7.1, we conclude that $\lim_{n \to \infty} \mathbb{E}\{J_{nh}\} = 0$.

Before completing the proof, there is still a gap to show the convergence for an arbitrary iteration number in the case of $\lim_{n \to \infty} \mathbb{E}\{J_{nh}\} = 0$. In fact, by noticing (7.59), it is a direct conclusion that $\lim_{k \to \infty} \mathbb{E}\{J_k\} = 0$. Moreover, as $\mathcal{Q}$ is a positive definite matrix, $\lim_{k \to \infty} \mathbb{E}\|\delta U_k\|^2 = 0$. This completes the proof of convergence in the mean square sense.

We can now demonstrate the almost sure convergence.

Theorem 7.5 *Consider system (7.1) and communication channel (7.4) with the data dropout being modeled by general Markov data dropout model, and assume Assumptions 7.1–7.7 hold. Then, the input sequence generated by Algorithm (7.7) converges to the desired input in the almost sure sense,*

$$\lim_{k \to \infty} u_k(t) = u_d(t) \ a.s., \quad \forall t \in \mathbf{Z}_0^{N-1}, \tag{7.61}$$

if L_t is designed such that $-L_t C_{t+1} B_t$ is stable for all $t = 0, \ldots, N-1$.

Proof To show the convergence, we divide the iteration numbers into h subsequences $\{nh + j, n = 0, 1, \ldots\}$, $j = 1, \ldots, h$. For each subsequence of the iteration numbers, we follow the derivation steps of (7.58) to obtain

$$\mathbb{E}\left\{J_{(n+1)h+i} \mid \mathscr{G}''_{nh+i}\right\}$$
$$\leq J_{nh+i} - c_{15} a_{nh+i} J_{nh+i} + c_{11} a_{nh+i}^2 J_{nh+i}$$
$$+ c_{13} \sum_{j=nh+i}^{(n+1)h-1+i} a_j^2 \bigg(c_{14} + \mathbb{E}\left\{J_j \mid \mathscr{G}''_{nh+i}\right\}$$
$$+ \mathbb{E}\left\{\|W_j\|^2 \mid \mathscr{G}''_{nh+i}\right\} + \mathbb{E}\left\{\|V_j\|^2 \mid \mathscr{G}''_{nh+i}\right\} \bigg),$$
$$\forall i = 1, 2, \ldots, h. \tag{7.62}$$

Noting that $\sum_{n=0}^{\infty} a_{nh+i}^2 < \infty$, $\sum_{n=0}^{\infty} \sum_{j=nh+i}^{(n+1)h-1+i} a_j^2 < \infty$ and using the result $\lim_{k \to \infty} \mathbb{E}\{J_k\} = 0$ or its corollary that $\mathbb{E}\{J_k\}$ is bounded, we obtain

$$\sum_{n=0}^{\infty} \mathbb{E}\left\{c_{11} a_{nh+i}^2 J_{nh+i}\right\} < \infty,$$

$$\sum_{n=0}^{\infty} \mathbb{E}\left\{c_{13} \sum_{j=nh+i}^{(n+1)h-1+i} a_j^2 \mathbb{E}\{J_j \mid \mathscr{G}_{nh+i}''\}\right\}$$

$$= c_{13} \sum_{n=0}^{\infty} \sum_{j=nh+i}^{(n+1)h-1+i} a_j^2 \mathbb{E}\{J_j\} < \infty,$$

$$\sum_{n=0}^{\infty} \mathbb{E}\left\{c_{13} \sum_{j=nh+i}^{(n+1)h-1+i} a_j^2 \mathbb{E}\{\|W_j\|^2 + \|V_j\|^2 \mid \mathscr{G}_{nh+i}''\}\right\}$$

$$= c_{13} \sum_{n=0}^{\infty} \sum_{j=nh+i}^{(n+1)h-1+i} a_j^2 \mathbb{E}\{\|W_j\|^2 + \|V_j\|^2\} < \infty.$$

Therefore, by Lemma 7.2, we have that J_{nh+i} converges almost surely as $n \to \infty$ and $\sum_{n=0}^{\infty} a_{nh+i} J_{nh+i} < \infty, i = 1, 2, \ldots, h$. From $\lim_{k \to \infty} \mathbb{E}\{J_k\} = 0$ it follows that $\lim_{n \to \infty} J_{nh+i} = 0, \forall i = 1, 2, \ldots, h$, which further implies that $\lim_{k \to \infty} J_k = 0$, a.s. Similarly to the proof of Theorem 7.2, using the positive definiteness of $\mathscr{Q}$, we obtain $\lim_{k \to \infty} u_k(t) = u_d(t), \forall t$, a.s. The proof is thus completed.

As in Corollary 7.1, we can apply Algorithm (7.47) (which varies the step size at different time instants) based on certain prior knowledge of the collection set $\varXi_G$. In such case, the convergence is summarized in the following corollary.

Corollary 7.2 *Consider system* (7.1) *and communication channel* (7.4) *with the data dropout being modeled by general Markov data dropout model, and assume Assumption 7.1–7.7 hold. Then, the input sequence generated by the algorithm* (7.47) *converges to the desired input in both the mean square and almost sure senses, if L_t is designed such that $-L_t C_{t+1} B_t$ is stable for all $t = 0, \ldots, N - 1$ and the step size sequences $\{a_k(t)\}$ additionally satisfy*

$$\max_{0 \le t, s \le N-1} |a_k(t) - a_k(s)| = o\left(\sum_{t=0}^{N-1} a_k(t)\right), \quad \text{as } k \to \infty. \tag{7.63}$$

Moreover, under GMDDM, we require that the union of all collections of $\varXi_G$ should be $\zeta^{(1)}$. One special case is $\zeta^{(1)} \in \varXi_G$ and $\min_{1 \le i \le \kappa} p_{i1} > 0$. In this special case, the outputs of the entire iteration have a positive probability to be all transmitted back. Consequently, the proofs of Theorems 7.1 and 7.2 can be employed to derive the following corollary.

7.4 ILC for General Markov Data Dropout Model Case

Corollary 7.3 *Consider system* (7.1) *and communication channel* (7.4) *and assume Assumption 7.1–7.4, 7.6 hold. Moreover, the set of all possible collections* Ξ_G *contains* $\zeta^{(1)}$. *Without loss of generality, we denote this set as* $\zeta_G^{(1)} = \zeta^{(1)}$ *and assume* $\min_{1 \leq i \leq \kappa} p_{i1} > 0$. *Then, the input sequence generated by Algorithm* (7.7) *converges to the desired input in both the mean square and almost sure senses if* L_t *is designed such that* $-L_t C_{t+1} B_t$ *is stable for all* $t = 0, \ldots, N - 1$.

Remark 7.7 The proofs show that under the classical Markov chain model or the special case in Corollary 7.3, the Lyapunov function decays during every iteration. Under the general Markov data dropout, it decays after a certain number of successive iterations. Thus, it seems that Algorithm (7.7) converges more quickly under MCM than under GMDDM. However, this intuition is not true. In fact, both models generally have the same convergence rate. The reason is that the decay rate of both cases generally is with the rate $1 - Ca_k$, where the value of C differs between the two cases. As we know, different C does not imply different convergence rate in essence. This phenomenon is also verified by simulations.

Remark 7.8 While we consider the homogeneous ergodic Markov chain in Assumption 7.5, the ergodic condition can be further relaxed to the case that the union of a certain irreducible closed subset of positive recurrent states is $\zeta^{(1)}$ and the other states are transient. In this case, the state space Ξ_G of m_t can be partitioned as a direct sum of two subsets $\Xi_G = \Xi_G' \bigcup \Xi_G''$, where Ξ_G' is the set of transient states and Ξ_G'' is the irreducible closed set of recurrent states. We assume that Ξ_G'' fulfill Assumption 7.5. Then, the initial state of the Markov chain belongs to either Ξ_G' or Ξ_G''. For the former case, m_t will finally reach the irreducible closed set after a finite number of iterations and remains in that set thereafter; for the latter case, m_t will never leave the irreducible closed set after the initial iteration. Consequently, by applying the results of Theorems 7.4 and 7.5, we can confirm the convergence properties of the proposed algorithms.

Remark 7.9 From another angle, our results provide an essential requirement for data scheduling under severe transmission conditions. That is, the data in the whole time interval should be transmitted back during a finite number of iterations. If the communication bandwidth is rather limited and the data in part of an iteration cannot be transmitted back, we can divide the entire iteration into several segments and transmit each segment periodically.

7.5 Illustrative Simulations

In this section, we apply the proposed algorithms to a permanent magnet linear motor (PMLM), which is described by the following discretized model [9]:

$$\begin{cases} x(t+1) = x(t) + v(t)\Delta \\ v(t+1) = v(t) - \Delta \frac{k_1 k_2 \psi_f^2}{Rm} v(t) + \Delta \frac{k_2 \psi_f}{Rm} u(t) \\ y(t) = v(t) + \varepsilon(t) \end{cases}$$

where x and v denote the motor position and rotor velocity, $\Delta = 10\,\text{ms}$ the sampling time interval, $R = 8.6\,\Omega$ the resistance of stator, $m = 1.635\,\text{kg}$ the rotor mass, and $\psi_f = 0.35\,\text{Wb}$ the flux linkage, $k_1 = \pi/\tau$ and $k_2 = 1.5\pi/\tau$, where $\tau = 0.031\,\text{m}$ is the pole pitch. The stochastic noise $\varepsilon(t)$ obeys a zero-mean normal distribution $N(0, \sigma^2)$ with $\sigma = 0.02$. In this simulation, we set the whole iteration length to be $1\,\text{s}$, i.e., $N = 100$.

We consider five possible collections of successful transmission time instants, $\zeta^{(1)} = \{1, \ldots, 80\}, \zeta^{(2)} = \{1, \ldots, 60, 81, \ldots, 100\}, \zeta^{(3)} = \{1, \ldots, 40, 61, \ldots, 100\}$, $\zeta^{(4)} = \{1, \ldots, 20, 41, \ldots, 100\}$, and $\zeta^{(5)} = \{21, \ldots, 100\}$. Therefore, the state space of the Markov chain m_k is $\{1, 2, 3, 4, 5\}$. This Markov chain drives the switches of the possible collections with the transition probability matrix

$$P = \begin{bmatrix} 0.2 & 0.21 & 0.19 & 0.21 & 0.19 \\ 0.21 & 0.2 & 0.19 & 0.19 & 0.21 \\ 0.19 & 0.21 & 0.2 & 0.19 & 0.21 \\ 0.21 & 0.21 & 0.19 & 0.2 & 0.19 \\ 0.19 & 0.21 & 0.19 & 0.21 & 0.2 \end{bmatrix}$$

for illustration. The transition probability matrix need not to be even. If m_k visit a certain state with a higher probability than the others, the convergence performance would be dominated by this state.

The measured output is transformed into a signal of the form $z_k(t) = \alpha_k(t) y_k(t) + \varepsilon_k(t)$, which is received by the controller. The fading factor $\alpha_k(t)$ follows a normal distribution $N(0.95, 0.1^2)$. The additive noise $\varepsilon_k(t)$ in the received signal follows the same distribution as $\varepsilon_k(t)$.

The desired reference is $y_d(t) = 2/3[\sin(t/20) + 1 - \cos(3t/20)]$, $0 \le t \le 100$. The initial state satisfies Assumption 7.1. The control input for the first iteration is simply set to 0. The learning gain is $L_t = 12$ and the decreasing sequence is set to be $a_k = 1.8/k^{2/3}$. Algorithm (7.7) is run for 100 iterations.

The actual tracking error is $e_k(t) = y_d(t) - y_k(t)$ and the correction error in the updating algorithm is $\widehat{e}_k(t) \triangleq y_d(t) - \bar{\alpha}_t^{-1} z_k(t)$. Figure 7.2 shows the averaged tracking error and correction error profiles along iteration axis, where both averaged errors are defined as $\sum_{t=1}^{N} |e_k(t)|/N$ and $\sum_{k=1}^{N} |\widehat{e}_k(t)|/N$, respectively. It can be seen from the figure that both errors decrease as the iteration number increases. Moreover, the actual tracking error outperforms the correction error, revealing that the tracking performance can be continuously improved even when the correction error in the input algorithm (7.7) deviates from the actual tracking error.

The specific tracking performance of the 2nd, 5th, 20th, and 100th iterations are plotted in Fig. 7.3. As can be seen, the output profile after two iterations largely deviates from the desired reference; but is already acceptable after 20 iterations.

7.5 Illustrative Simulations

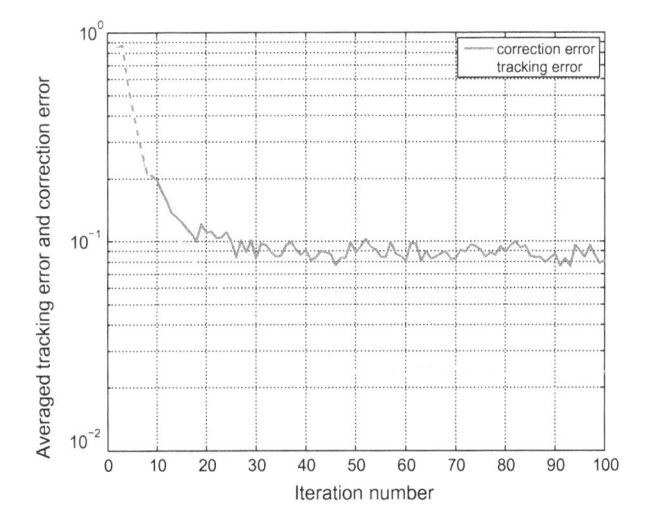

Fig. 7.2 Tracking error and correction error profiles along the iteration axis (correction error: solid line; tracking error: dashed line)

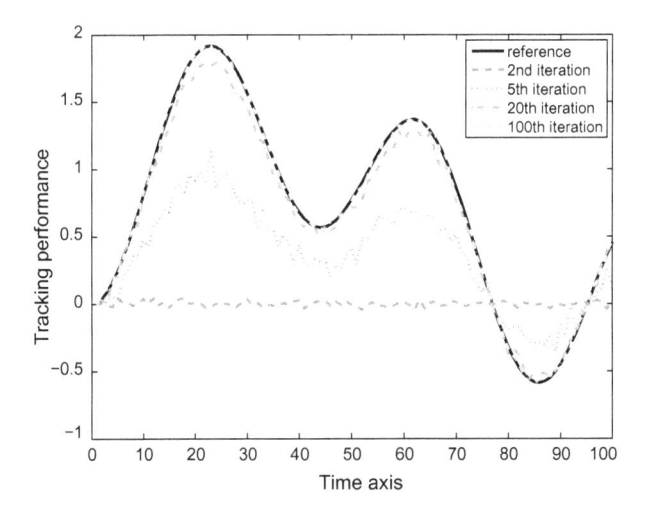

Fig. 7.3 Tracking performance of the 2nd, 5th, 20th, and 100th iterations

The distinct fluctuations in all output profiles arise from stochastic noises, random channel fading, and Markov data dropouts.

In addition, Fig. 7.4 displays the input profiles generated by Algorithm (7.7) at $t = 20, 40, 60, 80$, and 99. This figure demonstrates the essential effect of the step size sequence $\{a_k\}$. Note that $\{a_k\}$ enables stable convergence of the input sequence

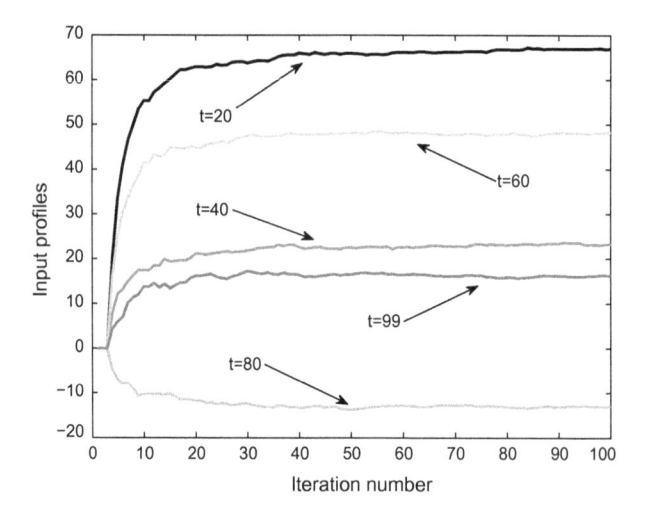

Fig. 7.4 Input profiles along the iteration axis at $t = 20, 40, 60, 80$, and 99.eps

generated by Algorithm (7.7). If the step size is fixed for all iterations, the input sequence will continue to fluctuate under the various uncertainties and randomness in the configuration. Moreover, by designing a flexible step size, we can also impose a mild condition on the selection of the learning gain matrix L_t (as shown in the theorems).

We simulate Algorithm (7.39) under the same conditions clarified above. Figure 7.5 shows the averaged tracking error and correction error profiles along iteration axis. Comparing with Fig. 7.2, we observe that Algorithm (7.39) achieves similar tracking performance to Algorithm (7.7); however, both averaged tracking error and correction error of Algorithm (7.39) converge more quickly than that of Algorithm (7.7). This observation provides the advantage of update-triggered step size.

In order to illustrate the influence of data dropouts and fading channel separately, we conduct the simulations for Algorithm (7.7) in absence of channel fading and data dropouts, respectively. On the one hand, when the communication channel only suffers data dropouts (that is, both multiplicative and additive randomness does not exist in the communication channel), the correction error actually is the tracking error. The averaged tracking error and the input profiles at $t = 20, 40, 60, 80$, and 99 are shown in Figs. 7.6 and 7.7, respectively. From Fig. 7.6 we observe that tracking precision and convergence rate are distinctly improved comparing with the general communication channel case (see Fig. 7.2). Moreover, comparing Fig. 7.7 with Fig. 7.4, we further find that the input profiles for the former case are much flatter than that for the latter case. In short, the channel fading condition introduces more effects on the tracking and learning performance. On the other hand, when all data packets can be successfully transmitted through the fading channel, we plot the corresponding averaged tracking error and correction error profiles in Fig. 7.8. It is seen from this figure that the convergence rate is much faster than Fig. 7.2 and ever a little faster

7.5 Illustrative Simulations

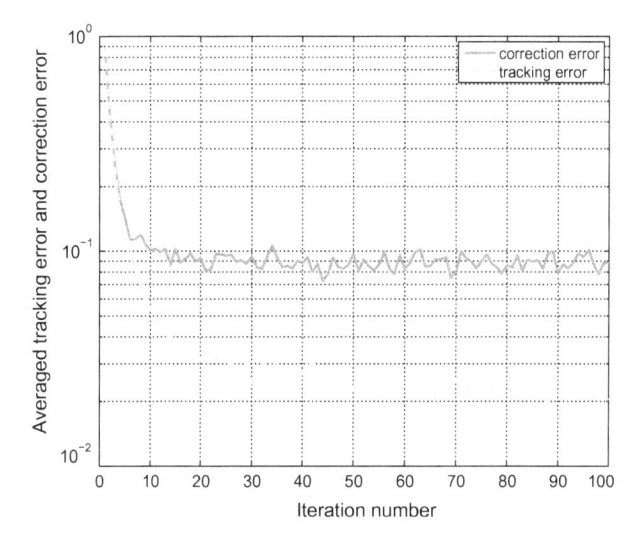

Fig. 7.5 Tracking error and correction error profiles along the iteration axis (correction error: solid line; tracking error: dashed line) for Algorithm (7.39)

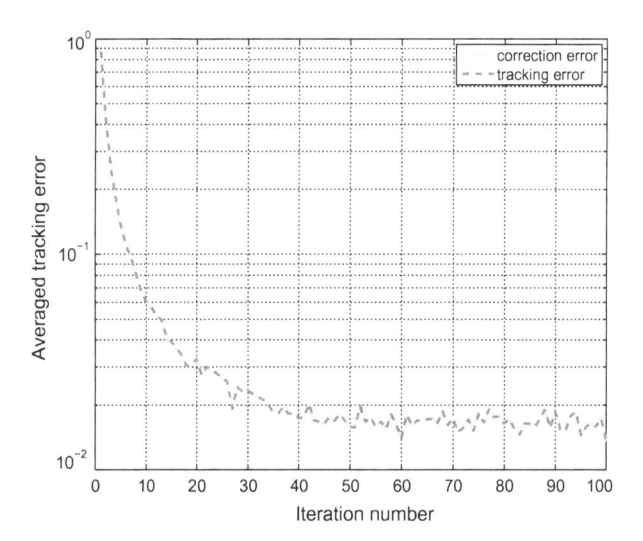

Fig. 7.6 Tracking error and correction error profiles along the iteration axis (correction error: solid line; tracking error: dashed line) for no fading case

than Fig. 7.5. In other words, the convergence rate is determined by both the learning gain sequence and the data dropout rate.

We note that different decreasing gain sequences $\{a_k\}$ imply different convergence performances of the proposed algorithm. To clearly show this intuition, we consider the decreasing gain as $a_k = 1.8/k^\rho$ with $\rho = 0.6, 0.7, 0.8, 0.9$, and 1. The averaged

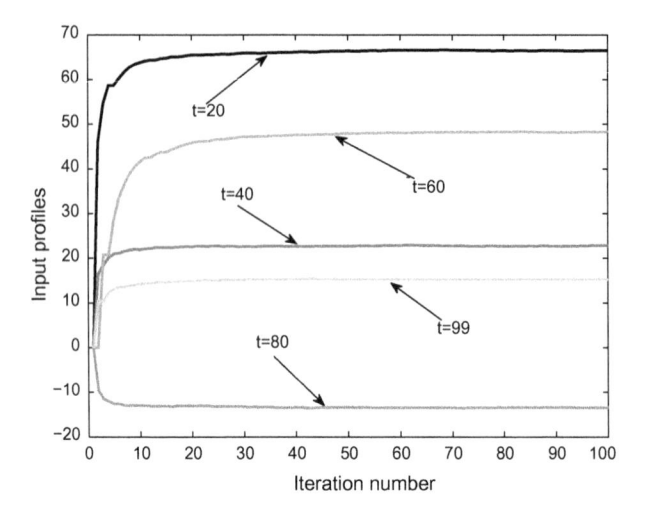

Fig. 7.7 Input profiles along the iteration axis at $t = 20, 40, 60, 80$, and 99 for no fading case

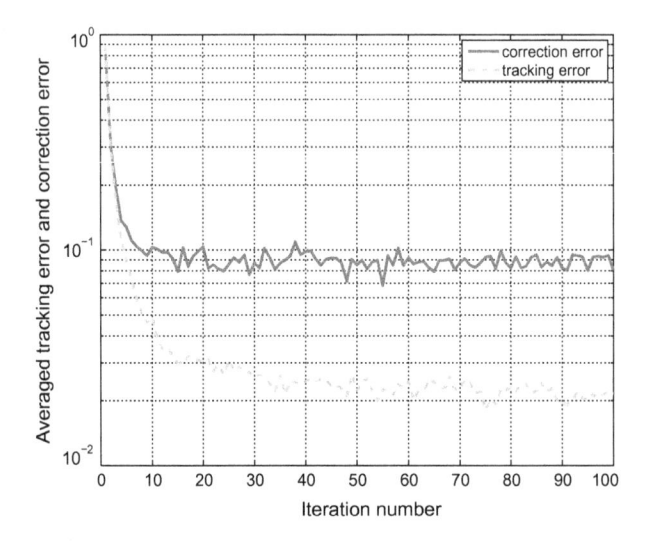

Fig. 7.8 Tracking error and correction error profiles along the iteration axis (correction error: solid line; tracking error: dashed line) for no data dropout case

tracking error profiles are shown in Fig. 7.9. Clearly, we observe that smaller ρ corresponds to the faster convergence rate. However, we should point out that the convergence performance also depends on the numerator value. Generally, if the numerator is too large, a contraction is not valid for first few iterations and then the tracking error profiles may arise for these iterations. As a result, the convergence performance can be much different.

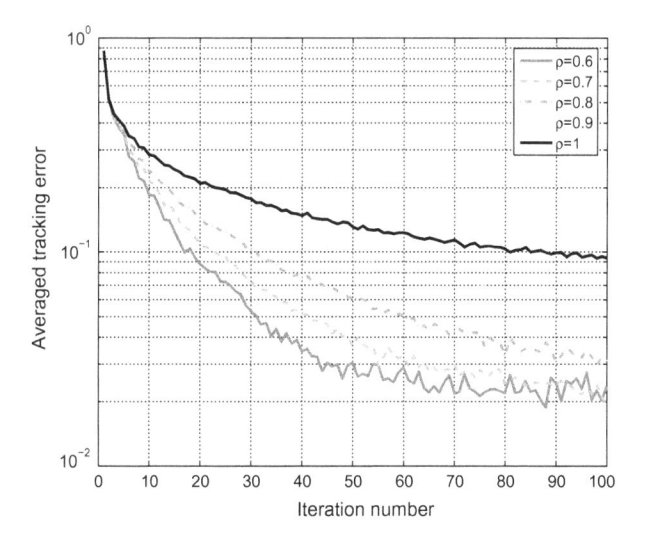

Fig. 7.9 Tracking error profiles along the iteration axis for different gain sequence

7.6 Summary

This chapter contributes to propose a systematic convergence analysis for linear systems with Markov data dropouts. We first provided a general Markov chain model for data dropouts under severe communication conditions, allowing dependence among the dropouts along the time and iteration axes. The mean square and almost sure convergence of the proposed algorithm were established for the conventional and the general Markov chain models in sequence. Both time-invariant and varying step sizes were discussed. The algorithm achieved high tracking performance even when part of information gathered over the entire iteration was not transmitted.

References

1. Saab, S.S.: A discrete-time stochastic learning control algorithm. IEEE Trans. Autom. Control **46**(6), 877–887 (2001)
2. Chen, H.F.: Almost sure convergence of iterative learning control for stochastic systems. Science in China (Series F) **46**(1), 67–79 (2003)
3. Huang, S.N., Tan, K.K., Lee, T.H.: Necessary and sufficient condition for convergence of iterative learning algorithm. Automatica **38**(7), 1257–1260 (2002)
4. Meng, D., Jia, Y., Du, J., Yu, F.: Necessary and sufficient stability condition of LTV iterative learning control systems using a 2-D approach. Asian J. Control **13**(1), 25–37 (2011)
5. Xiao, N., Xie, L., Qiu, L.: Feedback stabilization of discrete-time networked systems over fading channels. IEEE Trans. Autom. Control **57**(9), 2176–2189 (2012)
6. Dey, S., Leong, A.S., Evans, J.S.: Kalman filtering with faded measurements. Automatica **45**, 2223–2233 (2009)

7. Polyak, B.: Introduction to Optimization. Optimization Software Inc., New York (1987)
8. Goodwin, G., Sin, K.: Adaptive Filtering, Prediction and Control. Prentice-Hall, Englewood Cliffs, N.J. (1984)
9. Zhou, W., Yu, M., Huang, D.: A high-order internal model based iterative learning control scheme for discrete linear time-varying systems. Int. J. Autom. Comput. **12**(3), 330–336 (2015)

Part II
Two-Side Data Dropout

In this part, we concentrate on the case that data dropout occurs at both the measurement and actuator sides. In other words, the output of the plant may suffer random data dropouts during the transmission from the plant to the controller, and the generated input of the controller can be also lost during the transmission from the controller to the plant. In such case, the asynchronism between the control input generated by the controller and the actual input fed into the plant should be carefully analyzed.

Chapter 8
Two-Side Data Dropout for Linear Deterministic Systems

8.1 Problem Formulation

Consider the lifted causal system

$$y_k = Hu_k + y_k(0), \tag{8.1}$$

where $u_k \in \mathbf{R}^{pN}$ and $y_k \in \mathbf{R}^{qN}$ denote the lifted input vector and output vector of an iteration, respectively. $y_k(0)$ denotes the initial response of each iteration. k is the iteration index, $k = 1, 2, \ldots$, and N denotes the iteration length. p and q denote the dimensions of the input and output, respectively. $H \in \mathbf{R}^{qN \times pN}$ is the system matrix, which is a lower triangular block matrix for the causal relationship between the input and output. It is usually formulated as

$$H = \begin{bmatrix} H_{1,1} & 0 & 0 & \cdots & 0 \\ H_{2,1} & H_{2,2} & 0 & \cdots & 0 \\ H_{3,1} & H_{3,2} & H_{3,3} & \cdots & 0 \\ \vdots & \vdots & \vdots & \ddots & \vdots \\ H_{N,1} & H_{N,2} & H_{N,3} & \cdots & H_{N,N} \end{bmatrix}, \tag{8.2}$$

where $H_{i,j} \in \mathbf{R}^{q \times p}$ are Markov parameter matrices [1–3]. As a special case, for the linear time-invariant (LTI) system describe by state space representation (A, B, C) with the relative degree being one, the diagonal parameters $H_{i,i}$ is CB while the off-diagonal parameters $H_{i,j}$ are usually computed as $CA^{i-j}B$, $i > j$.

Remark 8.1 In this chapter, to make our idea clear to understand, we directly employ the lifted model of a discrete-time system, i.e., (8.1). This formulation would save the notations without loss of any generality compared with the traditional state space model. Such model has been used in many previous studies, especially when considering the LTI system. In addition, the proposed control laws can be applied to nonlinear systems but the convergence analysis would be quite different and thus will be detailed in Chap. 10.

The desired trajectory is given as y_d. The tracking error is denoted by $e_k \triangleq y_d - y_k$.

Assumption 8.1 For the desired trajectory, a unique input u_d exists such that $y_d = Hu_d + y_d(0)$.

Assumption 8.2 The system can be reset accurately for each iteration, i.e., $y_k(0) = y_d(0), \forall k$.

Remark 8.2 Assumption 8.1 is imposed mainly to guarantee the convergence of the input sequence generated by the proposed algorithms. If such assumption is not satisfied, it is only the convergence of the tracking error can be derived. Indeed, Assumption 8.1 is valid as long as H is of full-column rank. As an illustration, when considering the LTI system (A, B, C) with the relative degree being one, the above requirement is satisfied if the input–output coupling matrix CB is of full-column rank. If the relative degree is larger than one, Assumption 8.1 can be ensured by suitable adjustment to H similar to [2]. Assumption 8.2 is the requirement on initial condition. It is worth pointing out that many efforts have been made to remove or relax such condition; however, the progress is limited. In this chapter, our critical objective is to handle the general data dropouts problem and the involved asynchronism of inputs; thus, we simply assume Assumptions 8.1 and 8.2 to make our expressions concentrated.

In this chapter, the general network framework is considered; that is, both networks at the measurement and actuator sides would suffer from random data dropouts. The data dropouts are modeled by two random variables, σ_k and γ_k, subject to Bernoulli distribution for both sides, respectively. In other words, both σ_k and γ_k are equal to 1 if the corresponding data are transmitted successfully, and 0 otherwise. Moreover, $\mathbb{P}(\sigma_k = 1) = \bar{\sigma}$ and $\mathbb{P}(\gamma_k = 1) = \bar{\gamma}$, $0 < \bar{\sigma}, \bar{\gamma} < 1$. For clear explanation of the critical idea, the data of each iteration is assumed to be transmitted as one package so that the statements are concise, and the extension to the case that data dropout at different time instants is separately transmitted will be detailed later.

Remark 8.3 Generally, the condition $0 < \bar{\sigma}, \bar{\gamma} < 1$ implies that random data dropouts indeed exist but the network is not completely broken. If $\bar{\sigma} = 0$ or $\bar{\gamma} = 0$, the network at the measurement side or the actuator side would be completely broken. In such case, no data can be successfully transmitted, and thus, it is impossible to achieve a perfect tracking. If $\bar{\gamma} = 1$, the problem reduces to the traditional one-side data dropout case, which has been well addressed in Part I of this monograph. If $\bar{\sigma} = 1$, the network at the measurement side works well. Then, we only have to discuss the effect of the data dropout at the actuator side. The discussions below are still valid for this case.

Now, we can give our problem statement as follows.

Problem Statement: The objective of this chapter is to design a suitable compensation mechanism for the dropped data, reveal the inherent character of the asynchronism between the inputs generated by the learning controller and fed to the system, and show the almost sure convergence of the input sequence to the desired input.

Fig. 8.1 Block diagram of the proposed ILC framework

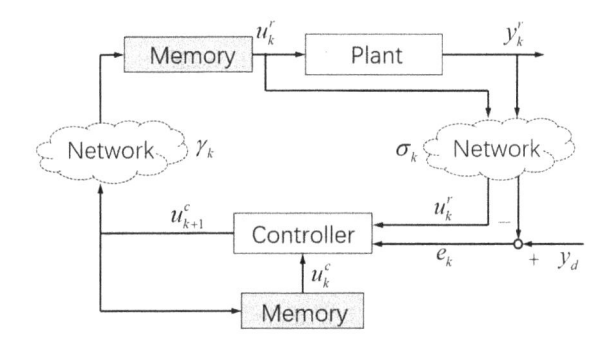

8.2 ILC Algorithms

The block diagram of the control structure is illustrated in Fig. 8.1. For the learning controller, if the data is successfully transmitted at the measurement side, then the algorithm would update its input signal; if the data is lost during transmission at the measurement side, then the algorithm would stop updating and retain the input signal of the previous iteration. At the actuator side, if the input signal is successfully transmitted, then the plant will use this new input signal; if the input signal is lost, then the plant will operate the process using the stored input signal of the previous iteration.

In Fig. 8.1, $u_k^c \in \mathbf{R}^{pN}$ and $u_k^r \in \mathbf{R}^{pN}$ denote the computed control signal and the actually used control signal at the kth iteration, respectively. In the following, u_k^c and u_k^r are labeled "*computed control*" and "*real control*", respectively. Then, the learning algorithm for the computed control is formulated as

$$u_{k+1}^c = \sigma_{k+1} u_k^r + (1 - \sigma_{k+1}) u_k^c + \sigma_{k+1} L e_k. \tag{8.3}$$

The actually used control signal for the $(k+1)$th iteration is

$$u_{k+1}^r = \gamma_{k+1} u_{k+1}^c + (1 - \gamma_{k+1}) u_k^r. \tag{8.4}$$

From (8.3), it is noticed that if $\sigma_{k+1} = 1$, i.e., the data is successfully transmitted, then the *computed control* is updated; otherwise if $\sigma_{k+1} = 0$, then the *computed control* copies the value of the previous iteration. It should be noted that such copy may retain for successive iterations. On the other hand, it is noticed from (8.4) that if $\gamma_{k+1} = 1$, the *real control* is successfully updated with the latest *computed control*; otherwise if $\gamma_{k+1} = 0$, then the *real control* retains its previous value and the operation is thus repeated.

Remark 8.4 The learning algorithm is updated entirely; that is, the input signal of the entire iteration is lifted and updated in (8.3). One may question the computation load of this algorithm when the iteration length N is large. We should explain that the learning algorithm (8.3) is formulated only for the convenience of the following technical analysis. For practical applications, the algorithm can be divided into a time domain-based form as we will design the leaning matrix L in diagonal form. Thus no computation problem is involved.

Remark 8.5 From (8.3) and (8.4), one observes that the data of each iteration is transmitted as one package. As a matter of fact, the data dropout variable can be appended to each time instant. In this case, the data dropout variables σ_k and γ_k are replaced by matrices Σ_k and Γ_k, defined by $\Sigma_k = \text{diag}\{\sigma_k^1, \sigma_k^2, \ldots, \sigma_k^N\}$ and $\Gamma_k = \text{diag}\{\gamma_k^1, \gamma_k^2, \ldots, \gamma_k^N\}$, where σ_k^i and γ_k^j describe the random data dropouts both at the measurement and the actuator sides, respectively. The following results are still valid, but they require more complex derivations. The detailed extension will be detailed at the end of Sect. 8.4.

8.3 Markov Chain Model of Input Evolution

In this section, we will give a new perspective for understanding the relationship between u_k^c and u_k^r. To this end, subtracting both sides of (8.3) from the desired input u_d yields

$$
\begin{aligned}
& u_d - u_{k+1}^c \\
={} & u_d - [\sigma_{k+1} u_k^r + (1 - \sigma_{k+1}) u_k^c + \sigma_{k+1} L e_k] \\
={} & \sigma_{k+1}(u_d - u_k^r) + (1 - \sigma_{k+1})(u_d - u_k^c) - \sigma_{k+1} L e_k \\
={} & \sigma_{k+1}(u_d - u_k^r) + (1 - \sigma_{k+1})(u_d - u_k^c) \\
& - \sigma_{k+1} L H(u_d - u_k^r) \\
={} & \sigma_{k+1}(I - LH)(u_d - u_k^r) + (1 - \sigma_{k+1})(u_d - u_k^c).
\end{aligned}
$$

Denote $\delta u_k^r \triangleq u_d - u_k^r$ and $\delta u_k^c \triangleq u_d - u_k^c$. Then, we have

$$\delta u_{k+1}^c = \sigma_{k+1}(I - LH)\delta u_k^r + (1 - \sigma_{k+1})\delta u_k^c. \tag{8.5}$$

Similarly, subtracting both sides of (8.4) yields

$$\delta u_{k+1}^r = \gamma_{k+1} \delta u_{k+1}^c + (1 - \gamma_{k+1})\delta u_k^r. \tag{8.6}$$

Then, substituting (8.5) into (8.6), we have

$$
\begin{aligned}
& \delta u_{k+1}^r \\
={} & \gamma_{k+1}[\sigma_{k+1}(I - LH)\delta u_k^r + (1 - \sigma_{k+1})\delta u_k^c] \\
& + (1 - \gamma_{k+1})\delta u_k^r \\
={} & [I - \gamma_{k+1} I + \gamma_{k+1}\sigma_{k+1}(I - LH)]\delta u_k^r \\
& + \gamma_{k+1}(1 - \sigma_{k+1})\delta u_k^c. \tag{8.7}
\end{aligned}
$$

8.3 Markov Chain Model of Input Evolution

Now, we first show the inherent Markov character of the sample behaviors of δu_k^c and δu_k^r as follows.

Lemma 8.1 *The input signals δu_k^c and δu_k^r generated by* (8.3) *and* (8.4) *form a Markov chain.*

Proof The computed control and the real control are evidently updated to be the same new input signal when $\sigma_k = \gamma_k = 1$. We first check the sample path behavior of the learning and updating progress. In the following, the path behavior is called "*synchronization*" if the computed control and real control are equal to each other; otherwise, it is called "*asynchronization*". Moreover, it is called a "*renewal*" if both computed control and real control are in a state of *synchronization* but different from their last *synchronization*.

We start from the kth iteration where $\sigma_k = \gamma_k = 1$ and therefore $\delta u_k^r = \delta u_k^c$. That is, the computed control and real control are in a state of *synchronization* at the kth iteration. Then, for the $(k+1)$th iteration, four possible outcomes exist.

- Case 1: $\sigma_{k+1} = 0$ and $\gamma_{k+1} = 1$.
  In this case, from (8.5) and (8.7) we have
  $$\delta u_{k+1}^c = \delta u_k^c = \delta u_k^r,$$
  $$\delta u_{k+1}^r = \delta u_{k+1}^c = \delta u_k^r.$$

  Thus, the computed control and the real control retain the same status as the kth iteration.

- Case 2: $\sigma_{k+1} = 0$ and $\gamma_{k+1} = 0$.
  In this case, it is obvious that
  $$\delta u_{k+1}^c = \delta u_k^c = \delta u_k^r, \quad \delta u_{k+1}^r = \delta u_k^r.$$

  That is, no change of both computed control and real control occurs.

- Case 3: $\sigma_{k+1} = 1$ and $\gamma_{k+1} = 1$.
  In this case, we find that
  $$\delta u_{k+1}^c = (I - LH)\delta u_k^r,$$
  $$\delta u_{k+1}^r = \delta u_{k+1}^c = (I - LH)\delta u_k^r.$$

  In other words, the computed control and the real control are updated simultaneously and are still equal to each other. In short, a *renewal* occurs.

- Case 4: $\sigma_{k+1} = 1$ and $\gamma_{k+1} = 0$.
  In this case, only the computed control is updated,

  $$\delta u_{k+1}^c = (I - LH)\delta u_k^r,$$
  $$\delta u_{k+1}^r = \delta u_k^r.$$

As a result, it becomes *asynchronization*.

The probabilities for the above four cases are $(1 - \overline{\sigma})\overline{\gamma}$, $(1 - \overline{\sigma})(1 - \overline{\gamma})$, $\overline{\sigma}\overline{\gamma}$, and $\overline{\sigma}(1 - \overline{\gamma})$, respectively. From the above discussions, we find that (a) the computed control and the real control stay in the state of *synchronization* except in the last case, and (b) a *renewal* occurs when no data dropouts happen at the measurement and the actuator sides simultaneously.

Therefore, we further discuss the last case; that is, we assume that the computed control and the real control become the last case at the $(k+1)$th iteration. Then, four possible outcomes exist for the $(k+2)$th iteration.

- Case 1': $\sigma_{k+2} = 0$ and $\gamma_{k+2} = 1$.
  In this case, the real control is updated,

  $$\delta u_{k+2}^c = \delta u_{k+1}^c = (I - LH)\delta u_k^r,$$
  $$\delta u_{k+2}^r = \delta u_{k+2}^c = (I - LH)\delta u_k^r.$$

  That is, the computed control and the real control achieve *synchronization*, and a *renewal* occurs.

- Case 2': $\sigma_{k+2} = 0$ and $\gamma_{k+2} = 0$.
  In this case, no change happens to both the computed control and the real control,

  $$\delta u_{k+2}^c = \delta u_{k+1}^c = (I - LH)\delta u_k^r,$$
  $$\delta u_{k+2}^r = \delta u_{k+1}^r = \delta u_k^r.$$

  Then, the computed control and real control are still in the state of *asynchronization*.

- Case 3': $\sigma_{k+2} = 1$ and $\gamma_{k+2} = 0$.
  In this case, only the computed control is updated,

  $$\delta u_{k+2}^c = (I - LH)\delta u_{k+1}^r = (I - LH)\delta u_k^r,$$
  $$\delta u_{k+2}^r = \delta u_{k+1}^r = \delta u_k^r.$$

  The values of the computed control and the real control remain the same to the $(k+1)$th iteration, and they are in the state of *asynchronization*

8.3 Markov Chain Model of Input Evolution

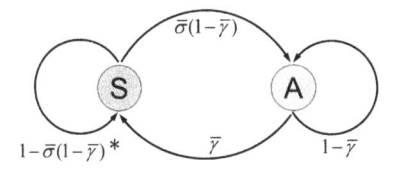

Fig. 8.2 Illustration of the Markov chain of *synchronization* and *asynchronization*. S: synchronization; A: asynchronization; ∗: there is a *renewal* with probability $\overline{\sigma}\overline{\gamma}$

- Case 4': $\sigma_{k+2} = 1$ and $\gamma_{k+2} = 1$.
  In this case, both the computed control and the real control are updated,

$$\delta u^c_{k+2} = (I - LH)\delta u^r_{k+1} = (I - LH)\delta u^r_k,$$
$$\delta u^r_{k+2} = \delta u^c_{k+2} = (I - LH)\delta u^r_k.$$

As a result, the computed control and the real control become *synchronization* again, and a *renewal* occurs.

The probabilities of the four outcomes, i.e., Cases 1'–4', being the same to Cases 1–4 above. The analysis indicates that (a) from the *asynchronization* state, the computed control and the real control will either remain unchanged or become *synchronization* again, and (b) a *renewal* occurs whenever the state changes to *synchronization*.

Thus, we can conclude that the relationship between both computed control and real control have two states, namely, *synchronization* and *asynchronization*, respectively. Moreover, the two states would switch between each other following a Markov chain, as shown in Fig. 8.2. Specifically, from Fig. 8.2, we can observe the transition probability distribution of the Markov chain. That is, from the state of *synchronization*, the probability of the inputs retaining *synchronization* is $1 - \overline{\sigma}(1 - \overline{\gamma})$, whereas the probability of the inputs switching to *asynchronization* is $\overline{\sigma}(1 - \overline{\gamma})$. From the state of *asynchronization*, the probabilities of retaining *asynchronization* and switching to *synchronization* are $1 - \overline{\gamma}$ and $\overline{\gamma}$, respectively. Note that although the transition probability from *synchronization* to *synchronization* is $1 - \overline{\sigma}(1 - \overline{\gamma})$, there exists a *renewal* of the inputs with probability $\overline{\sigma}\overline{\gamma}$. All these probabilities are computed above.

Note that we have shown that the switching between *synchronization* and *asynchronization* only depends on the state of the last iteration. Thus, the proof is completed.

8.4 Convergence Analysis

In this section, we will give the convergence proof with the help of the Markov chain property specified in the last section. To this end, let us first design the learning gain matrix and then propose the convergence results. Note that the system matrix H is a block lower triangular matrix and the diagonal Markov parameter is $H_{i,i}$, $1 \leq i \leq N$. Thus, we could design the learning gain matrix L as a block diagonal matrix, i.e., $L = \mathrm{diag}\{L_1, \ldots, L_N\}$. Then, all eigenvalues of $I - L_i H_{i,i}$, $1 \leq i \leq N$ evidently lie in the range $(0, 1)$ whenever the learning gain matrix L is designed such that $0 < \rho(I - L_i H_{i,i}) < 1$, $1 \leq i \leq N$, where $\rho(\cdot)$ denotes the spectral radius.

The main result of this chapter is given in the following theorem.

Theorem 8.1 *Consider the linear system* (8.1) *and assume Assumptions 8.1 and 8.2 hold. The learning update laws* (8.3) *and* (8.4) *guarantee the zero-error convergence of the output to any desired trajectory* y_d *asymptotically as the iteration number goes to infinity, if the learning gain matrix* L_i *satisfies* $0 < \rho(I - L_i H_{i,i}) < 1$, $1 \leq i \leq N$.

Proof From Fig. 8.2, the transition matrix of the Markov chain is formulated as

$$P = \begin{bmatrix} p_{11} & p_{12} \\ p_{21} & p_{22} \end{bmatrix} = \begin{bmatrix} 1 - \overline{\sigma}(1 - \overline{\gamma}) & \overline{\sigma}(1 - \overline{\gamma}) \\ \overline{\gamma} & 1 - \overline{\gamma} \end{bmatrix}, \tag{8.8}$$

where $p_{11} \triangleq \mathbb{P}(\tau_{k+1} = S | \tau_k = S)$, $p_{12} \triangleq \mathbb{P}(\tau_{k+1} = A | \tau_k = S)$, $p_{21} \triangleq \mathbb{P}(\tau_{k+1} = S | \tau_k = A)$, and $p_{22} \triangleq \mathbb{P}(\tau_{k+1} = A | \tau_k = A)$ with τ_k being the state of the kth iteration, S the *synchronization* state, and A the *asynchronization* state. Note that $0 < \overline{\sigma}, \overline{\gamma} < 1$; thus, P is irreducible, aperiodic, and recurrent, which further means P is ergodic. The stationary distribution π of this Markov chain through solving the equation $\pi P = \pi$ is given as

$$\pi \triangleq \left[\frac{\overline{\gamma}}{\overline{\gamma} + \overline{\sigma} - \overline{\sigma} \cdot \overline{\gamma}}, \frac{\overline{\sigma} - \overline{\sigma} \cdot \overline{\gamma}}{\overline{\gamma} + \overline{\sigma} - \overline{\sigma} \cdot \overline{\gamma}} \right]. \tag{8.9}$$

Note that the *renewal* only occurs when the state changes to *synchronization*. Specifically, the occurrence probability of *renewal* is $\overline{\sigma} \cdot \overline{\gamma}$ for the case that the state switches from *synchronization* to *synchronization*, and such probability is $\overline{\gamma}$ for the case that the state switches from *asynchronization* to *synchronization*. Thus, the probability of *renewal* along the iteration axis can be calculated

$$\begin{aligned} \mathbb{P}(\text{renewal}) &= \frac{\overline{\gamma}}{\overline{\gamma} + \overline{\sigma} - \overline{\sigma} \cdot \overline{\gamma}} \cdot \overline{\sigma} \cdot \overline{\gamma} + \frac{\overline{\sigma} - \overline{\sigma} \cdot \overline{\gamma}}{\overline{\gamma} + \overline{\sigma} - \overline{\sigma} \cdot \overline{\gamma}} \cdot \overline{\gamma} \\ &= \frac{\overline{\sigma} \cdot \overline{\gamma}}{\overline{\gamma} + \overline{\sigma} - \overline{\sigma} \cdot \overline{\gamma}}. \end{aligned} \tag{8.10}$$

In addition, whenever a *renewal* occurs, both the computed control and the real control will improve. To be specific, with the help of the above analysis, we can introduce a random variable λ_k to denote whether a renewal happens or not, i.e., $\lambda_k = 1$ if a renewal happens, and 0 otherwise. Moreover, λ_k obeys the Bernoulli distribution, $p_1 \triangleq \mathbb{P}(\lambda_k = 1) = \frac{\overline{\sigma} \cdot \overline{\gamma}}{\overline{\gamma} + \overline{\sigma} - \overline{\sigma} \cdot \overline{\gamma}}$ and $p_2 \triangleq \mathbb{P}(\lambda_k = 0) = 1 - \mathbb{P}(\lambda_k = 1)$. Then, the recursion of the real input error could be formulated as

$$\delta u_k^r = \lambda_k (I - LH) \delta u_{k-1}^r + (1 - \lambda_k) \delta u_{k-1}^r. \tag{8.11}$$

8.4 Convergence Analysis

The last recursion can be regarded as a switched system

$$\delta u_k^r = \Gamma_k \delta u_{k-1}^r, \tag{8.12}$$

where $\Gamma_k = I - LH$ when $\lambda_k = 1$ and $\Gamma_k = I$ when $\lambda_k = 0$. Let us denote $\overline{\Gamma}_k = \Gamma_k \Gamma_{k-1} \cdots \Gamma_1$ and $\mathscr{S}^k = \{\overline{\Gamma}_k$: taken over all sample paths$\}$. We have the following claim.

Claim 1: The mean of $\mathscr{S}^k$, denoted by M_k, is defined recursively by

$$M_k = (p_1(I - LH) + p_2 I) M_{k-1}. \tag{8.13}$$

The proof for this claim is given by direct calculations. According to the definition of mean, $M_k = \sum_{\overline{\Gamma}_k \in \mathscr{S}^k} \mathbb{P}(\overline{\Gamma}_k) \overline{\Gamma}_k$. Then,

$$\begin{aligned}
M_k &= \sum_{\overline{\Gamma}_k \in \mathscr{S}^k} \mathbb{P}(\overline{\Gamma}_k) \overline{\Gamma}_k \\
&= \sum_{\overline{\Gamma}_{k-1} \in \mathscr{S}^{k-1}} \mathbb{P}(\overline{\Gamma}_{k-1})(p_1(I - LH)\overline{\Gamma}_{k-1} + p_2 \overline{\Gamma}_{k-1}) \\
&= (p_1(I - LH) + p_2 I) \sum_{\overline{\Gamma}_{k-1} \in \mathscr{S}^{k-1}} \mathbb{P}(\overline{\Gamma}_{k-1}) \overline{\Gamma}_{k-1} \\
&= (p_1(I - LH) + p_2 I) M_{k-1}.
\end{aligned}$$

The claim is proved. To show the convergence of δu_k^r in mathematical expectation sense, (8.12) indicates that

$$\begin{aligned}
\mathbb{E}\delta u_k^r &= \mathbb{E}(\overline{\Gamma}_k \delta u_0^r) \\
&= (p_1(I - LH) + p_2 I)^k \mathbb{E}\delta u_0^r.
\end{aligned}$$

Thus, it suffices to show $\rho(p_1(I - LH) + p_2 I) < 1$. On the one hand, $I - LH$ is a block lower triangular matrix with its diagonal elements being $I - L_i H_{i,i}$, $1 \le i \le N$ and I is the identity matrix. On the other hand, $0 < p_1, p_2 < 1$ and $p_1 + p_2 = 1$. Verifying $\rho(p_1(I - LH) + p_2 I) < 1$ thus requires little effort.

In addition, given that $\rho(I - LH) < 1$, a suitable induced matrix norm exists such that $0 < p_1 \|I - LH\| + p_2 \|I\| < 1$. Then, a positive constant $0 < \mu < 1$ exists such that

$$0 < p_1 \|I - LH\| + p_2 \|I\| < \mu.$$

Then, the recursion (8.12) leads to

$$\mathbb{E}\|\delta u_k^r\| \leq \mathbb{E}\|\overline{\Gamma}_k\|\mathbb{E}\|\delta u_0^r\|$$
$$\leq (\mathbb{E}\|\Gamma_k\|)^k \mathbb{E}\|\delta u_0^r\|$$
$$= (p_1\|I - LH\| + p_2\|I\|)^k \mathbb{E}\|\delta u_0^r\|$$
$$\leq \mu^k \mathbb{E}\|\delta u_0^r\|.$$

Consequently,

$$\sum_{k=1}^{\infty} \mathbb{E}\|\delta u_k^r\| \leq \sum_{k=1}^{\infty} \mu^k \mathbb{E}\|\delta u_0^r\|$$
$$= \frac{\mu}{1-\mu}\mathbb{E}\|\delta u_0^r\| < \infty.$$

Then by Markov inequality, for any $\varepsilon > 0$ we have

$$\sum_{k=1}^{\infty} \mathbb{P}(\|\delta u_k^r\| > \varepsilon) \leq \sum_{k=1}^{\infty} \frac{\mathbb{E}\|\delta u_k^r\|}{\varepsilon} < \infty. \tag{8.14}$$

Therefore, $\mathbb{P}(\|\delta u_k^r\| > \varepsilon$, infinitely often$) = 0$ is concluded by Borel–Cantelli lemma, $\forall \varepsilon > 0$, and $\mathbb{P}(\lim_{k\to\infty}\|\delta u_k^r\| = 0) = 1$ is obtained further. The zero-error of input convergence is thus proved. From the relationship $e_k = H\delta u_k^r$, the proof is completed.

Remark 8.6 Define the function $f(\overline{\sigma}, \overline{\gamma}) = \mathbb{P}(renewal)$. Evidently, $f(\overline{\sigma}, \overline{\gamma}) = f(\overline{\gamma}, \overline{\sigma})$. Moreover, through simple calculations, one has $\frac{\partial f(\overline{\sigma},\overline{\gamma})}{\partial \overline{\sigma}} = \frac{\overline{\gamma}^2}{(\overline{\gamma}+\overline{\sigma}-\overline{\sigma}\cdot\overline{\gamma})^2} > 0$. This condition means that a larger successful transmission rate corresponds to more renewals, and thus, faster algorithm convergence. This result coincides with our intuitive knowledge.

Now, let us further discuss the extension to time-dependent update case. That is, as discussed in Remark 8.5, we introduce σ_k^t and γ_k^t to denote the data dropout occurring at time instant t for the updating of the kth iteration. According to Remark 8.4, the control law could run time instant by time instant. In other words, it is not mandatory to use the lifted forms (8.3) and (8.4). The update algorithms for such general data dropout case take the following forms:

$$u_{k+1}^c(t) = \sigma_{k+1}^t u_k^r(t) + (1 - \sigma_{k+1}^t)u_k^c(t)$$
$$+ \sigma_{k+1}^t L_0 e_k(t+1), \tag{8.15}$$

$$u_{k+1}^r(t) = \gamma_{k+1}^t u_{k+1}^c(t) + (1 - \gamma_{k+1}^t)u_k^r(t). \tag{8.16}$$

In such case, the equivalent switching process of the input error, i.e., a counterpart to (8.11), is formulated as

$$\delta u_k^r = \Lambda_k(I - LH)\delta u_k^r + (I - \Lambda_k)\delta u_k^r, \tag{8.17}$$

8.4 Convergence Analysis

where $\Lambda_k = \text{diag}\{\lambda_k^0, \ldots, \lambda_k^{N-1}\}$ with λ_k^t being defined similar to λ_k in (8.11). Therefore, different from (8.11) where only two cases are involved, there are 2^N cases in (8.17) for the switching.

Note that the Markov chain established in Lemma 8.1 is in terms of the synchronization and asynchronization of the computed control and real control of the entire iteration. Thus, when considering the time-dependent data dropout case, it is easy to find that the computed control and real control for arbitrary time instant also follow the same path behaviors. That is, for arbitrary given t, the synchronous and asynchronous states of δu_k^c and δu_k^r form a two-state Markov chain. The combinations of the states of all time instants will no longer be described by two states only, i.e., *synchronization* and *asynchronization*. Instead, such combination consists of 2^N states because the controls at different time instants switch independently. Consequently, the combination of N independent Markov chains still behaves as a Markov chain with 2^N states. Among the 2^N states, two special cases should be pointed out. The first one is that both the computed control and real control at all time instants achieve *synchronization* simultaneously, and the other one is that both controls at all time instants achieve *asynchronization* simultaneously. Moreover, the corresponding switched recursion (8.12) now becomes a switched system with 2^N cases of Γ_k, which includes two cases, i.e., $I - LH$ and I, as its special cases according to the above special states. Specifically, the diagonal blocks of Γ_k are defined as follows: the tth diagonal block matrix of Γ_k is $I - L_t H_{t,t}$ if the computed control and real control at time instant t achieve *synchronization*; otherwise, it becomes $I_{p \times p}$. Then, the convergence of the time-dependent data dropouts case can be verified following the same steps of Theorem 8.1.

Remark 8.7 Taking (8.16) as an example, we give a further remark on the compensation mechanism. In (8.16), if the corresponding data packet $u_{k+1}^c(t)$ is dropped during the transmission, then the available latest data $u_k^r(t)$ is used to compensate for the data. Thus, this compensation mechanism admits successive random data dropouts along both the iteration and time axes. This is different from the existing compensation mechanisms in [4–7]. Under the time-axis-based compensation mechanism of [4, 5], the update law (8.16) should be formulated as $u_{k+1}^r(t) = \gamma_{k+1}^t u_{k+1}^c(t) + (1 - \gamma_{k+1}^t) u_{k+1}^c(t-1)$. Then, the data packet at adjacent time instants is not allowed to be dropped simultaneously. On the other hand, when applying the iteration-axis-based compensation mechanism of [6, 7], the update law (8.16) should be formulated as $u_{k+1}^r(t) = \gamma_{k+1}^t u_{k+1}^c(t) + (1 - \gamma_{k+1}^t) u_k^c(t)$. Then, the data dropouts for adjacent iterations are excluded.

Remark 8.8 Note that the learning process is essentially iteration-dependent (cf. (8.15) and (8.16)). It should be pointed out that the dependence is generated by the random data dropouts; that is, the updating of the input would be different with the data dropouts occurring or not. However, from Theorem 8.1 it is observed that the convergence condition only depends on the system matrix, which is essential iteration-independent. This is because we adopt a simple holding mechanism when data dropouts occur. Therefore, the essential system dynamics is repetitive and the inherent improvement of the input sequence is determined by the system information only, whereas the data dropout rate mainly affect the learning speed. This fact is verified in the next section.

8.5 Illustrative Simulations

In this section, we first verify our theoretical results using a numerical simulation of linear time-varying (LTV) system. Then, a case study on an industrial robot is also provided to demonstrate the effectiveness. It should be emphasized that the general time-dependent data dropout problem is considered and the algorithms (8.15), (8.16) are applied in these simulations.

Example 8.1 (*Numerical Example*) In this example, a numerical system is given to verify the theoretical analysis. To show the effectiveness of the proposed algorithm, consider the LTV system (A_t, B_t, C_t) with A_t, B_t, and C_t being

$$A_t = \begin{bmatrix} 0.2\exp(-t/100) & -0.60 & 0 \\ 0 & 0.50 & \sin(t) \\ 0 & 0 & 0.70 \end{bmatrix},$$

$$B_t = [0 \ 0.3\sin(t) \ 1]^T,$$

$$C_t = [0 \ 0.1 + 0.05\cos(t) \ 0.8].$$

The iteration length is $N = 100$. The initial condition is given as $y_k(0) = y_d(0) = 0$ for all k. The control input for the first iteration is set $u_0^r = u_0^c = 0$. The desired trajectory is $y_d = 0.5\sin(\pi t/20) + 0.25\sin(\pi t/10)$. The lifted model H can be calculated directly. The learning gains are selected as $L_i = 0.4$, $1 \le i \le 100$.

According to Remarks 8.4 and 8.5, we simulate the general case. That is, the algorithms used here are time-dependent (8.15) and (8.16). Meanwhile, the data dropout is introduced to each time instant separately rather than to the entire iteration. The algorithms are run for 150 iterations.

We define data dropout rate (DDR) as the probability $\mathbb{P}(\sigma_k^t = 0)$ or $\mathbb{P}(\gamma_k^t = 0)$. It is noticed that DDR indicates the average ratio of lost transmission among all iterations. In the simulation, for experiments simplicity, the DDR at the measurement side is set to equal to the one at the actuator side. In addition, five cases of DDR are simulated, namely, DDR = 0, 10, 20, 30, and 40%.

Figure 8.3 shows the output tracking profiles at the 2nd, 5th, and 8th iterations as well as the desired trajectory for the case that the DDRs at the measurement and actuator sides are both 10, 20, 30, and 40%, respectively. The figures show that the output converges to the reference after several iterations.

Figure 8.4 further displays the influence of data dropouts on convergence performance, where the maximal tracking error profiles are compared for DDR = 0, 10, 20, 30, and 40%. Two facts are observed. First, the convergence speed decreases

8.5 Illustrative Simulations

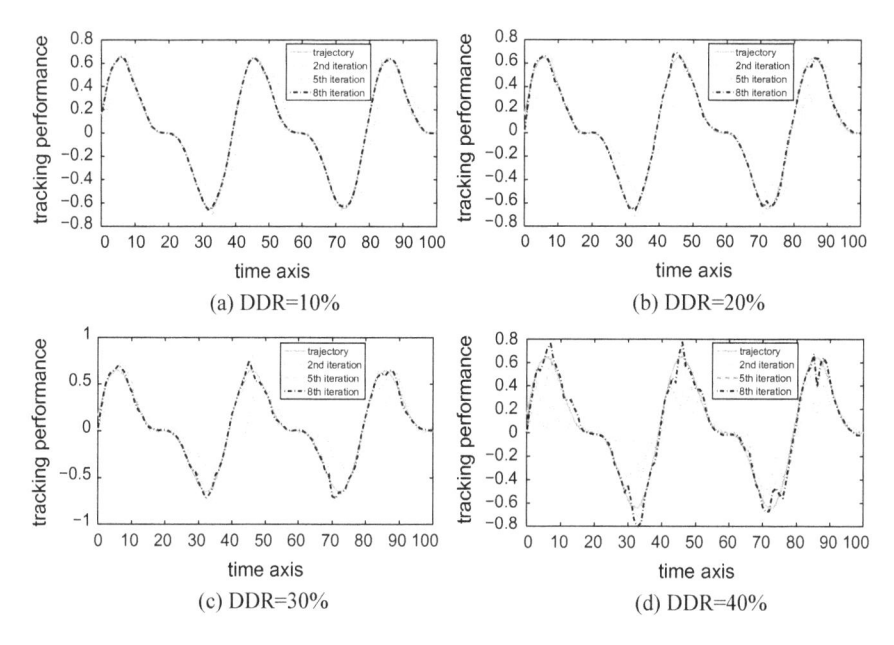

Fig. 8.3 Output profiles at the 2nd, 5th, and 8th iterations and the desired trajectory

as the DDR increases. This observation coincides with our intuitive recognition as larger DDR means less updating iterations. Second, all the maximal tracking error profiles are approximate lines in logarithm axis, thereby implying an exponential convergence speed. This observation verifies the theoretical results given in the last section.

In addition, to demonstrate the asynchronization of the computed input signal and the real input signal, we introduce a counter $\tau_k(t)$ for any given time instant t, denoting the amount number up to the kth iteration of the case that the computed input signal is not equal to the real input signal. That is, the counter value increases only when both the computed and real input signals are not equal to each other. The profiles for all time instants are displayed in Fig. 8.5 for four cases, i.e., DDR = 10, 20, 30, and 40%, where the profiles rise as the iteration number goes up. This fact illustrates that the asynchronization occurs randomly along the iteration axis independently for different time instants. Moreover, the larger DDR, the larger average value of $\tau_k(t)$ at the last iteration.

Example 8.2 (*A Case Study on Industrial Robot*) In this example, the algorithms are applied to an industrial robot. The nominal model of the closed-loop joint is given as follows [8]:

$$G_p(s) = \frac{948}{s^2 + 42s + 948}. \tag{8.18}$$

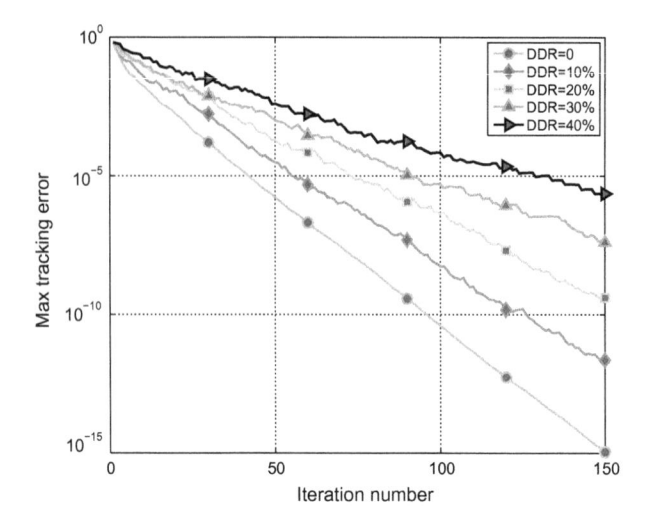

Fig. 8.4 Maximal error profile along iteration axis for DDR = 0, 10, 20, 30, and 40%

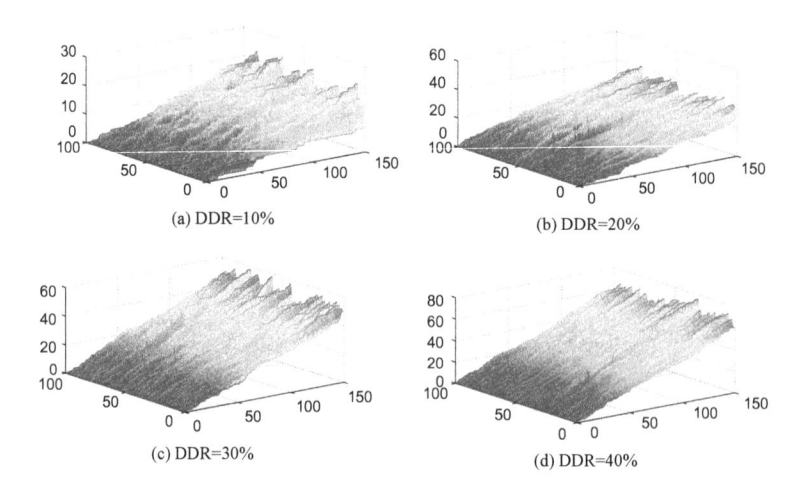

Fig. 8.5 Asynchronization of the computed and real input signals

The desired trajectory is given as follows $y_d(t) = \sin(\pi t/20) + 0.3\cos(\pi t/10)$. The operation length is 2 s and the sampling frequency is 100 Hz. Then, the system is discretized with iteration length $N = 200$. The learning gains used in this example are $L_i = 0.0625$, $1 \leq i \leq 200$. The algorithms are run for 100 iterations. Similar to the last example, we also consider five cases of DDR, i.e., DDR = 0, 10, 20, 30, and 40%.

Figure 8.6 shows the output tracking profiles at the 2nd, 10th, and 50th iterations as well as the desired trajectory for the cases that the DDRs at the measurement and actuator sides are both 10, 20, 30, and 40%, respectively. It is evident that the output

8.5 Illustrative Simulations

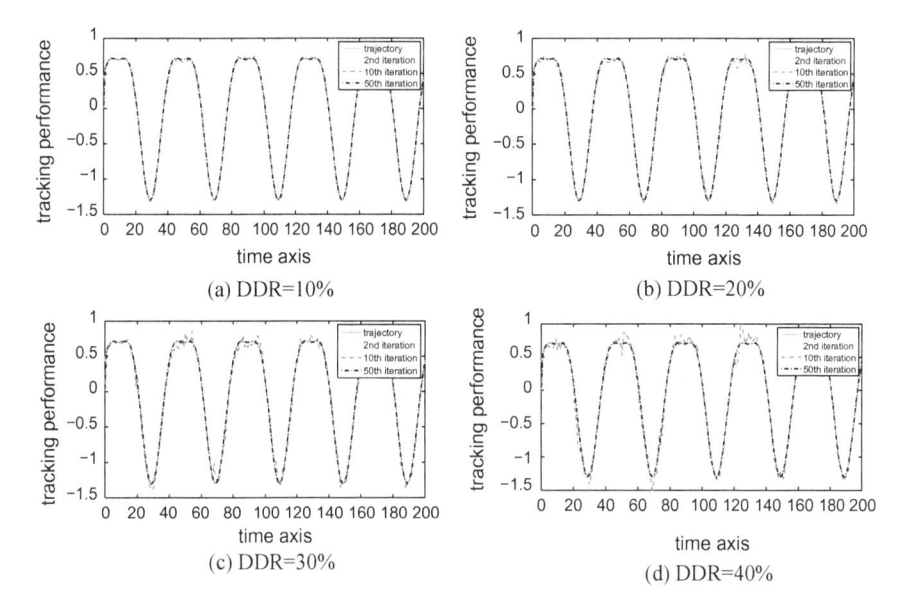

Fig. 8.6 Output profiles at the 2nd, 10th, and 50th iterations and the desired trajectory

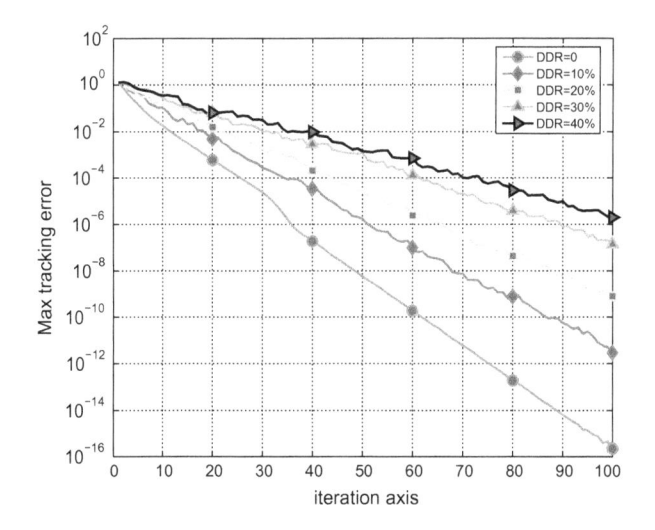

Fig. 8.7 Maximal error profile along iteration axis for DDR = 0, 10, 20, 30, and 40%

profiles converge to the desired trajectory asymptotically. The maximal tracking error profiles are illustrated in Fig. 8.7. It can be seen from this figure that the exponential convergence speed is guaranteed under general data dropout environments. This fact verifies the effectiveness of the proposed compensation mechanism and its associated analysis results. The asynchronization phenomena between the computed control and the real control is also simulated. The results are same to Fig. 8.5 and thus are omitted for saving space.

8.6 Summary

This chapter proposes the first convergence analysis of ILC for a linear system with data dropouts at both measurement and actuator sides. The proof is carried out by carefully analyzing the sample path behavior and using Markov chain techniques. While the results are carried out for the classic P-type update law, it can be generalized to other update laws such as PD-type. The results in this chapter are mainly based on [9].

References

1. Son, T.D., Pipeleers, G., Swevers, J.: Robust monotonic convergent iterative learning control. IEEE Trans. Autom. Control **61**(4), 1063–1068 (2016)
2. Pipeleers, G., Moore, K.L.: Unified analysis of iterative learning and repetitive controllers in trial domain. IEEE Trans. Autom. Control **59**(4), 953–965 (2014)
3. Oh, S.-K., Lee, J.M.: Stochastic iterative learning control for discrete linear time-invariant system with batch-varying reference trajectories. J. Process Control **36**, 64–78 (2015)
4. Bu, X., Yu, F., Hou, Z., Wang F.: Terative learning control for a class of nonlinear systems with random packet losses. Nonlinear Anal. Real World Appl. **14**(1), 567–580 (2013)
5. Pan, Y.-J., Marquez, H.J., Chen, T., Sheng, L.: Effects of network communications on a class of learning controlled non-linear systems. Int. J. Syst. Sci. **40**(7), 757–767 (2009)
6. Huang, L.-X., Fang, Y.: Convergence analysis of wireless remote iterative learning control systems with dropout compensation. Math. Prob. Eng. **2013**, 609284 (2013)
7. Liu, J., Ruan, X.: Networked iterative learning control approach for nonlinear systems with random communication delay. Int. J. Syst. Sci. **47**(16), 3960–3969 (2016)
8. Zhang, B., Ye, Y., Zhou, K., Wang, D.: Case studies of filtering techniques in multirate iterative learning control. Control Eng. Pract. **26**, 116–124 (2014)
9. Shen, D., Jin, Y., Xu, Y.: Learning control for linear systems under general data dropouts at both measurement and actuator sides: a Markov chain approach. J. Franklin Inst. **354**(13), 5091–5109 (2017)

Chapter 9
Two-Side Data Dropout for Linear Stochastic Systems

9.1 Problem Formulation

Consider the following linear time-varying stochastic system:

$$\begin{aligned} x_k(t+1) &= A_t x_k(t) + B_t u_k(t) + w_k(t), \\ y_k(t) &= C_t x_k(t) + v_k(t), \end{aligned} \quad (9.1)$$

where $t = 0, 1, \ldots, N$ denotes the time index, N is the iteration length, and $k = 1, 2, \ldots$, denotes the iteration index. $u_k(t) \in \mathbf{R}^p$, $y_k(t) \in \mathbf{R}^q$, and $x_k(t) \in \mathbf{R}^n$ are the system input, output, and state, respectively. The system matrices A_t, B_t, and C_t are with appropriate dimensions. $w_k(t)$ and $v_k(t)$ are system noises and measurement noises, respectively.

The desired reference is $y_d(t)$, which satisfies the following formulation:

$$\begin{aligned} x_d(t+1) &= A_t x_d(t) + B_t u_d(t), \\ y_d(t) &= C_t x_d(t). \end{aligned} \quad (9.2)$$

The following assumptions are required for further analysis.

Assumption 9.1 The input–output coupling matrix $C_{t+1} B_t \in \mathbf{R}^{q \times p}$ is of full-column rank, $t = 0, \ldots, N - 1$ and therefore $q \geq p$.

Assumption 9.2 The initial state $x_k(0)$ is precisely reset, i.e., $x_k(0) = x_d(0)$.

Remark 9.1 The initial state reset is a critical issue in ILC field. Assumption 9.2 is the well-known identical initialization condition (i.i.c.), which has been used in many ILC papers. Efforts have been dedicated to the relaxation of i.i.c., but they require further information on the system or additional control mechanism. However, the extension of i.i.c. is out of our scope, thus we use Assumption 9.2 for simplicity.

© Springer Nature Singapore Pte Ltd. 2018
D. Shen, *Iterative Learning Control with Passive Incomplete Information*,
https://doi.org/10.1007/978-981-10-8267-2_9

Denote the increasing σ-algebra $\mathscr{F}_k \triangleq \sigma\{x_j(t), w_j(t), v_j(t), 1 \leq j \leq k, t = 0, 1, \ldots, N\}$ generated by the information from the first iteration up to the kth iteration. We give the following assumption on stochastic noises.

Assumption 9.3 The stochastic noises $w_k(t)$ and $v_k(t)$ are independent for different time instants and they are independent with each other. For each t, $\mathbb{E}\{w_k(t)|\mathscr{F}_{k-1}\} = 0$, $\mathbb{E}\{v_k(t)|\mathscr{F}_{k-1}\} = 0$, $\sup_k \mathbb{E}\{w_k^2(t)|\mathscr{F}_{k-1}\} < \infty$, $\sup_k \mathbb{E}\{v_k^2(t)|\mathscr{F}_{k-1}\} < \infty$. Here, $\mathbb{E}(\cdot)$ denotes the mathematical expectation operator.

The control objective of this chapter is to design a learning algorithm such that the generated input sequence could track the desired trajectory asymptotically along the iteration axis under data dropouts and stochastic noises conditions. Because of the existence of stochastic noises, the actual output could not precisely track the desired trajectory. Thus, our objective is to minimize the following performance index:

$$V_t = \lim_{n \to \infty} \frac{1}{n} \sum_{k=1}^{n} \|y_d(t) - y_k(t)\|^2. \tag{9.3}$$

As has been proved in previous chapters, in order to minimize the above performance index, it is sufficient to show that the input sequence $u_k(t) \to u_d(t)$, $t = 0, 1, \ldots, N - 1$. This is the objective of subsequent analysis. In addition, it is noticed from Lemma 2.1 that the minimum of the above performance index (9.3) is a linear combination of the upper bounds of covariance of both system and measurement noises $w_k(t)$, $v_k(t)$. In other words, the ultimate tracking performance is determined by the stochastic noises.

In this chapter, the general networked framework is considered; that is, both networks at the measurement and actuator sides would suffer random data dropouts. The data dropouts are modeled by two random variables, $\sigma_k(t)$ and $\gamma_k(t)$, subject to Bernoulli distribution. Specifically, both $\sigma_k(t)$ and $\gamma_k(t)$ are equal to 1 if the corresponding data is transmitted successfully, and 0 otherwise. Moreover, $\mathbb{P}(\sigma_k(t) = 1) = \bar{\sigma}$ and $\mathbb{P}(\gamma_k(t) = 1) = \bar{\gamma}$, $0 < \bar{\sigma}, \bar{\gamma} < 1$, where $\mathbb{P}(\cdot)$ denotes the probability of the indicted event. Both $\sigma_k(t)$ and $\gamma_k(t)$ are independent for different time instant t and iteration index k.

The block diagram of the framework is illustrated in Fig. 9.1. In this framework, the update of the input follows the intermittent type. In other words, if the data is successfully transmitted at the measurement side, then the algorithm would update its input signal; while if the data is lost during transmission at the measurement side, then the algorithm would stop updating and retain the stored input signal of the previous iteration. At the actuator side, if the input signal is successfully transmitted, then the plant will use this new input signal; if the input signal is lost, then the plant will operate with the stored input signal of the previous iteration. The data dropouts at measurement and actuator sides occur independently.

In this framework, we denote the control signal generated by the learning controller, called the *computed control*, as $u_k^c(t)$, and the real control signal fed to the plant, called the *real control*, as $u_k^r(t)$. The workflow of Fig. 9.1 is as follows: when

9.1 Problem Formulation

Fig. 9.1 Block diagram of the proposed ILC framework

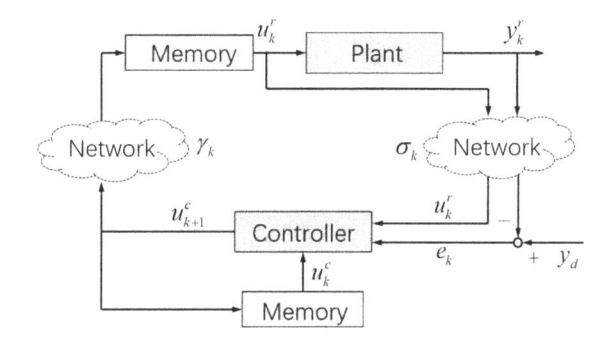

the system finishes one batch, all the data are transmitted back to the controller, then the controller computes the control signal for the next batch, and the computed control would then be transmitted to the plant so that the system could run for the next batch. Then the updating law in the controller is formulated as

$$u^c_{k+1}(t) = \sigma_{k+1}(t)u^r_k(t) + (1 - \sigma_{k+1}(t))u^c_k(t) \\ + \sigma_{k+1}(t)a_k L_t e_k(t+1), \tag{9.4}$$

while the actually used control signal for the $(k+1)$th iteration is

$$u^r_{k+1}(t) = \gamma_{k+1}(t)u^c_{k+1}(t) + (1 - \gamma_{k+1}(t))u^r_k(t), \tag{9.5}$$

where $e_k(t) \triangleq y_d(t) - y_k(t)$, L_t is the learning gain matrix and a_k is a decreasing sequence to ensure zero-error tracking. The sequence $\{a_k\}$ should satisfy $a_k > 0$, $\sum_{k=1}^{\infty} a_k = \infty$, and $\sum_{k=1}^{\infty} a_k^2 < \infty$.

Remark 9.2 The decreasing sequence $\{a_k\}$ used in (9.4) is a technical means to handle the stochastic noises. If the stochastic noises are eliminated from the system, the term a_k can be removed from (9.4). It is well known that an appropriate decreasing gain for the correction term in updating processes is a necessary requirement to ensure convergence in the recursive computation for optimization, identification, and tracking of stochastic systems [1, 2].

Remark 9.3 The random variables $\sigma_{k+1}(t)$ and $\gamma_{k+1}(t)$ are defined independently along both iteration and time axes. Thus, it is apparent that successive data dropouts in time axis are allowed in this formulation. Moreover, from (9.4) it is noticed that if $\sigma_{k+1}(t) = 1$, i.e., the data is successfully transmitted, then the *computed control* is updated; otherwise if $\sigma_{k+1}(t) = 0$, then the *computed control* copies its corresponding value of the previous iteration. However, in the latter case, the corresponding *computed control* of the previous iteration may likewise copy the value of its previous iteration. Consequently, successive data dropouts in iteration axis are also allowed. In addition, no extra storage beyond one batch size is required by the memory array because only the latest data needs to be stored, as shown in the updating laws (9.4) and (9.5) through two random variables $\sigma_{k+1}(t)$ and $\gamma_{k+1}(t)$. In other words, at each time instant, if the input is updated, then the updated value replaces the stored data; otherwise, the stored input keeps unchanged.

9.2 Markov Chain of Input Evolution

In this section, we will establish the Markov chain of the input sequences. To make our idea clear, we first consider the case for an arbitrary time instant and then generalize it to the whole iteration.

Now, for arbitrary fixed time instant t, $0 \leq t \leq N - 1$, let us consider the sample path behaviors of update algorithms (9.4) and (9.5), where a sample path means an arbitrary sequence with respect to iteration k. It is noticed that the computed control and the real control are the same whenever $\gamma_k = 1$. In the following, the sample path behavior is called "*synchronization*" if the computed control and real control are equal to each other; otherwise, it is called "*asynchronization*". Moreover, it is called a "*renewal*" if both computed control and real control are in the state of *synchronization* but different from their last *synchronization*. The following lemma shows that the sample path behavior actually is a Markov chain in terms of *synchronization* and *asynchronization*.

Lemma 9.1 *Consider the updating laws (9.4) and (9.5). The updating of the values of $u_k^r(t)$ and $u_k^c(t)$ forms a Markov chain.*

Proof We start from the kth iteration where $u_k^r(t) = u_k^c(t)$. That is, the computed control and real control are in the state of *synchronization* at the kth iteration. Then, for the $(k + 1)$th iteration, four possible outcomes exist.

- Case 1: $\sigma_{k+1}(t) = 0$ and $\gamma_{k+1}(t) = 1$.
  The probability of this case is $\mathbb{P}(\sigma_{k+1}(t) = 0)\mathbb{P}(\gamma_{k+1}(t) = 1) = (1 - \bar{\sigma})\bar{\gamma}$. In this case, from (9.4) and (9.5) we have
  $$u_{k+1}^c(t) = u_k^c(t) = u_k^r(t),$$
  $$u_{k+1}^r = u_{k+1}^c(t) = u_k^r(t).$$

  Thus, the computed control and the real control retain the same as the kth iteration.

- Case 2: $\sigma_{k+1}(t) = 0$ and $\gamma_{k+1}(t) = 0$.
  The probability of this case is $\mathbb{P}(\sigma_{k+1}(t) = 0)\mathbb{P}(\gamma_{k+1}(t) = 0) = (1 - \bar{\sigma})(1 - \bar{\gamma})$. In this case, it is obvious that
  $$u_{k+1}^c(t) = u_k^c(t) = u_k^r(t),$$
  $$u_{k+1}^r(t) = u_k^r(t).$$

  That is, no change in computed control nor in real control occurs.

9.2 Markov Chain of Input Evolution

- **Case 3:** $\sigma_{k+1}(t) = 1$ and $\gamma_{k+1}(t) = 1$.
  The probability of this case is $\mathbb{P}(\sigma_{k+1}(t) = 1)\mathbb{P}(\gamma_{k+1}(t) = 1) = \overline{\sigma}\overline{\gamma}$. In this case, we find that

$$u_{k+1}^c(t) = u_k^r(t) + a_k L_t e_k(t+1),$$
$$u_{k+1}^r(t) = u_{k+1}^c(t) = u_k^r(t) + a_k L_t e_k(t+1).$$

In other words, the computed control and the real control are updated simultaneously and are still equal to each other. In short, a *renewal* occurs.

- **Case 4:** $\sigma_{k+1}(t) = 1$ and $\gamma_{k+1}(t) = 0$.
  The probability of this case is $\mathbb{P}(\sigma_{k+1}(t) = 1)\mathbb{P}(\gamma_{k+1}(t) = 0) = \overline{\sigma}(1-\overline{\gamma})$. In this case, only the computed control is updated,

$$u_{k+1}^c(t) = u_{k+1}^c(t) = u_k^r(t) + a_k L_t e_k(t+1),$$
$$u_{k+1}^r(t) = u_k^r(t).$$

As a result, the state becomes *asynchronization*.

From the above discussions, we find that (a) the computed control $u_{k+1}^c(t)$ and the real control $u_{k+1}^r(t)$ stay in the state of *synchronization* except in the last case; and (b) a *renewal* occurs when no data dropouts happen at the measurement and the actuator sides with a probability $\overline{\sigma}\overline{\gamma}$.

Therefore, we further discuss the last case, that is, we assume that the computed control and the real control become the last case at the $(k+1)$th iteration. Then, four possible outcomes exist for the $(k+2)$th iteration with probabilities of the four outcomes being the same to cases 1–4 above.

- **Case 1':** $\sigma_{k+2}(t) = 0$ and $\gamma_{k+2}(t) = 1$.
  In this case, the real control is updated,

$$u_{k+2}^c(t) = u_{k+1}^c(t) = u_k^r(t) + a_k L_t e_k(t+1),$$
$$u_{k+2}^r(t) = u_{k+2}^c(t) = u_k^r(t) + a_k L_t e_k(t+1).$$

That is, the computed control and the real control achieve *synchronization*, and a *renewal* occurs.

- **Case 2':** $\sigma_{k+2}(t) = 0$ and $\gamma_{k+2}(t) = 0$.
  In this case, no change happens to both the computed control and the real control,

$$u_{k+2}^c(t) = u_{k+1}^c(t) = u_k^r(t) + a_k L_t e_k(t+1),$$
$$u_{k+2}^r(t) = u_{k+1}^r(t) = u_k^r(t).$$

Then the computed control and the real control are still in the state of *asynchronization*.

- **Case 3':** $\sigma_{k+2}(t) = 1$ and $\gamma_{k+2}(t) = 1$.
  In this case, both the computed control and the real control are updated,

$$u^c_{k+2}(t) = u^r_{k+1}(t) + a_{k+1}L_t e_{k+1}(t+1),$$
$$u^r_{k+2}(t) = u^c_{k+2}(t).$$

As a result, the computed control and the real control become *synchronization* again, and a *renewal* occurs.

- Case 4': $\sigma_{k+2}(t) = 1$ and $\gamma_{k+2}(t) = 0$.
  In this case, only the computed control is updated,

$$u^c_{k+2}(t) = u^r_{k+1}(t) + a_{k+1}L_t e_{k+1}(t+1),$$
$$u^r_{k+2}(t) = u^r_{k+1}(t).$$

However, the real control remains the same to the $(k+1)$th iteration and therefore, the computed control retains the same state to the $(k+1)$th iteration. Thus, the computed control and real control are in the state of *asynchronization*.

The analysis indicates that (a) from the *asynchronization* state, the computed control $u^c_{k+2}(t)$ and the real control $u^r_{k+2}(t)$ will either remain unchanged state or become *synchronization* again; and (b) a *renewal* occurs whenever the state changes into *synchronization*.

As a result, we can conclude that the computed control and the real control have two states, i.e., *synchronization* and *asynchronization*, respectively. Moreover, from the state of *synchronization*, the probability of the inputs retaining *synchronization* is $1 - \bar{\sigma}(1 - \bar{\gamma})$, while the probability of the inputs switching to *asynchronization* is $\bar{\sigma}(1 - \bar{\gamma})$. From the state of *asynchronization*, the probabilities of retaining *asynchronization* and switching to *synchronization* are $1 - \bar{\gamma}$ and $\bar{\gamma}$, respectively. Therefore, the two states would switch between each other following a Markov chain, as shown in Fig. 9.2. The proof of this lemma is completed.

Remark 9.4 If $\bar{\gamma} = 1$, which means there is no data dropout at the actuator side, then the matrix P is singular as the second column being zero. This implies that the computed control and real control would always be in the state of *synchronization*. This special case has been discussed in many papers. On the other hand, if $\bar{\sigma} = 1$, which means that there is no data dropout at the measurement side, then P is also singular as its first row coincides with the last row, and the Markov chain degrades

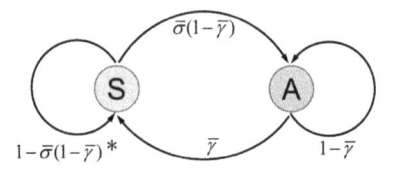

Fig. 9.2 Illustration of the Markov chain of *synchronization* and *asynchronization*. S: synchronization; A: asynchronization; ∗: there is a *renewal* with probability $\bar{\sigma}\bar{\gamma}$

9.2 Markov Chain of Input Evolution

to be a simple Bernoulli sequence. The convergence of such special case could be easy to established.

Next we verify that the sample path behaviors for the whole iteration also form a Markov chain. Lemma 9.1 indicates that, for arbitrary time instant t, the states of *synchronization* and *asynchronization* between $u_k^c(t)$ and $u_k^r(t)$ form a Markov chain. From the analysis steps of Lemma 9.1, it can be seen that the switching of such two states are only determined by the data dropout variables at both sides, i.e., $\sigma_k(t)$ and $\gamma_k(t)$. In other words, the Markov property of the sample path behavior of $u_k^c(t)$ and $u_k^r(t)$ is irrelevant with their specific values.

Moreover, notice that the random variables $\sigma_k(t)$ and $\gamma_k(t)$ modeling the data dropouts at both sides are independent for different time instants. Owing to such independence among different time instants, the combination of the N Markov chains, generated by the computed control and the real control for each time instant t, $0 \le t \le N - 1$, is also a Markov chain.

Specifically speaking, let us introduce a new notation $\tau_k(t)$ to describe the states of *synchronization* and *asynchronization*. That is, we let $\tau_k(t) = 1$ if the computed control and the real control achieve *synchronization*; otherwise, $\tau_k(t) = 0$ if they achieve *asynchronization*. As we have explained above, the states of *synchronization* or *asynchronization* are determined only by the data dropout variables $\sigma_k(t)$ and $\gamma_k(t)$. Therefore, $\tau_k(i)$ is independent of $\tau_k(j)$ for any $i \ne j$. Meanwhile, the evolution of $\tau_k(t)$ is a Markov chain. To show the overall behaviors of all time instants, we further introduce a vector $\varphi_k \triangleq [\tau_k(0), \ldots, \tau_k(N-1)]^T$, which is a stack of $\tau_k(t)$. Since each variable $\tau_k(t)$ is binary, the vector φ_k has 2^N possible outcomes, i.e., $[0, 0, \ldots, 0]^T, [1, 0, \ldots, 0]^T, \ldots, [1, 1, \ldots, 1]^T$. We denote the set of all possible values of φ_k as $\mathscr{S}$. The Markov property of φ_k can be proved directly by the definition of Markov chain. Note that, $\{\tau_k(t)\}$ is a Markov chain, $\forall t$, that is, $\mathbb{P}\{\tau_k(t) = i_k | \tau_{k-1} = i_{k-1}, \ldots, \tau_1(t) = i_1\} = \mathbb{P}\{\tau_k(t) = i_k | \tau_{k-1} = i_{k-1}\}$, $\forall t$, $i_k \in \{0, 1\}$. Therefore, $\mathbb{P}\{\varphi_k = \theta_k | \varphi_{k-1} = \theta_{k-1}, \ldots, \varphi_1 = \theta_1\} = \mathbb{P}\{\varphi_k = \theta_k | \varphi_{k-1} = \theta_{k-1}\}$, where $\theta_i \in \mathscr{S}$. As a consequence, the switching of the inputs for the general case is also a Markov chain of 2^N states.

9.3 Convergence Analysis

In this section, the convergence of the input sequences in both mean square and almost sure senses is established. To this end, the original algorithms (9.4) and (9.5) are first transformed as a switching system whose random matrix switches as a Markov chain.

We first consider the algorithms for arbitrary fixed time instant t. From Fig. 9.2, it is observed that the transition matrix of the Markov chain is

$$P = \begin{bmatrix} p_{11} & p_{12} \\ p_{21} & p_{22} \end{bmatrix} = \begin{bmatrix} 1 - \overline{\sigma}(1 - \overline{\gamma}) & \overline{\sigma}(1 - \overline{\gamma}) \\ \overline{\gamma} & 1 - \overline{\gamma} \end{bmatrix}, \tag{9.6}$$

where

$$p_{11} \triangleq \mathbb{P}(\tau_{k+1}(t) = 1 | \tau_k(t) = 1),$$
$$p_{12} \triangleq \mathbb{P}(\tau_{k+1}(t) = 0 | \tau_k(t) = 1),$$
$$p_{21} \triangleq \mathbb{P}(\tau_{k+1}(t) = 1 | \tau_k(t) = 0),$$
$$p_{22} \triangleq \mathbb{P}(\tau_{k+1}(t) = 0 | \tau_k(t) = 0),$$

with p_{ij} being the element of P at the ith row and jth column and $\tau_k(t)$ denoting the state at iteration k, $\forall t$ (in particular, $\tau_k(t) = 1$ and $\tau_k(t) = 0$ denote the states of *synchronization* and *asynchronization*, respectively). Note that $0 < \overline{\sigma}, \overline{\gamma} < 1$; thus, P is irreducible, aperiodic, and recurrent, which further means P is ergodic. In addition, $p_{ij} > 0, i, j = 1, 2$.

Note that *renewal* can occur both at the state of *synchronization* and *asynchronization*. Moreover, whenever a *renewal* occurs, both the computed control and the real control are improved. In other words, the real control is updated if a *renewal* occurs; otherwise, it remains unchanged. Therefore, it is concluded that updating of the real control also follows a Markov jump way. We could further introduce a random variable $\lambda_k(t)$ to denote whether a renewal happens or not, i.e., $\lambda_k(t) = 1$ if a renewal happens, and 0 otherwise. Recalling Fig. 9.2, we find that the occurrence probability of *renewal* depends on its state of the last iteration. That is, the evolution of $\lambda_k(t)$ is also a irreducible, aperiodic, recurrent, and ergodic Markov chain.

The update of the real control has the following two formulations. That is, when $\lambda_k(t) = 0$,

$$u_{k+1}^r(t) = u_k^r(t) = u_k^r(t) + a_k \mathbf{0}_{p \times q} e_k(t+1), \tag{9.7}$$

and when $\lambda_k = 1$,

$$u_{k+1}^r(t) = u_k^r(t) + a_k L_t e_k(t+1), \tag{9.8}$$

where $\mathbf{0}_{i \times j}$ denotes zero matrix with appropriate dimensions. We can unify these two cases into the following one

$$u_{k+1}^r(t) = u_k^r(t) + a_k \lambda_k(t) L_t e_k(t+1), \tag{9.9}$$

where $\lambda_k(t)$ values 0 or 1 subject to a two-state Markov chain.

In order to show the convergence, we now lift the above recursion along the time axis. Specifically, denote

$$Y_k = [y_k(1), \ldots, y_k(N)]^T \in \mathbf{R}^{qN},$$
$$U_k^r = [u_k^r(0), \ldots, u_k^r(N-1)]^T \in \mathbf{R}^{pN},$$
$$Y_k(0) = [C_1 A_0 x_k(0), \ldots, C_N A_{N-1,0} x_k(0)]^T \in \mathbf{R}^{qN},$$

9.3 Convergence Analysis

and $\mathscr{H}$ is

$$\mathscr{H} = \begin{bmatrix} C_1 B_0 & 0 & 0 & \cdots & 0 \\ C_2 A_1 B_0 & C_2 B_1 & 0 & \cdots & 0 \\ C_3 A_{2,1} B_0 & C_3 A_2 B_1 & C_3 B_2 & \cdots & 0 \\ \vdots & \vdots & \vdots & \ddots & \vdots \\ C_N A_{N-1,1} B_0 & C_N A_{N-1,2} B_1 & C_N A_{N-1,3} B_2 & \cdots & C_N B_{N-1} \end{bmatrix}$$

with $A_{i,j} \triangleq A_i A_{i-1} \cdots A_j, i \geq j$. The stochastic noise term ε_k is expressed as

$$\varepsilon_k = \begin{bmatrix} v_k(1) + C_1 w_k(0) \\ v_k(2) + C_2 w_k(1) + C_2 A_1 w_k(0) \\ \vdots \\ v_k(N) + \sum_{j=1}^{N} C_N A_{N-1,j} w_k(j-1) \end{bmatrix}.$$

The lifted system is

$$Y_k = \mathscr{H} U_k + \varepsilon_k + Y_k(0),$$

while the desired reference $y_d(t)$ and associated desired input $u_d(t)$ can be lifted in similar formulations,

$$Y_d = \mathscr{H} U_d + Y_d(0).$$

From Assumptions 9.2, 9.3, one has that $Y_k(0) = Y_d(0)$ and $\mathbb{E}\{\varepsilon_k | \mathscr{F}_{k-1}\} = 0$, $\mathbb{E}\{\|\varepsilon_k\|^2 | \mathscr{F}_{k-1}\} < \infty$. In addition, the lifted tracking error is

$$E_k \triangleq Y_d - Y_k = \mathscr{H}(U_d - U_k^r) - \varepsilon_k.$$

Then, the lifted form of the recursion (9.9) is

$$\begin{aligned} U_{k+1}^r &= U_k^r + a_k \Lambda_k \mathscr{L} E_k \\ &= U_k^r + a_k \Lambda_k \mathscr{L} \mathscr{H} \Delta U_k^r - a_k \Lambda_k \varepsilon_k, \end{aligned} \quad (9.10)$$

where $\Lambda_k = \text{diag}\{\lambda_k(0), \ldots, \lambda_k(N-1)\} \otimes I_{p \times p}$, $\mathscr{L} = \text{diag}\{L_0, \ldots, L_{N-1}\} \in \mathbf{R}^{pN \times qN}$, and $\Delta U_k^r = U_d - U_k^r$. Subtracting both sides of the last equation (9.10) from U_d leads to

$$\Delta U_{k+1}^r = \Delta U_k^r - a_k \Lambda_k \mathscr{L} \mathscr{H} \Delta U_k^r + a_k \Lambda_k \varepsilon_k. \quad (9.11)$$

It is worth pointing out that Λ_k is a random matrix with 2^N possible outcomes because all its diagonal entries are binary and the switching of Λ_k among its outcomes follows a irreducible, aperiodic, recurrent, and ergodic Markov chain. In particular, there are two special cases of Λ_k, namely, $I_{pN \times pN}$ and $\mathbf{0}_{pN \times pN}$ denoting that the inputs at all the N time instants are renewed and unchanged, respectively.

Next, it is sufficient to show the zero-error convergence of the recursion (9.11) under mild design conditions. That is, we can move to design the learning gain matrix $\mathscr{L}$ and propose the convergence results. Note that $\mathscr{H}$ is a lower triangular block matrix with its diagonal blocks being $C_{t+1}B_t$. Thus, we could design the learning gain matrix L_t such that $L_t C_{t+1} B_t$ are with positive eigenvalues. Before presenting the main theorem, we remind the technical Lemmas 9.1 and A.1 for further analysis.

Theorem 9.1 *Consider the linear time-varying system (9.1) and assume Assumptions 9.1–9.3 hold. The learning update laws (9.4) and (9.5) guarantee that the generated input sequence converges to the desired input both in mean square sense and almost sure sense if the learning gain matrix L_t satisfies that all eigenvalues of $L_t C_{t+1} B_t$ are positive constants, $t = 0, 1, \ldots, N - 1$. As a result, the desired reference $y_d(t)$ is asymptotically tracked according to the index (9.3).*

Proof From Lemma 9.1 and the transformations, we have (9.11). Now we show that ΔU_k^r converges to zero both in mean square sense and almost sure sense.

Note that all eigenvalues of the matrix $\mathscr{L}\mathscr{H}$ are positive constants, thus there exists a positive definite matrix $\mathscr{Q}$ such that

$$(\mathscr{L}\mathscr{H})^T \mathscr{Q} + \mathscr{Q}\mathscr{L}\mathscr{H} = I.$$

Moreover, according to the form of Λ_k, we have

$$(\mathscr{L}\mathscr{H})^T \Lambda_k^T \mathscr{Q} + \mathscr{Q}\Lambda_k \mathscr{L}\mathscr{H} \geq 0.$$

Then, we define a weighted norm for ΔU_k^r as $\|\Delta U_k^r\|_{\mathscr{Q}}^2 \triangleq (\Delta U_k^r)^T \mathscr{Q} \Delta U_k^r$, which can be regarded as a Lyapunov function. Now, we take the weighted norm to both sides of (9.11),

$$\begin{aligned}
\|\Delta U_{k+1}^r\|_{\mathscr{Q}}^2 = &\|\Delta U_k^r\|_{\mathscr{Q}}^2 + a_k^2 \|\Lambda_k \mathscr{L}\mathscr{H} \Delta U_k^r\|_{\mathscr{Q}}^2 + a_k^2 \|\Lambda_k \varepsilon_k\|_{\mathscr{Q}}^2 \\
&- a_k (\Delta U_k^r)^T ((\mathscr{L}\mathscr{H})^T \Lambda_k^T \mathscr{Q} + \mathscr{Q}\Lambda_k \mathscr{L}\mathscr{H}) \Delta U_k^r \\
&+ 2 a_k (\Delta U_k^r)^T \mathscr{Q} \Lambda_k \varepsilon_k \\
&- 2 a_k (\Delta U_k^r)^T (\mathscr{L}\mathscr{H})^T \Lambda_k^T \mathscr{Q} \Lambda_k \varepsilon_k. \quad (9.12)
\end{aligned}$$

Define a new increasing σ-algebra $\mathscr{F}_k' \triangleq \sigma\{x_j(t), w_j(t), v_j(t), \sigma_j(t), \gamma_j(t), 1 \leq j \leq k-1, t = 0, \ldots, N\}$. In view of (9.4)–(9.5), it is evident that $U_k^r \in \mathscr{F}_k'$; that is, U_k^r is adapted to $\mathscr{F}_k'$. From Assumption 9.3, we have $\mathbb{E}\{\varepsilon_k | \mathscr{F}_k'\} = 0$. As a result,

$$\mathbb{E}\{2a_k (\Delta U_k^r)^T \mathscr{Q} \Lambda_k \varepsilon_k | \mathscr{F}_k'\} = 0, \quad (9.13)$$

$$\mathbb{E}\{2a_k (\Delta U_k^r)^T (\mathscr{L}\mathscr{H})^T \Lambda_k^T \mathscr{Q} \Lambda_k \varepsilon_k | \mathscr{F}_k'\} = 0. \quad (9.14)$$

9.3 Convergence Analysis

Therefore, it is straightforward to have that

$$\mathbb{E}\{\|\Delta U_{k+1}^r\|_{\mathscr{Q}}^2|\mathscr{F}_k'\}$$
$$= \|\Delta U_k^r\|_{\mathscr{Q}}^2 + c_0 a_k^2(\|\Delta U_k^r\|_{\mathscr{Q}}^2 + \mathbb{E}\{\|\varepsilon_k\|_{\mathscr{Q}}^2|\mathscr{F}_k'\})$$
$$- a_k(\Delta U_k^r)^T \mathbb{E}\{(\mathscr{L}\mathscr{H})^T \Lambda_k^T \mathscr{Q} + \mathscr{Q}\Lambda_k \mathscr{L}\mathscr{H}|\mathscr{F}_k'\}\Delta U_k^r, \tag{9.15}$$

where $c_0 = \max\{\|LH\|^2, 1\}$ as $\|\Lambda_k\| \leq 1$. Note that Λ_k is a diagonal matrix, and the evolution of Λ_k is a irreducible and ergodic Markov chain. Thus, there is a positive probability for Λ_k to be the identity matrix $I_{pN \times pN}$. In addition, all the eigenvalues of the remaining $2^N - 1$ possible diagonal-matrix of Λ_k are either 1 or 0. Consequently, there exists some constant $c_1 > 0$ such that

$$\mathbb{E}\{(\mathscr{L}\mathscr{H})^T \Lambda_k^T \mathscr{Q} + \mathscr{Q}\Lambda_k \mathscr{L}\mathscr{H}|\mathscr{F}_k'\} \geq c_1 I. \tag{9.16}$$

By Assumption 9.3 we have $\mathbb{E}\{\|\varepsilon_k\|_{\mathscr{Q}}^2|\mathscr{F}_k'\} < c_2$ where $c_2 > 0$ is a suitable constant.

Now we move to show the mean square convergence. Denote $\xi_k \triangleq \mathbb{E}\|\Delta U_k^r\|_{\mathscr{Q}}^2$. Then, taking mathematical expectation of both sides of (9.15) and using (9.16) lead to

$$\xi_{k+1} \leq (1 - c_1 c_3 a_k)\xi_k + c_0 a_k^2(c_2 + \xi_k).$$

where c_3 is a positive constant such that $I \geq c_3 \mathscr{Q}$. Then by Lemma A.1, we have that $\mathbb{E}\|\Delta U_k^r\|_{\mathscr{Q}}^2 \to 0$, implying that $\mathbb{E}\|\Delta U_k^r\|^2 \to 0$. That is, the zero-error convergence of ΔU_k^r in mean square sense is proved.

Next, we proceed to show the almost sure convergence of ΔU_k^r. Denote $\eta_k \triangleq \|\Delta U_k^r\|_{\mathscr{Q}}^2$. Substituting (9.16) into (9.15), we have that

$$\mathbb{E}\{\eta_{k+1}|\mathscr{F}_k'\} \leq (1 - c_1 c_3 a_k)\eta_k + c_0 a_k^2(\mathbb{E}\{\|\varepsilon_k\|^2|\mathscr{F}_k'\} + \eta_k)$$
$$\leq \eta_k + c_0 a_k^2(\mathbb{E}\{\|\varepsilon_k\|^2|\mathscr{F}_k'\} + \eta_k). \tag{9.17}$$

Note that η_k and $c_0 a_k^2(\mathbb{E}\{\|\varepsilon_k\|^2|\mathscr{F}_k'\} + \eta_k)$ correspond to $X(n)$ and $Z(n)$ in Lemma A.2, respectively. Moreover, it is evident that

$$\sum_{k=1}^{\infty} \mathbb{E}\left[c_0 a_k^2(\mathbb{E}\{\|\varepsilon_k\|^2|\mathscr{F}_k'\} + \eta_k)\right] = \sum_{k=1}^{\infty} c_0 a_k^2(\mathbb{E}\|\varepsilon_k\|^2 + \xi_k) < \infty,$$

where the convergence of ξ_k is used. In other words, (A.5) and (A.6) in Lemma A.2 are fulfilled. Therefore, it follows that $\eta_k = \|\Delta U_k^r\|_{\mathscr{Q}}^2$ converges almost surely as $k \to \infty$. On the other hand, we have shown that ΔU_k^r converges to zero in mean square, thus ΔU_k^r converges to zero almost surely from probability theory. The proof is completed.

Remark 9.5 In this chapter, the data dropouts are modeled by random variables subject to Bernoulli distribution, which is a widely used model in this research area. The major reason for such assumption is to allow the successive data dropouts with arbitrary length. In addition, such assumption also helps us to make an explicit convergence proof for the general data dropout problem. Note that the critical technique for establishing the convergence is the Markov property of the sample path behavior, thus the proposed method given in this chapter can be applied to the Markovian data dropout case. On the other hand, the inherent update mechanism is the occurrence of *renewal*, which implies that the essential convergence of the proposed algorithms only requires that the data is not completely lost for each time instant.

Remark 9.6 If no noise is involved in the system, that is, both $w_k(t)$ and $v_k(t)$ are eliminated, then the decreasing gain a_k could be removed from the algorithms. In this case, the recursion of input error (9.11) reduces to

$$\Delta U_{k+1}^r = \Delta U_k^r - \Lambda_k \mathscr{L} \mathscr{H} \Delta U_k^r, \tag{9.18}$$

and an exponential convergence speed can be then obtained. Moreover, the design of learning gain matrix L_t would be different with or without a_k. For the case with a_k, the condition on L_t is that all eigenvalues of $L_t C_{t+1} B_t$ are positive real numbers. Roughly speaking, this condition can be relaxed to the one that all eigenvalues of $L_t C_{t+1} B_t$ are with positive real parts. On the other hand, for the case without a_k, the matrix L_t should satisfy that the spectral radius of $I - L_t C_{t+1} B_t$ is less than one to ensure convergence. The latter design condition can be derived following the traditional contraction mapping method. Therefore, the introduction of decreasing sequence $\{a_k\}$ also relaxes the design range of L_t.

9.4 Discussions on Convergence Speed

In this section, we give a brief description on the convergence speed of the proposed algorithms. As a matter of fact, the convergence speed depends on two individual factors, namely, the *renewal* frequency and the decreasing gain sequence $\{a_k\}$. The former factor reflects the influence of data dropout on ILC, while the latter factor is a design factor originating from the stochastic approximation algorithm. In the traditional ILC problem where no data dropout occurs, the convergence speed is only determined by the designed decreasing gain a_k.

Now let us check how the influence of data dropouts on the convergence speed is. To this end, we give an explicit description on the *renewal* frequency. Recalling Fig. 9.2 and its transition matrix (9.6), the associated stationary distribution π of the Markov chain can be calculated from $\pi P = \pi$ and is given as

$$\pi \triangleq \left[\frac{\overline{\gamma}}{\overline{\gamma} + \overline{\sigma} - \overline{\sigma} \cdot \overline{\gamma}}, \frac{\overline{\sigma} - \overline{\sigma} \cdot \overline{\gamma}}{\overline{\gamma} + \overline{\sigma} - \overline{\sigma} \cdot \overline{\gamma}} \right]. \tag{9.19}$$

9.4 Discussions on Convergence Speed

Note that *renewal* can occur both at the state of *synchronization* and *asynchronization*. From Fig. 9.2, it is observed that the probability of occurrence of *renewal* at the state of *synchronization* is $\overline{\sigma}\overline{\gamma}$, while the probability at the state of *asynchronization* is $\overline{\gamma}$. Therefore, the probability of *renewal* along the iteration axis can be calculated as

$$\mathbb{P}(renewal) = \frac{\overline{\gamma}}{\overline{\gamma} + \overline{\sigma} - \overline{\sigma} \cdot \overline{\gamma}} \cdot \overline{\sigma} \cdot \overline{\gamma} + \frac{\overline{\sigma} - \overline{\sigma} \cdot \overline{\gamma}}{\overline{\gamma} + \overline{\sigma} - \overline{\sigma} \cdot \overline{\gamma}} \cdot \overline{\gamma}$$
$$= \frac{\overline{\sigma} \cdot \overline{\gamma}}{\overline{\gamma} + \overline{\sigma} - \overline{\sigma} \cdot \overline{\gamma}}. \tag{9.20}$$

In this chapter, for clear expression, we simply assume that the probability distributions for different time instants are the same. Thus, the above probability of *renewal* is the average of the whole iteration. This probability describes the *renewal* frequency of the learning algorithms under the data dropout environment. It is noticed that the probability is determined by the sum and product of the successful transmission probability at both sides, i.e., $\overline{\sigma} + \overline{\gamma}$ and $\overline{\sigma}\overline{\gamma}$. As a consequence, the *renewal* frequency is neither determined by the worst side nor the simple sum of both sides. Two facts are observed as follows:

(1) Define the function $f(\overline{\sigma}, \overline{\gamma}) = \mathbb{P}(renewal)$. Evidently, $f(\overline{\sigma}, \overline{\gamma}) = f(\overline{\gamma}, \overline{\sigma})$. Moreover, through simple calculations, we have

$$\frac{\partial f(\overline{\sigma}, \overline{\gamma})}{\partial \overline{\sigma}} = \frac{\overline{\gamma}^2}{(\overline{\gamma} + \overline{\sigma} - \overline{\sigma} \cdot \overline{\gamma})^2} > 0.$$

This condition means that a large successful transmission rate corresponds to increased number of renewals, and thus, faster convergence speed.

(2) It is well known that $\overline{\sigma}\overline{\gamma} \leq \left(\frac{\overline{\sigma}+\overline{\gamma}}{2}\right)^2$ where the equality holds if and only if $\overline{\sigma} = \overline{\gamma}$. This implies that, when the sum $\overline{\sigma} + \overline{\gamma}$ is fixed, the closer $\overline{\sigma}$ approaches to $\overline{\gamma}$, the larger the product $\overline{\sigma}\overline{\gamma}$ is and so is the probability $\mathbb{P}(renewal)$. In other words, the convergence speed increases.

9.5 Illustrative Simulations

In this section, we apply the proposed algorithms to a permanent magnet linear motor (PMLM), which is described by the following discretized model [3]

$$\begin{cases} x(t+1) = x(t) + v(t)\Delta + \varepsilon_1(t+1) \\ v(t+1) = v(t) - \Delta \frac{k_1 k_2 \psi_f^2}{Rm} v(t) + \Delta \frac{k_2 \psi_f}{Rm} u(t) + \varepsilon_2(t+1) \\ y(t) = v(t) + \varepsilon(t) \end{cases}$$

where x and v denote the motor position and rotor velocity, $\Delta = 10$ ms the sampling time interval, $R = 8.6\,\Omega$ the resistance of stator, $m = 1.635$ kg the rotor mass, and $\psi_f = 0.35$ Wb the flux linkage, $k_1 = \pi/\tau$ and $k_2 = 1.5\pi/\tau$, where $\tau = 0.031$ m is the pole pitch. The stochastic noises $\varepsilon_1(t), \varepsilon_2(t)$, and $\varepsilon(t)$ obey zero-mean distribution $N(0, \sigma^2)$ with $\sigma = 0.03$.

In this simulation, we set the whole iteration length as 1 s, i.e., $N = 100$. The desired reference is $y_d(t) = 1/3[\sin(t/20) + 1 - \cos(3t/20)]$, $0 \le t \le 100$. The initial state satisfies Assumption 9.2. The control input for the first iteration is simply set to be 0. The learning gain $L_t = 50$ and the decreasing sequence is set to be $a_k = 1/k$. The algorithms are run for 150 iterations.

The general algorithms (9.4) and (9.5) are used. The random variables $\gamma_k(t)$ and $\sigma_k(t)$ for data dropouts are defined separately for different time instants rather than a unified variable for the entire iteration. We introduce data dropout rate (DDR) as the probability $\mathbb{P}(\sigma_k(t) = 0)$ or $\mathbb{P}(\gamma_k(t) = 0)$. In other words, DDR denotes the percentage of lost packages over the total packages. For simplicity, the DDRs for both the measurement and actuator sides are equal in the following. In this example, four cases are simulated, that is, DDR $= 10\%$, 20%, 30%, and 40%, respectively. The averaged tracking error profiles along iteration axis for all cases are shown in Fig. 9.3, where the averaged tracking error is defined as $\overline{e}_k \triangleq \frac{\sum_{t=1}^{N} |e_k(t)|}{N}$. As can be seen, the convergence speed slows as the DDR increases.

As has been shown in Theorem 9.1, the condition on learning gain L_t is that all eigenvalues of $L_t C_{t+1} B_t$ are positive. Moreover, a faster convergence speed can be achieved when L_t is designed such that the eigenvalues are with larger magnitude; however, such case would lead to bad transient performance such as overshot before

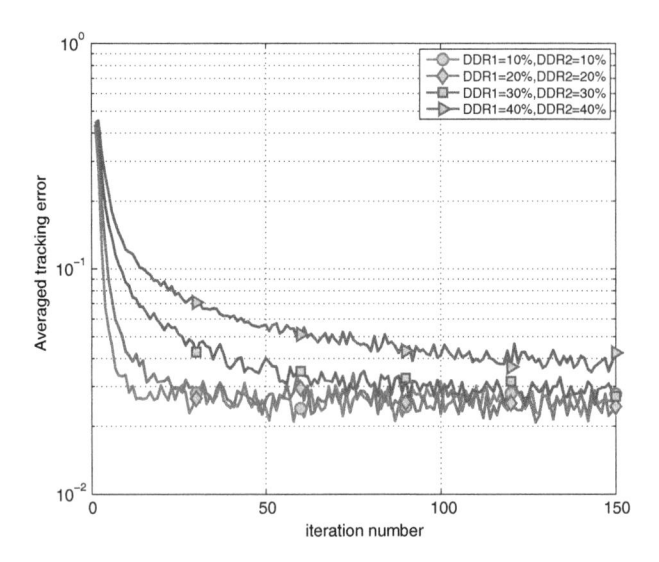

Fig. 9.3 Tracking error profiles along iteration axis with different DDRs: Noise case (DDR1: DDR at the measurement side; DDR2: DDR at the actuator side)

9.5 Illustrative Simulations

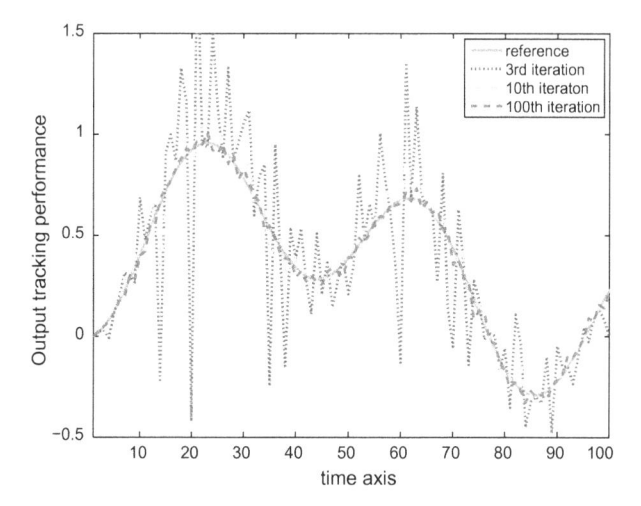

Fig. 9.4 Tracking performance for illustrated iterations

convergence. Thus, there is a trade-off between the convergence speed and transient performance.

The tracking performance is plotted for the 3rd, 10th, and 100th iterations as illustrations. It is seen that the system output at the 3rd iteration is far away from the reference; however, the system output at the 100th iteration is already acceptable (Fig. 9.4).

If there is no noise involved in the system, i.e., the noises $w_k(t)$ and $v_k(t)$ are removed from the system (9.1), then we can also delete the decreasing sequence a_k from the update laws (9.4) as mentioned in Remark 9.2. In this case, an exponential convergence speed is achieved, as shown in Fig. 9.5, where the learning gain L_t is selected as $L_t = 8$ due to the previous value does not satisfy the condition given in Remark 9.6. In the figure, all profiles are approximate lines in the logarithmic coordinate, which imply the exponential convergence property.

In addition, Fig. 9.5 also verifies the relationship between convergence speed and DDR rate. For the above three lines, the sum of DDRs at both sides are identical. It can be found that the fastest convergence speed belongs to the case DDR1 =DDR2. Moreover, although the worst DDR in the fourth line is 40%, it still behaves faster than the third line where the worst DDR is 30%. Such observations coincide with the discussions in Sect. 9.4.

To demonstrate the effect of the decreasing sequence a_k for stochastic systems, we display the input profiles for (9.4) with and without a_k in Fig. 9.6. As can be seen, the introduction of a_k enables a stable convergence of input, while the input keeps fluctuating if such sequence is removed from the update law. This verifies the necessity of decreasing term in algorithms for stochastic systems. However, it should be pointed out that the decreasing sequence may make the learning controller

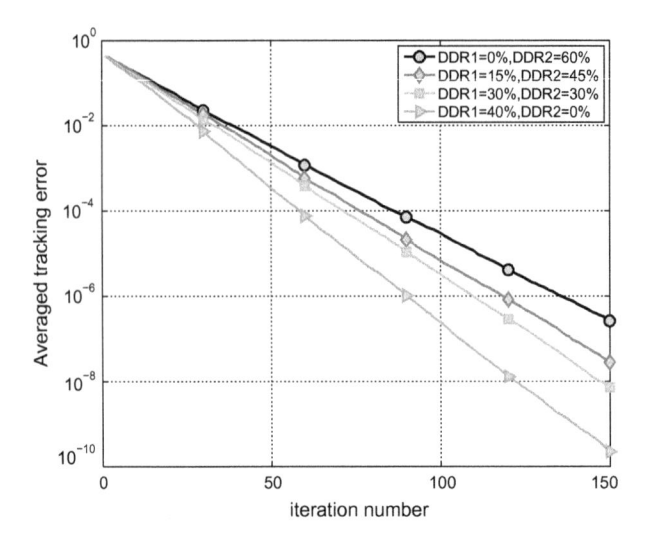

194   9 Two-Side Data Dropout for Linear Stochastic Systems

Fig. 9.5 Tracking error profiles along iteration axis with different DDRs: Noise free case (DDR1: DDR at the measurement side; DDR2: DDR at the actuator side)

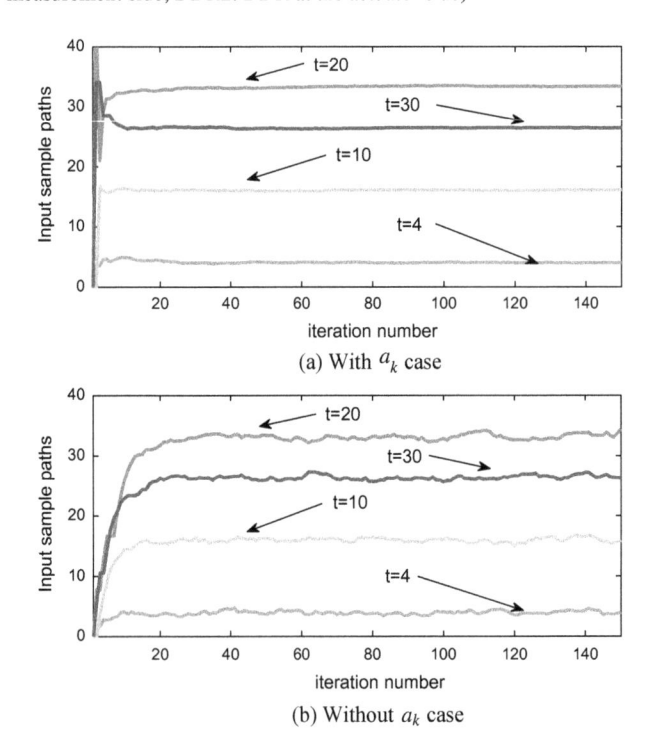

(a) With a_k case

(b) Without a_k case

Fig. 9.6 Input sample paths along iteration axis with different DDRs

unsuitable if large changes occur to the desired reference after several iterations. As a consequence, the design of learning laws depends on the practical application requirements.

9.6 Summary

ILC under general data dropout environments is explored in this chapter. The data dropouts are allowed to occur randomly at both the measurement and actuator sides. As a result, the control update process consists of two parts, i.e., the computed control and the real control. A novel convergence analysis framework is proposed in this chapter. To be specific, the update process is first proved to be a Markov chain by directly analyzing its sample path behavior. Then, the convergence in both mean square and almost sure senses is established strictly. In addition, the chapter also demonstrates the effectiveness and robustness of conventional P-type update law against random factors. The results in this chapter are mainly based on [4].

References

1. Benveniste, A., Métivier, M., Priouret, P.: Adaptive Algorithms and Stochastic Approximations. Springer, New York (1990)
2. Caines, P.E.: Linear Stochastic Systems. Wiley, New York (1988)
3. Zhou, W., Yu, M., Huang, D.: A high-order internal model based iterative learning control scheme for discrete linear time-varying systems. Int. J. Autom. Comput. **12**(3), 330–336 (2015)
4. Shen, D., Xu, J.-X.: A novel Markov chain based ILC analysis for linear stochastic systems under general data dropouts environments. IEEE Trans. Autom. Control **62**(11), 5850–5857 (2017)

Chapter 10
Two-Side Data Dropout for Nonlinear Systems

10.1 Problem Formulation

Consider the following affine nonlinear system:

$$\begin{aligned} x_k(t+1) &= f(t, x_k(t)) + B(t)u_k(t), \\ y_k(t) &= C(t)x_k(t), \end{aligned} \quad (10.1)$$

where k is the iteration number, $k = 1, 2, \ldots$, t denotes the time instant, $t = 0, 1, 2, \ldots, N$, and N is the iteration length. The variables $x_k(t) \in \mathbf{R}^n$, $u_k(t) \in \mathbf{R}^p$, and $y_k(t) \in \mathbf{R}^q$ denote the system state, input, and output, respectively. $f(\cdot, \cdot)$ is a nonlinear continuous function. $C(t)$ and $B(t)$ are unknown time-varying matrices with appropriate dimensions. For brevity, we denote $C^+B(t) \triangleq C(t+1)B(t)$. To simplify the convergence analysis, we assume that $C^+B(t)$ is of full-column rank.

Let $y_d(t), t \in \{0, 1, 2, \ldots, N\}$ be the desired reference. For the suitable initial state $x_d(0)$ such that $y_d(0) = C(0)x_d(0)$, there always exists a unique desired input $u_d(t)$ that can generate the reference signal $y_d(t)$. Specifically, the desired input $u_d(t)$ is recursively defined as follows:

$$\begin{aligned} u_d(t) &= [(C^+B(t))^T C^+B(t)]^{-1}(C^+B(t))^T \\ &\quad \times (y_d(t+1) - C(t+1)f(t, x_d(t))), \\ x_d(t+1) &= f(t, x_d(t)) + B(t)u_d(t). \end{aligned}$$

With this control signal, it is apparent that the following equations for the desired reference are satisfied; that is, the input $u_d(t)$ computed above drives the plant to generate the desired reference $y_d(t)$,

$$\begin{aligned} x_d(t+1) &= f(t, x_d(t)) + B(t)u_d(t), \\ y_d(t) &= C(t)x_d(t). \end{aligned} \quad (10.2)$$

Define the tracking error as

$$e_k(t) = y_d(t) - y_k(t). \tag{10.3}$$

The following assumptions are required for the technical analysis.

Assumption 10.1 For all time instants $t \in \{0, \cdots, N\}$, the nonlinear continuous function $f(t, \cdot) : \mathbf{R}^n \to \mathbf{R}^n$ satisfies the globally Lipschitz condition, that is, $\forall x_1, x_2 \in \mathbf{R}^n$,

$$\|f(t, x_1) - f(t, x_2)\| \le k_f \|x_1 - x_2\|, \tag{10.4}$$

where $k_f > 0$ is the Lipschitz constant.

This assumption is made mainly for the technical analysis because a modified λ-norm technique is employed to derive the convergence of tracking error in the next section. For the extension from globally Lipschitz condition to locally Lipschitz condition, a possible way is to adopt similar techniques in Chaps. 2 and 3. However, this chapter aims to provide a novel convergence proof for ILC under data dropouts at both measurement and actuator sides, of which the data dropout condition is rather relaxed, thus we assume the globally Lipschitz condition for a concise proof.

Assumption 10.2 The initial state of the system is reset to be $x_d(0)$ at every iteration, i.e., $x_k(0) = x_d(0)$, $\forall k \ge 1$.

This assumption is the well-known identical initialization condition (i.i.c.), one of the fundamental issues in ILC. It has been used in many ILC papers as repetition is the basic premise of ILC. If i.i.c. is not satisfied, then the perfect tracking is hard to achieve by learning algorithms, at least for the initial position/portion of the desired reference. Many papers have dedicated to extending i.i.c. by introducing additional mechanisms such as initial rectifying mechanism [1] or initial learning mechanism [2]. Such mechanisms can be combined with the results given in this chapter to deal with the initial resetting issue. Besides, if the initial state is not identically reset but locates in a bounded range around $x_d(0)$, then one can obtain that the tracking error converges to a small zone around zero.

In this chapter, we consider a general formulation of the networked ILC framework, in which the plant and the learning controller are connected by the wired or wireless networks as shown in Fig. 10.1. In this framework, two networks exist both from the plant to the learning controller, namely, at the measurement side, and from the learning controller to the plant, namely, at the actuator side. Moreover, both networks would suffer random data dropouts. To model this point, we introduce two random variables $\sigma_k(t)$ and $\gamma_k(t)$ subjected to 0 − 1 Bernoulli distribution for both sides, respectively. In other words, both $\sigma_k(t)$ and $\gamma_k(t)$ are equal to 1 if the corresponding data is successfully transmitted, and 0 otherwise. In addition, $\mathbb{P}(\sigma_k(t) = 1) = \overline{\sigma}(t)$ and $\mathbb{P}(\gamma_k(t) = 1) = \overline{\gamma}(t)$ where $0 < \overline{\sigma}(t), \overline{\gamma}(t) < 1$. Note that both networks work individually, thus it is rational to assume that $\sigma_k(t)$ is independent of $\gamma_k(t)$.

The control objective of this chapter is to design a suitable updating scheme such that the generated input sequence ensures zero-error convergence with probability

10.1 Problem Formulation

Fig. 10.1 Block diagram of the networked ILC framework

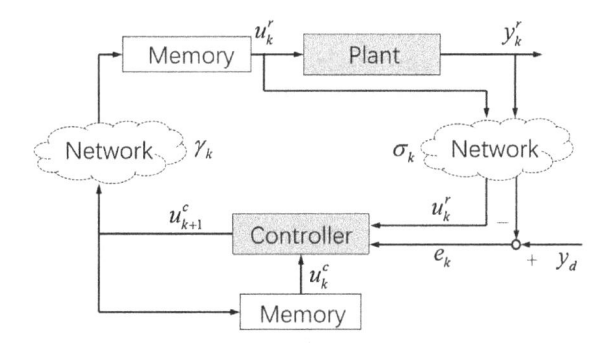

one for nonlinear systems with data dropouts. Moreover, the system output driven by such updating scheme can track the desired reference asymptotically as the iteration number goes to infinity.

To achieve the control objective, the controller update law follows the basic holding strategy. To be specific, if the data transmits successfully at the measurement side, then the learning controller would update its input signal. Otherwise, if the data is lost during the transmission at the measurement side, then the learning controller stops updating and retains the previous input signal. On the other hand, if the input signal is successfully transmitted at the actuator side, then the plant would use the newly arrived input signal. Otherwise, if the input signal is lost during the transmission, then the plant would retain the previous input signal stored in the memory. To make the following expressions concise, hereafter we call the input generated by the learning controller as computed input signal, denoted by $u_k^c(t)$, and the input used for the plant as real input signal, denoted by $u_k^r(t)$, respectively. Then, the computed input signal is updated as

$$u_{k+1}^c(t) = \sigma_{k+1}(t)u_k^r(t) + [1 - \sigma_{k+1}(t)]u_k^c(t) \\ + \sigma_{k+1}(t)L_t e_k(t+1), \tag{10.5}$$

where L_t is the learning gain matrix to be designed later. Moreover, the real input signal used for the plant is given as

$$u_{k+1}^r(t) = \gamma_{k+1}(t)u_{k+1}^c(t) + [1 - \gamma_{k+1}(t)]u_k^r(t). \tag{10.6}$$

Remark 10.1 Note that the random data dropouts occur independently at both measurement and actuator sides, thus the update of both computed and real input signals might be asynchronous. That is, the computed input is updated when the data is successfully transmitted back from the plant. However, this latest input may fail to be transmitted to the plant so that the real input signal retains the previous one. In this case, the asynchronization between the computed and real input signals arises. Moreover, it is worth pointing out that the update of both inputs is also asynchronous along the time axis as the random variables $\sigma_k(t)$ and $\gamma_k(t)$ are independent for different time instants. In addition, it should be noted that the ILC scheme given in Fig. 10.1 requires no transient growth problem existing when output dropouts occur, because in this case, when large transient errors occur, the controller may have no information about them due to the data dropouts and thus cannot stop the transient growth.

10.2 Convergence Analysis of ILC Algorithms

In this section, the convergence of the proposed algorithms (10.5) and (10.6) for both the computed and real input signals to the desired input $u_d(t)$ with probability one is proved and then the output of the system (10.1) would track the desired reference $y_d(t)$ asymptotically as iteration number goes to infinity.

As remarked in the last section, there exists asynchronization in the updating of the computed and real input signals. Such asynchronization makes it nontrivial to establish the convergence proof. To this end, we first derive the expressions for both input errors and then build an augmented regression model, so that the asynchronization can be treated as internal randomness (see Lemma 10.1). The property of the newly introduced random matrix in the regression model of the augmented input errors is then analyzed (see Lemma 10.2). By applying a modified λ-norm technique according to the random asynchronization, the contraction mapping of the input errors is strictly established to show the convergence (see Theorem 10.1).

We first state the auxiliary lemmas. Denote $\delta u_k^c \triangleq u_d(t) - u_k^c(t)$ and $\delta u_k^r \triangleq u_d(t) - u_k^r(t)$ as the errors of the computed and real inputs, respectively. Define the augmented input error

$$\delta u_k(t) = [(\delta u_k^c(t))^T, (\delta u_k^r(t))^T]^T. \tag{10.7}$$

Then we have the following characterization of this augmented input error.

Lemma 10.1 *For the augmented input error given in (10.7), the following regression holds,*

$$\delta u_{k+1}(t) = P_k(t)\delta u_k(t) \\ - Q_k(t)[f(t, x_d(t)) - f(t, x_k(t))], \tag{10.8}$$

where

$$P_k(t) = \begin{bmatrix} [1 - \sigma_{k+1}(t)]I & \sigma_{k+1}(t)[I - L_t C^+ B(t)] \\ \gamma_{k+1}(t)[1 - \sigma_{k+1}(t)]I & * \end{bmatrix}, \tag{10.9}$$

$$Q_k(t) = \begin{bmatrix} \sigma_{k+1}(t) L_t C(t+1) \\ \gamma_{k+1}(t)\sigma_{k+1}(t) L_t C(t+1) \end{bmatrix}, \tag{10.10}$$

with the expression in the position marked by "" being*

10.2 Convergence Analysis of ILC Algorithms

$$* \triangleq [1 - \gamma_{k+1}(t)]I + \gamma_{k+1}(t)\sigma_{k+1}(t)[I - L_t C^+ B(t)].$$

Proof Subtracting both sides of (10.5) from $u_d(t)$ leads to

$$\begin{aligned}\delta u_{k+1}^c &= u_d(t) - u_{k+1}^c(t)\\ &= u_d(t) - \{\sigma_{k+1}(t)u_k^r(t) + [1 - \sigma_{k+1}(t)]u_k^c(t)\\ &\quad + \sigma_{k+1}(t)L_t e_k(t+1)\}\\ &= \sigma_{k+1}(t)\delta u_k^r(t) + (1 - \sigma_{k+1}(t))\delta u_k^c(t)\\ &\quad - \sigma_{k+1}(t)L_t e_k(t+1).\end{aligned} \qquad (10.11)$$

Similarly, subtracting both sides of (10.6) from $u_d(t)$ yields

$$\delta u_{k+1}^r = \gamma_{k+1}(t)\delta u_{k+1}^c(t) + [1 - \gamma_{k+1}(t)]\delta u_k^r(t), \qquad (10.12)$$

where $\delta u_k^c \triangleq u_d(t) - u_k^c(t)$ and $\delta u_k^r \triangleq u_d(t) - u_k^r(t)$ denote errors for the computed and real input signals, respectively.

Moreover, from the system formulation we have

$$\begin{aligned}\delta x_k(t+1) &= [f(t, x_d(t)) - f(t, x_k(t))]\\ &\quad + B(t)\delta u_k^r(t),\end{aligned} \qquad (10.13)$$

where $\delta x_k(t) \triangleq x_d(t) - x_k(t)$. Meanwhile, the tracking error is $e_k(t) = C(t)\delta x_k(t)$. Thus,

$$\begin{aligned}e_k(t+1) &= C^+[f(t, x_d(t)) - f(t, x_k(t))]\\ &\quad + C^+ B(t)\delta u_k^r(t),\end{aligned} \qquad (10.14)$$

where $C^+ = C(t+1)$ for short. Substituting (10.14) into (10.11) yields

$$\begin{aligned}\delta u_{k+1}^c &= \sigma_{k+1}(t)[I - L_t C^+ B(t)]\delta u_k^r(t)\\ &\quad + \sigma_{k+1}(t)L_t C^+[f(t, x_d(t)) - f(t, x_k(t))]\\ &\quad + [1 - \sigma_{k+1}(t)]\delta u_k^c(t).\end{aligned} \qquad (10.15)$$

Further, substituting (10.15) into (10.12) leads to

$$\begin{aligned}\delta u_{k+1}^r &= [1 - \gamma_{k+1}(t)]\delta u_k^r(t) + \gamma_{k+1}(t)[1 - \sigma_{k+1}(t)]\delta u_k^c(t)\\ &\quad + \gamma_{k+1}(t)\sigma_{k+1}(t)L_t C^+[f(t, x_d(t)) - f(t, x_k(t))]\\ &\quad + \gamma_{k+1}(t)\sigma_{k+1}(t)[I - L_t C^+ B(t)]\delta u_k^r(t).\end{aligned} \qquad (10.16)$$

Based on (10.15) and (10.16), noting the augmented input error $\delta u_k(t)$ and associated matrices $P_k(t)$ and $Q_k(t)$, the regression model (10.8) holds obviously. This completes the proof.

This lemma characterizes the random asynchronization between the computed and real inputs, demonstrated by the random matrix $P_k(t)$. It is clear that $P_k(t)$ depends on both k and t, which reflects the asynchronization in iteration-domain and time-domain, respectively. Note that $\sigma_{k+1}(t)$ is independent of $\gamma_{k+1}(t)$ and both of them take value of 0 or 1. Thus, $P_k(t)$ has four possible outcomes, in which one case implies the asynchronization state between the two inputs ($\sigma_{k+1}(t) = 1$ and $\gamma_{k+1}(t) = 0$), two cases imply the synchronization state ($\gamma_{k+1}(t) = 1$), and one case implies the maintenance of the previous state ($\sigma_{k+1}(t) = \gamma_{k+1}(t) = 0$).

For the regression model (10.8), the contraction mapping property of the matrix $P_k(t)$ is important for the convergence analysis. This property is clarified in the following lemma.

Lemma 10.2 *If the learning gain matrix L_t in (10.5) satisfies $\|I - L_t C^+ B(t)\|_\infty < 1$, then we have*

$$\sup_t \mathbb{E}\|P_k(t)\|_\infty < 1. \tag{10.17}$$

Proof It is seen that $P_k(t)$ is a stochastic matrix with two random variables $\sigma_{k+1}(t)$ and $\gamma_{k+1}(t)$, which has four possible situations as follows.

Case 1: $\sigma_{k+1}(t) = 1$, $\gamma_{k+1}(t) = 1$.

$$P_k^1(t) = \begin{bmatrix} 0 & I - L_t C^+ B(t) \\ 0 & I - L_t C^+ B(t) \end{bmatrix}.$$

Case 2: $\sigma_{k+1}(t) = 1$, $\gamma_{k+1}(t) = 0$.

$$P_k^2(t) = \begin{bmatrix} 0 & I - L_t C^+ B(t) \\ 0 & I \end{bmatrix}.$$

Case 3: $\sigma_{k+1}(t) = 0$, $\gamma_{k+1}(t) = 1$.

$$P_k^3(t) = \begin{bmatrix} I & 0 \\ I & 0 \end{bmatrix}.$$

Case 4: $\sigma_{k+1}(t) = 0$, $\gamma_{k+1}(t) = 0$.

$$P_k^4(t) = \begin{bmatrix} I & 0 \\ 0 & I \end{bmatrix}.$$

Then, we could introduce four binary random variables μ_i, $1 \leq i \leq 4$, such that $\mu_i \in \{0, 1\}$ and $\mu_1 + \mu_2 + \mu_3 + \mu_4 = 1$. Note that these four μ_i are dependent, since whenever any one is equal to 1, all the others have to be 0. The random variable μ_i is used to describe the occurrence of $P_k^i(t)$ for $P_k(t)$; that is, if $P_k(t)$ values $P_k^i(t)$, then $\mu_i = 1$. Recalling the formulation of $\sigma_k(t)$ and $\gamma_k(t)$ in Sect. 10.1, we have that

10.2 Convergence Analysis of ILC Algorithms

$$p_1 = \mathbb{P}(\mu_1 = 1) = \overline{\sigma}(t)\overline{\gamma}(t),$$
$$p_2 = \mathbb{P}(\mu_2 = 1) = \overline{\sigma}(t)[1 - \overline{\gamma}(t)],$$
$$p_3 = \mathbb{P}(\mu_3 = 1) = [1 - \overline{\sigma}(t)]\overline{\gamma}(t),$$
$$p_4 = \mathbb{P}(\mu_4 = 1) = [1 - \overline{\sigma}(t)][1 - \overline{\gamma}(t)].$$

Then we can obtain that

$$\mathbb{E}\|P_k(t)\|_\infty$$
$$= \mathbb{E}\|\mu_1 P_k^1(t) + \mu_2 P_k^2(t) + \mu_3 P_k^3(t) + \mu_4 P_k^4(t)\|_\infty$$
$$= \sum_{i=1}^{4} \mathbb{P}(\mu_i = 1) \left\| \sum_{j=1}^{4} \mu_j P_k^j(t) \right\|_\infty$$
$$= \sum_{i=1}^{4} \mathbb{P}(\mu_i = 1)\|P_k^i(t)\|_\infty. \tag{10.18}$$

Noticing the form of $P_k^i(t)$, $1 \leq i \leq 4$, and definition of ∞-norm, we have that $\|P_k^i(t)\|_\infty = 1$, $i = 2, 3, 4$. While for $P_k^1(t)$, it is apparent that $\|P_k^1(t)\|_\infty < 1$ as long as L_t is designed satisfying that $\|I - L_t C^+ B(t)\|_\infty < 1$. As long as the networks at both measurement and actuator sides are not completely broken, we must have $p_1 > 0$, and then $\mathbb{E}\|P_k(t)\|_\infty < 1$, $\forall t$. This further results in that $\sup_t \mathbb{E}\|P_k(t)\|_\infty < 1$. The proof is completed.

Now, the main theorem is given as follows.

Theorem 10.1 *Consider the nonlinear system* (10.1) *and assume Assumptions 10.1–10.2 hold. If the learning gain matrix L_t in* (10.5) *satisfies* $\|I - L_t C^+ B(t)\|_\infty < 1$, *then both the computed and real input sequences generated by the algorithms* (10.5) *and* (10.6) *converge to the desired input $u_d(t)$ given in* (10.2) *with probability one as $k \to \infty$. That is, $u_k^c(t) \to u_d(t)$, $u_k^r(t) \to u_d(t)$, $\forall t$, with probability one as $k \to \infty$. Consequently, the actual tracking error $e_k(t) \to 0$ with probability one as $k \to \infty$.*

Proof Taking ∞-norm to both sides of the regression model for the augmented input error (10.8) yields that

$$\|\delta u_{k+1}(t)\|_\infty \leq \|P_k(t)\delta u_k(t)\|_\infty + \|Q_k(t)[f(t, x_d(t)) - f(t, x_k(t))]\|_\infty$$
$$\leq \|P_k(t)\|_\infty \|\delta u_k(t)\|_\infty$$
$$+ \|Q_k(t)\|_\infty \|[f(t, x_d(t)) - f(t, x_k(t))]\|_\infty$$
$$\leq \|P_k(t)\|_\infty \|\delta u_k(t)\|_\infty + k_f \|Q_k(t)\|_\infty \|\delta x_k(t)\|_\infty, \tag{10.19}$$

where Assumption 10.1 is applied to the last inequality.

Noticing the independence property of the involved variables, we take mathematical expectation to (10.19) and obtain that

$$\mathbb{E}\|\delta u_{k+1}(t)\|_\infty \leq \mathbb{E}\|P_k(t)\|_\infty \cdot \mathbb{E}\|\delta u_k(t)\|_\infty$$
$$+ k_f \cdot \mathbb{E}\|Q_k(t)\|_\infty \cdot \mathbb{E}\|\delta x_k(t)\|_\infty, \quad (10.20)$$

because all terms in (10.19) are positive and the inequality holds by the order preservation property of mathematical expectation for random variables.

Noticing system (10.1) and the desired reference model (10.2) as well as the control framework in Fig. 10.1, we have

$$\delta x_k(t+1) = [f(t, x_d(t)) - f(t, x_k(t))]$$
$$+ B(t)\delta u_k^r(t), \quad (10.21)$$

where $\delta x_k(t) \triangleq x_d(t) - x_k(t)$. Then taking ∞-norm to both sides of (10.21) leads to

$$\|\delta x_k(t+1)\|_\infty$$
$$\leq \|[f(t, x_d(t)) - f(t, x_k(t))]\|_\infty + \|B(t)\delta u_k^r(t)\|_\infty$$
$$\leq k_f \|\delta x_k(t)\|_\infty + \|B(t)\|_\infty \|\delta u_k^r(t)\|_\infty$$
$$\leq k_f \|\delta x_k(t)\|_\infty + k_b \|\delta u_k^r(t)\|_\infty, \quad (10.22)$$

where $k_b \geq \max_t \|B(t)\|_\infty$. We further take mathematical expectation to the last inequality, where all variables are positive,

$$\mathbb{E}\|\delta x_k(t+1)\|_\infty \leq k_f \mathbb{E}\|\delta x_k(t)\|_\infty + k_b \mathbb{E}\|\delta u_k^r(t)\|_\infty. \quad (10.23)$$

Backward iterating this inequality along the time axis further leads to

$$\mathbb{E}\|\delta x_k(t+1)\|_\infty \leq k_f^2 \mathbb{E}\|\delta x_k(t-1)\|_\infty + k_b \mathbb{E}\|\delta u_k^r(t)\|_\infty$$
$$+ k_f k_b \mathbb{E}\|\delta u_k^r(t-1)\|_\infty$$
$$\leq k_b \sum_{i=0}^{t} k_f^{t-i} \mathbb{E}\|\delta u_k^r(i)\|_\infty, \quad (10.24)$$

where Assumption 10.2 (i.e., $\delta x_k(0) = 0$) is applied. Consequently, we have

$$\mathbb{E}\|\delta x_k(t)\|_\infty \leq k_b \sum_{i=0}^{t-1} k_f^{t-1-i} \mathbb{E}\|\delta u_k^r(i)\|_\infty. \quad (10.25)$$

Now substituting (10.25) into (10.20) leads to

$$\mathbb{E}\|\delta u_{k+1}(t)\|_\infty \leq \mathbb{E}\|P_k(t)\|_\infty \cdot \mathbb{E}\|\delta u_k(t)\|_\infty$$
$$+ k_b \mathbb{E}\|Q_k(t)\|_\infty \sum_{i=0}^{t-1} k_f^{t-i} \mathbb{E}\|\delta u_k^r(i)\|_\infty. \quad (10.26)$$

10.2 Convergence Analysis of ILC Algorithms

Because $\delta u_k^r(t)$ is part of $\delta u_k(t)$, we have $\|\delta u_k^r(t)\|_\infty \leq \|\delta u_k(t)\|_\infty$ for all t. Thus, from (10.26) it follows

$$\mathbb{E}\|\delta u_{k+1}(t)\|_\infty \leq \mathbb{E}\|P_k(t)\|_\infty \cdot \mathbb{E}\|\delta u_k(t)\|_\infty$$
$$+ k_b \mathbb{E}\|Q_k(t)\|_\infty \sum_{i=0}^{t-1} k_f^{t-i} \mathbb{E}\|\delta u_k(i)\|_\infty. \qquad (10.27)$$

Now the classical λ-norm technique can be used. Specifically, multiply both sides of last inequality with $\alpha^{-\lambda t}$ where $\alpha > 1$ and $\lambda > 1$ are defined later, and then take supremum according to all time instants t,

$$\sup_t (\alpha^{-\lambda t} \mathbb{E}\|\delta u_{k+1}(t)\|_\infty)$$
$$\leq \sup_t \mathbb{E}\|P_k(t)\|_\infty \cdot \sup_t (\alpha^{-\lambda t} \mathbb{E}\|\delta u_k(t)\|_\infty)$$
$$+ k_b \sup_t \mathbb{E}\|Q_k(t)\|_\infty \times \sup_t \alpha^{-\lambda t} \left(\sum_{i=0}^{t-1} k_f^{t-i} \mathbb{E}\|\delta u_k(i)\|_\infty \right). \qquad (10.28)$$

Let $\alpha > k_f$, then it is observed that

$$\sup_t \alpha^{-\lambda t} \left(\sum_{i=0}^{t-1} k_f^{t-i} \mathbb{E}\|\delta u_k(i)\|_\infty \right)$$
$$\leq \sup_t \alpha^{-\lambda t} \left(\sum_{i=0}^{t-1} \alpha^{t-i} \mathbb{E}\|\delta u_k(i)\|_\infty \right)$$
$$\leq \sup_t \left(\sum_{i=0}^{t-1} \alpha^{-(\lambda-1)t-i} \mathbb{E}\|\delta u_k(i)\|_\infty \right)$$
$$\leq \sup_t \left(\sum_{i=0}^{t-1} \alpha^{-\lambda i} \mathbb{E}\|\delta u_k(i)\|_\infty \cdot \alpha^{-(\lambda-1)(t-i)} \right)$$
$$\leq \sup_t \left(\alpha^{-\lambda i} \mathbb{E}\|\delta u_k(i)\|_\infty \right) \sup_t \left(\sum_{i=0}^{t-1} \alpha^{-(\lambda-1)(t-i)} \right)$$
$$\leq \sup_t \left(\alpha^{-\lambda i} \mathbb{E}\|\delta u_k(i)\|_\infty \right) \frac{1 - \alpha^{-(\lambda-1)t}}{\alpha^{\lambda-1} - 1}. \qquad (10.29)$$

Define a new λ-norm of $\delta u_k(t)$ as

$$\|\delta u_k(t)\|_\lambda \triangleq \sup_t \left(\alpha^{-\lambda t} \mathbb{E}\|\delta u_k(t)\|_\infty \right).$$

Substituting (10.29) into (10.28) yields that

$$\|\delta u_{k+1}(t)\|_\lambda \leq \left(\rho + k_b\varphi \frac{1-\alpha^{-(\lambda-1)t}}{\alpha^{\lambda-1}-1}\right)\|\delta u_k(t)\|_\lambda, \quad (10.30)$$

where ρ and φ are defined as

$$\rho = \sup_t \mathbb{E}\|P_k(t)\|_\infty, \quad \varphi = \sup_t \mathbb{E}\|Q_k(t)\|_\infty.$$

Note that $P_k(t)$ and $Q_k(t)$ depend on $\sigma_{k+1}(t)$ and $\gamma_{k+1}(t)$ only, while the latter are identically and independently distributed with respect to k and t. Thus, both ρ and φ are independent of iteration index k as the mathematical expectation operator $\mathbb{E}$ is involved.

From Lemma 10.2, we find $\rho < 1$. Let $\alpha > \max\{1, k_f\}$, then there always exists a sufficiently large λ such that $0 < k_b\varphi \frac{1-\alpha^{-(\lambda-1)t}}{\alpha^{\lambda-1}-1} < 1-\rho$. From this observation, we further get

$$\overline{\rho} \triangleq \rho + k_b\varphi \frac{1-\alpha^{-(\lambda-1)t}}{\alpha^{\lambda-1}-1} < 1. \quad (10.31)$$

Thus, from (10.30) we have $\lim_{k\to\infty}\|\delta u_k(t)\|_\lambda = 0$, $\forall t$. The time instant t is finite, then $\lim_{k\to\infty}\mathbb{E}\|\delta u_k(t)\|_\infty = 0$, $\forall t$. Further, noting $\|\delta u_k(t)\|_\infty \geq 0$, it is clear that

$$\lim_{k\to\infty}\|\delta u_k(t)\|_\infty = 0, \quad \forall t$$

with probability one.

Thus, it is apparent that $\lim_{k\to\infty}\|\delta u_k^c(t)\|_\infty = 0$ and $\lim_{k\to\infty}\|\delta u_k^r(t)\|_\infty = 0$. Furthermore, by (10.24) we know $\lim_{k\to\infty}\|\delta x_k(t)\|_\infty = 0$ and then $\lim_{k\to\infty} e_k(t) = 0$, $\forall t$. This completes the proof.

Remark 10.2 In the proof, the classical λ-norm is modified by introducing a mathematical expectation operator to the associated variables. Roughly speaking, this modification can effectively handle the newly introduced randomness (or asynchronization), which is generated by the random data dropouts at both measurement and actuator sides. This technique can be applied to deal with other similar random factors in ILC such as randomly iteration-varying lengths (see Chap. 13).

Remark 10.3 One may argue the conservativeness of the λ-norm technique, which has been discussed in some papers. However, it is worth pointing out that the λ-norm is only used to pave the way for convergence analysis. The intrinsic convergence property of the proposed algorithms is independent of the analysis technique. That is, the conservative analysis technique does not imply that the updating algorithms are conservative. Indeed, the P-type update law has remarkable tracking performance and thus it is therefore believed that the proposed algorithms behave well under general random data dropouts environments. The tracking performance of the proposed algorithms is illustrated in Sect. 10.4.

Remark 10.4 In the proof, the monotonic convergence in the λ-norm sense is shown in (10.30). However, One may interest in monotonic convergence in the vector norm sense. To this end, we can lift the augmented input into a super-vector form $U_k = [\mathbb{E}\|\delta u_k(0)\|_\infty^T, \mathbb{E}\|\delta u_k(1)\|_\infty^T, \cdots, \mathbb{E}\|\delta u_k(N-1)\|_\infty^T]^T$ and derive the associated matrix Γ from (10.26) as a block lower triangular matrix with its elements being the parameters of (10.26). Then, we have $\|U_{k+1}\|_\infty \leq \|\Gamma\|_\infty \|U_k\|_\infty$. Consequently, the input error converges to zero monotonically if one can design L_t satisfying $\|\Gamma\|_\infty < 1$. However, this condition requires additional system information, which may restrict the applicability.

10.3 Extensions to Non-affine Nonlinear Systems

In this section, we consider the following discrete-time non-affine nonlinear system

$$x_k(t+1) = g(t, x_k(t), u_k(t)), \\ y_k(t) = C(t)x_k(t), \quad (10.32)$$

where the notations have the same meaning to (10.1) except the nonlinear function $g(t, x_k(t), u_k(t))$. Here, $\forall t$, assume that $g(t, \cdot, \cdot) : \mathbf{R}^n \times \mathbf{R}^q \to \mathbf{R}^n$ are continuously differential with respect to its arguments x and u. To be specific, denote $D_{1,k}(t) \triangleq \frac{\partial g}{\partial x}\big|_{x_k^*(t)}$ and $D_{2,k}(t) \triangleq \frac{\partial g}{\partial u}\big|_{u_k^*(t)}$ where $x_k^*(t)$ denotes the vector that lies between $x_d(t)$ and $x_k(t)$ and $u_k^*(t)$ lies between $u_d(t)$ and $u_k(t)$.

The following assumptions are for the analysis.

Assumption 10.3 For the suitable initial state $x_d(0)$, there exists a unique $u_d(t)$ such that

$$x_d(t+1) = g(t, x_d(t), u_d(t)), \\ y_d(t) = C(t)x_d(t). \quad (10.33)$$

Assumption 10.4 For any time instant $t \in \{0, \cdots, N\}$, the globally Lipschitz condition holds for the nonlinear function $g(t, x, u)$ in the sense that $\|g(t, x_1, u_1) - g(t, x_2, u_2)\| \leq k_g \|x_1 - x_2\| + k_b \|u_1 - u_2\|$. Without loss of any generality, assume that $D_{2,k}(t)$ is non-singular. Moreover, $\forall k, t, \|D_{1,k}(t)\| \leq k_g, \|D_{2,k}(t)\| \leq k_b$.

Theorem 10.2 *Consider the nonlinear system (10.32) and assume Assumptions 10.2–10.4 hold. If the learning gain matrix L_t in (10.5) satisfies $\|I - L_t C^+ D_{2,k}(t)\|_\infty < 1$, then both the computed and real input sequences generated by the algorithms (10.5) and (10.6) converge to the desired input $u_d(t)$ given in (10.33) with probability one as $k \to \infty$. That is, $u_k^c(t) \to u_d(t)$, $u_k^r(t) \to u_d(t)$, $\forall t$, with probability one as $k \to \infty$. Consequently, the actual tracking error $e_k(t) \to 0$ with probability one as $k \to \infty$.*

Proof The proof can be performed similarly to that of Theorem 10.1. Thus, here we mainly provide the major revisions according to the general formulations. Based on (10.32) and (10.33), the state difference becomes

$$\delta x_k(t+1) = g(t, x_d(t), u_d(t)) - g(t, x_k(t), u_k^r(t))$$
$$= D_{1,k}(t)\delta x_k(t) + D_{2,k}(t)\delta u_k^r(t). \quad (10.34)$$

The error dynamics is then replaced by

$$e_k(t+1) = C(t+1)\delta x_k(t+1)$$
$$= C^+ D_{1,k}(t)\delta x_k(t) + C^+ D_{2,k}(t)\delta u_k^r(t), \quad (10.35)$$

where $C^+ \triangleq C(t+1)$. Comparing (10.35) with (10.14), we can observe the analogy with $B(t)$ being replaced by $D_{2,k}(t)$ and the associated matrix $P_k(t)$ now turns into

$$P_k(t) = \begin{bmatrix} [1-\sigma_{k+1}(t)]I & \sigma_{k+1}(t)[I - L_t C^+ D_{2,k}(t)] \\ \gamma_{k+1}(t)[1-\sigma_{k+1}(t)]I & * \end{bmatrix}, \quad (10.36)$$

where the expression in the position marked by "*" is

$$* \triangleq [1 - \gamma_{k+1}(t)]I + \gamma_{k+1}(t)\sigma_{k+1}(t)[I - L_t C^+ D_{2,k}(t)].$$

This further yields

$$\delta u_{k+1}(t) = P_k(t)\delta u_k(t) - Q_k(t)D_{1,k}(t)\delta x_k(t). \quad (10.37)$$

Thus, taking the ∞-norm first and then taking mathematical expectation to (10.37) yields

$$\mathbb{E}\|\delta u_{k+1}(t)\|_\infty \leq \mathbb{E}\|P_k(t)\|_\infty \cdot \mathbb{E}\|\delta u_k(t)\|_\infty$$
$$+ k_g \mathbb{E}\|Q_k(t)\|_\infty \cdot \mathbb{E}\|\delta x_k(t)\|_\infty, \quad (10.38)$$

where Assumption 10.4 is applied to the last inequality. From (10.34) and Assumption 10.4, backward iterating the state difference similarly to (10.24) we have

$$\mathbb{E}\|\delta x_k(t)\|_\infty \leq k_b \sum_{i=1}^{t-1} k_g^{t-1-i} \mathbb{E}\|\delta u_k(i)\|_\infty. \quad (10.39)$$

Similar to (10.28), apply the λ-norm to both sides of the inequality (10.38) and combine with (10.39), then,

10.3 Extensions to Non-affine Nonlinear Systems

$$\sup_t \left(\alpha^{-\lambda t} \mathbb{E} \|\delta u_{k+1}(t)\|_\infty \right)$$
$$\leq \sup_t \mathbb{E} \|P_k(t)\|_\infty \cdot \sup_t \left(\alpha^{-\lambda t} \mathbb{E} \|\delta u_k(t)\|_\infty \right)$$
$$+ k_b \sup_t \mathbb{E} \|Q_k(t)\|_\infty$$
$$\times \sup_t \alpha^{-\lambda t} \left(\sum_{i=0}^{t-1} k_g^{t-i} \mathbb{E} \|\delta u_k(i)\|_\infty \right). \tag{10.40}$$

According to the changes from (10.34) to (10.40) and following the similar steps to the proof of Theorem 10.1, we have

$$\|\delta u_{k+1}(t)\|_\lambda \leq \left(\rho + k_b \varphi \frac{1 - \alpha^{-(\lambda-1)t}}{\alpha^{\lambda-1} - 1} \right) \|\delta u_k(t)\|_\lambda. \tag{10.41}$$

Then, using the condition $\|I - LC^+ D_{2,k}(t)\|_\infty < 1$, it is easy to obtain $0 < \rho < 1$ following similar proof of Lemma 10.2. Hence, by choosing a sufficient large λ, it follows that (10.31) is valid for this case. Then, the proof can be completed following routine derivations.

Remark 10.5 In this section, the results are extended to the non-affine nonlinear system. One may argue that the condition on $D_{2,k}(t)$ is conservative because both time- and iteration-varying factors are taken into simultaneously. However, the condition are widely satisfied in practical applications as the system runs around the equilibrium or the desired state $x_d(t)$. Then, the partial derivative matrices in the neighborhood would ensure the validity of the condition and thus guarantee the convergence of the proposed algorithms. Such convergence, in turn, contributes the validity of the condition. In addition, following similar steps, we can also extend the linear output equation to the nonlinear case. This case is omitted in this chapter for brevity.

Remark 10.6 The problem formulations and updating laws in [3, 4] are much similar to those in this chapter. The major differences between [3, 4] and this chapter lie in three aspects: convergence analysis techniques, design of learning gain matrix, and conditions on data dropouts. First, [3, 4] established the convergence based on the limit analysis of series, while we formulate the asynchronism between the computed and real inputs by randomly switching matrices and show the convergence based on a modified contraction mapping method. Second, the selection of learning gain matrix in [3, 4] depends on not only the system information but also the data dropout rate, while in this chapter it only depends on the input/output coupling matrix. Last but not least, additional conditions on data dropout is imposed in [3, 4], while we only require that the transmission networks are not completely broken down.

10.4 Illustrative Simulations

To show the effectiveness of the proposed ILC algorithms, let us consider the following non-affine nonlinear system,

$$x_k^{(1)}(t+1) = -0.75\sin(t)\sin(x_k^{(1)}(t)) + 0.1x_k^{(1)}(t)\cos(x_k^{(2)}(t))$$
$$+ \left(0.5 + 0.1\cos\left(\frac{x_k^{(2)}(t) + u_k(t)}{5}\right)\right)u_k(t),$$
$$x_k^{(2)}(t+1) = -0.5\cos(t)\cos(x_k^{(2)}(t)) + 0.2\sin(t)\cos(x_k^{(1)}(t))$$
$$+ (1 + 0.1\sin(u_k(t)/10))u_k(t),$$
$$y_k(t) = 0.1x_k^{(1)}(t) + 0.02t^{1/3}x_k^{(2)}(t),$$

where $x_k(t) = [x_k^{(1)}(t)\ x_k^{(2)}(t)]^T$ denotes the state. The iteration length is $N = 50$.

The desired reference is $y_d(t) = 0.5\sin(\pi t/20) + 0.25\sin(\pi t/10)$. The initial state is set $x_k(0) = x_d(0) = 0$. Without loss of any generality, the initial input is set to be $u_0(t) = 0, \forall t$. The learning gain L_t is selected as 0.9, which satisfies the design condition given in Theorem 10.1; that is, $0 < 1 - L_t C^+ B(t) < 1$. The proposed algorithms (10.5) and (10.6) are run for 150 iterations.

To model the random data dropouts occurring at both measurement and actuator sides, in the simulation, we generate random variables $\sigma_k(t)$ and $\gamma_k(t)$ independently for different iterations and different time instants. In addition, $\sigma_k(t)$ is also independent of $\gamma_k(t)$. Both $\sigma_k(t)$ and $\gamma_k(t)$ are binary Bernoulli random variables with the expectation $\bar{\sigma}(t)$ and $\bar{\gamma}(t)$. Note that both $\bar{\sigma}(t)$ and $\bar{\gamma}(t)$ are also the probabilities of successful transmission. Then, the values $1 - \bar{\sigma}(t)$ and $1 - \bar{\gamma}(t)$ denote the average rate that the data is lost during the transmission. Thus, we called this value as data dropout rate (DDR) in the rest of this section.

In order to demonstrate the effectiveness of the learning algorithms under general data dropouts conditions, three scenarios are considered in this simulation. For simplicity, we let DDR at the measurement side is equal to that at the actuator side.

Case 1: DDR $= 15\%$ at both measurement and actuator sides. That is, $\bar{\sigma}(t) = \bar{\gamma}(t) = 0.85$ or $\mathbb{P}(\sigma_k(t) = 1) = \mathbb{P}(\gamma_k(t) = 1) = 0.85$.

Case 2: DDR $= 30\%$ at both measurement and actuator sides. That is, $\bar{\sigma}(t) = \bar{\gamma}(t) = 0.70$ or $\mathbb{P}(\sigma_k(t) = 1) = \mathbb{P}(\gamma_k(t) = 1) = 0.70$.

Case 3: DDR $= 45\%$ at both measurement and actuator sides. That is, $\bar{\sigma}(t) = \bar{\gamma}(t) = 0.55$ or $\mathbb{P}(\sigma_k(t) = 1) = \mathbb{P}(\gamma_k(t) = 1) = 0.55$.

The tracking performance of the system output at the 20th, 50th, and 150th iterations are illustrated in Fig. 10.2. As can be observed from this figure, the proposed algorithms ensure a convergence of the system output to the desired reference. At the 20th iteration, the outputs of three cases are deflected from the reference; while at the 150th iteration, all outputs achieve satisfactory tracking precision. Thus, the proposed algorithms have good behavior against general data dropouts conditions.

On the other hand, comparing Fig. 10.2a and c, it is seen that the tracking precision at the 50th iteration of the former case is better than that of the latter case. This

10.4 Illustrative Simulations

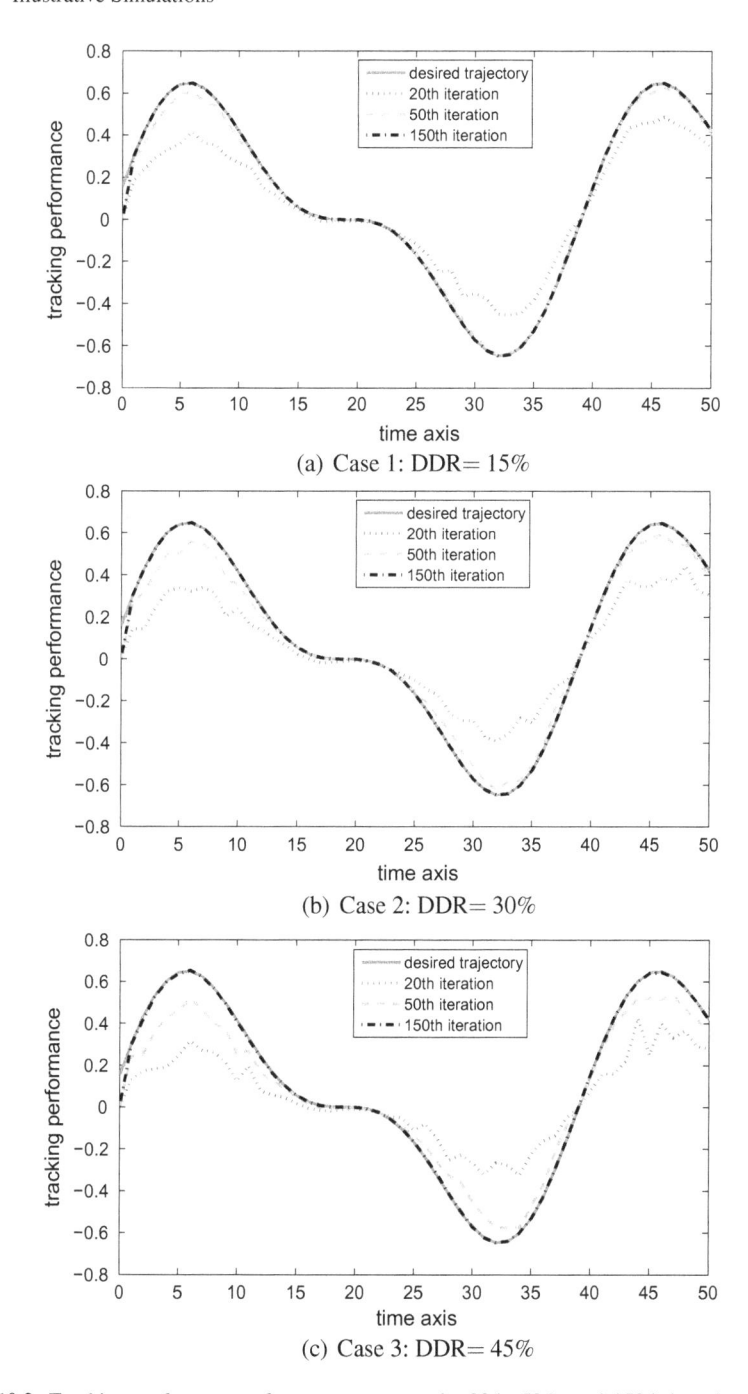

(a) Case 1: DDR= 15%

(b) Case 2: DDR= 30%

(c) Case 3: DDR= 45%

Fig. 10.2 Tracking performance of system output at the 20th, 50th, and 150th iterations under general data dropouts for three cases

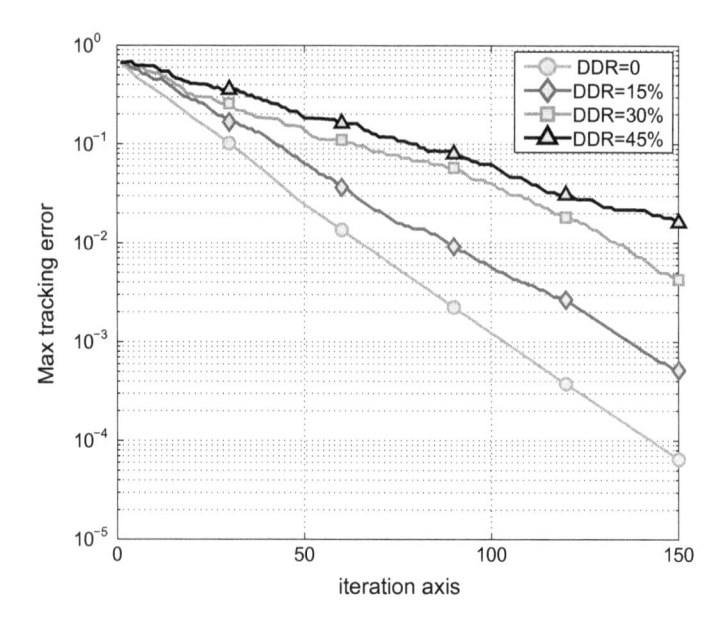

Fig. 10.3 Maximal tracking error profiles

observation implies that large DDR would slow the convergence speed. To further show this point, the maximal tracking error (MTE) profiles are displayed in Fig. 10.3 where the MTE is defined as $\max_t |e_k(t)|$ for the kth iteration. In Fig. 10.3, four lines are plotted with different markers, denoting the cases DDR = 0, 15, 30, and 45%, respectively. Two facts can be seen from the figure: the first one is that the larger the DDR, the slower the convergence speed (coinciding with Fig. 10.2); the other is that all lines decrease fast in the semilogarithmic coordinates, which shows the effectiveness of the proposed algorithms.

Moreover, to demonstrate the asynchronization between the computed input signal and the real input signal, we introduce a counter $\tau_k(t)$ for any given time instant t, denoting the amount number up to the kth iteration of the case that the computed input signal is not equal to the real input signal. That is, the counter value increases only when both computed input and real input achieve an asynchronous state. In other words, if $u_k^c(t) = u_k^r(t)$, then the counter $\tau_k(t)$ is unchanged; otherwise, if $u_k^c(t) \neq u_k^r(t)$, then the counter $\tau_k(t)$ increases one integer. The profiles for all time instants are plotted in Fig. 10.4, in which all profiles rise as the iteration number goes up. This figure illustrates that the asynchronization occurs randomly along the iteration axis and independently for different time instants. Moreover, the average value of $\tau_k(t)$ at the last iteration approximates the product of iteration number and the DDRs for all three cases. To be specific, when DDR = 15, 30, and 45%, the product (the expected amount of asynchronization states) is $150 \times 15\% = 22.5$, $150 \times 30\% = 45$, and $150 \times 45\% = 67.5$, respectively. To see this point, we provide the statistical results of Fig. 10.4 in the Table 10.1, where the ideal number denotes

10.4 Illustrative Simulations

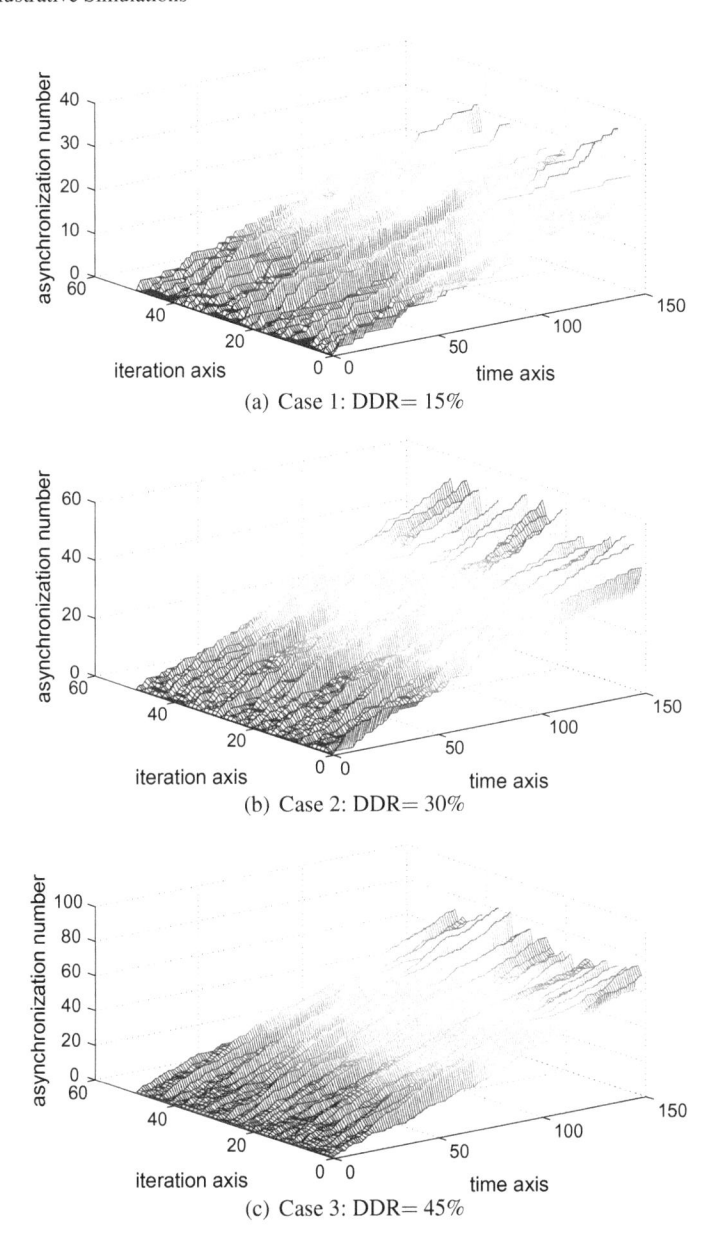

(a) Case 1: DDR= 15%

(b) Case 2: DDR= 30%

(c) Case 3: DDR= 45%

Fig. 10.4 Asynchronization of the computed and real input signals: $\tau_k(t)$

Table 10.1 Statistics of the asynchronization number, iteration maximum = 150

DDR (%)	Ideal number	Total number	Average number
15	22.5	1123	22.46
30	45	2220	44.40
45	67.5	3269	65.38

products of the iteration number and the DDR (first column), the total number denotes the amount occurrence of asynchronization for each case (second column), and the average number is computed by dividing the total number by the time length N (last column). It can be seen from the table that the average number almost equals to the ideal number for each case.

10.5 Summary

This chapter addresses the ILC problem for nonlinear discrete-time systems with data dropouts occurring at both measurement and actuator sides. Both updating laws are proposed for the computed input signal and the real input signal, whence the asynchronization between the two input signals are allowed. The zero-error convergence with probability one of the system output to the desired reference is strictly proved. In addition, the results show that the simple compensating mechanism has good tracking performance and robustness against random factors. Numerical simulations verify the effectiveness of the proposed algorithms. The results in this chapter are mainly based on [5].

References

1. Sun, M., Wang, D.: Iterative learning control with initial rectifying action. Automatica **38**(7), 1177–1182 (2002)
2. Chen, Y.Q., Wen, C., Gong, Z., Sun, M.: An iterative learning controller with initial state learning. IEEE Trans. Autom. Control **44**(2), 371–376 (1999)
3. Liu, J., Ruan, X.: Networked iterative learning control for discrete-time systems with stochastic packet dropouts in input and output channels. Adv. Differ. Equ. (2017). https://doi.org/10.1186/s13662-017-1103-8
4. Liu, J., Ruan, X.: Synchronous-substitution-type iterative learning control for discrete-time networked control systems with Bernoulli-type stochastic packet dropouts. IMA J. Math. Control Inf. (2017). https://doi.org/10.1093/imamci/dnx008
5. Jin, Y., Shen, D.: Iterative learning control for nonlinear systems with data dropouts at both measurement and actuator sides. Asian J. Control (2018). https://doi.org/10.1002/asjc.1656

Part III
General Incomplete Information Conditions

In this part, we concentrate on general incomplete information conditions such as data dropouts, random communication delays, packet disorder, finite memory, and randomly iteration-varying lengths. For these problems, the kernel issue is to investigate the inherent relationship between the incomplete information and corresponding algorithms design and analysis.

Chapter 11
Multiple Communication Conditions and Finite Memory

11.1 Problem Formulation

Consider the following single-input–single-output (SISO) nonlinear system:

$$\begin{aligned} x_k(t+1) &= f(t, x_k(t)) + \mathbf{b}(t, x_k(t))u_k(t), \\ y_k(t) &= \mathbf{c}(t)x_k(t) + v_k(t), \end{aligned} \quad (11.1)$$

where the subscript $k = 1, 2, \ldots$ denotes different iterations. The argument $t \in \{0, 1, \ldots, N\}$ labels the time instants in an iteration of the process, with N being the length of the iteration. The system input, state, and output are $u_k(t) \in \mathbf{R}$, $x_k(t) \in \mathbf{R}^n$, and $y_k(t) \in \mathbf{R}$, respectively, and $v_k(t)$ denotes random measurement noise. Both $f(t, x_k(t))$ and $\mathbf{b}(t, x_k(t))$ are continuous functions, where the argument t indicates that the functions are time-varying, and $\mathbf{c}(t)$ is the output coupling coefficient.

The setup of the control system is illustrated in Fig. 11.1, where the plant and learning controller are located separately and communicate via networks. To make our main idea intuitively understandable, the communication constraints are considered for the output side only. In other words, the random communication constraints occur only on the network from the measurement output to the buffer, whereas the network from the learning controller to the control plant is assumed to work well. If the network at the actuator side suffers from communication constraints, an asynchronism would arise between the control generated by the learning controller and the one fed to the plant. This asynchronism would require more steps to establish the convergence. Indeed, such an extension could be accomplished by incorporating the path analysis techniques similar to Chaps. 8–10. In this chapter, the data transmission of the measurement outputs might encounter multiple random factors, such as data dropouts, communication delays, and packet transmission disordering. Thus, as shown in Fig. 11.1, a buffer is required to allow the learning controller to provide a correction mechanism and ensure smooth running. The mechanism will be detailed in the next section.

Fig. 11.1 Block diagram of networked control system

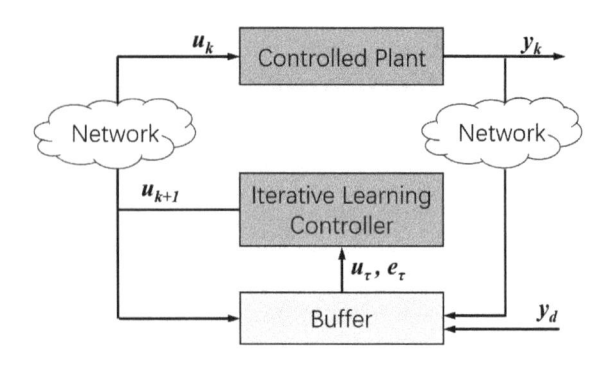

For system (11.1), we need the following assumptions:

Assumption 11.1 The desired reference $y_d(t)$, $t \in \{0, 1, \ldots, N\}$ is realizable, i.e., there exist a suitable initial state $x_d(0)$ and input $u_d(t)$ such that

$$x_d(t+1) = f(t, x_d(t)) + \mathbf{b}(t, x_d(t))u_d(t), \quad (11.2)$$
$$y_d(t) = \mathbf{c}(t)x_d(t).$$

Assumption 11.2 The real number $\mathbf{c}(t+1)\mathbf{b}(t, \cdot)$ that couples the input and output is unknown and nonzero. Its sign, which characterizes the control direction, is assumed to be known in advance. Without loss of generality, it is simply assumed that $\mathbf{c}(t+1)\mathbf{b}(t, \cdot) > 0$ for all iterations.

Assumption 11.3 For any t, the measurement noise $\{v_k(t)\}$ is an independent sequence along the iteration axis with zero-mean and finite second moment, i.e., $\mathbb{E}v_k(t) = 0$, $\mathbb{E}v_k^2(t) < \infty$, and $\limsup_{n \to \infty} \frac{1}{n} \sum_{k=1}^{n} v_k^2(t) = R_v^t$, a.s., where R_v^t is unknown.

Assumption 11.4 The initial values can be asymptotically reset precisely in the sense that $x_k(0) \to x_d(0)$ as $k \to \infty$ where $x_d(0)$ is given in Assumption 11.1.

Here, we make some remarks about these assumptions. Assumption 11.1 relates to the desired reference, which, if not realizable, means that no such input exists that satisfies (11.2). In that case, we would redefine the problem statement as one that achieves the best approximation of the reference. Assumption 11.2 requires the control direction to be known. However, if the direction is not known a priori, we can employ techniques similar to those proposed in Chap. 4 to regulate the control direction adaptively. This assumption also implies that the relative degree of system (11.1) is one. In addition, it is worth pointing out that the choice of an SISO system here is only to make the algorithm and analysis concise and easy to follow. The results in this chapter could be extended to a multi-input multi-output (MIMO) affine system by modifying the ILC update laws slightly; a gain matrix should multiply the tracking error term for regulating the control direction. The independence condition is

11.1 Problem Formulation

required in Assumption 11.3 along the iteration axis, but this is rational for practical applications because the process is repeatable. It is clear that common Gaussian white noise satisfies this assumption. Assumption 11.4 means that the desired initial state can be asymptotically achievable. This assumption is relaxed compared with the conventional identical initial condition for the initial state. The initial learning or rectifying mechanism given in [1, 2] can be incorporated in the following analysis to further deal with the initial shift problem. However, this is beyond the present scope and thus is omitted. In addition, we do not impose the conventional globally Lipschitz condition on the nonlinear functions.

11.2 Communication Constraints

In this chapter, three types of communication constraints are taken into consideration: data dropouts, communication delays, and transmission disordering. In this section, we discuss these random factors briefly and propose a unified description of the multiple communication constraints. In addition, a mechanism is provided to regulate the arriving packets.

Figure 11.2 shows the three communication constraints along the iteration axis for any fixed time instant. An orange square box denotes a data packet coming from the output side of the control plant, whereas a green square box denotes possible storage of the buffer. For brevity, we assume throughout that the data is packed and transmitted according to the time label, and in Fig. 11.2 we plot only the packets with the same time label. Thus, different square boxes denote data in different iterations. Focusing on the colored box in Fig. 11.2a, the packets before and after it would be successfully transmitted, whereas the colored one might be dropped during transmission. A communication delay is illustrated in Fig. 11.2b; adjacent colored boxes arrive at the buffer nonadjacently, which results in the second colored box being delayed. Figure 11.2c displays the disordering case, in which the second colored box arrives at the buffer ahead of the first. All these random communication conditions would make the data packets in the buffer chaotic.

For practicality and to reduce control costs, we limit the storage capacity of the buffer, which means that there will usually be insufficient storage for all the data coming from the output. In some cases, the available space may only accommodate the data of one iteration, which is the minimum buffer capacity with which to

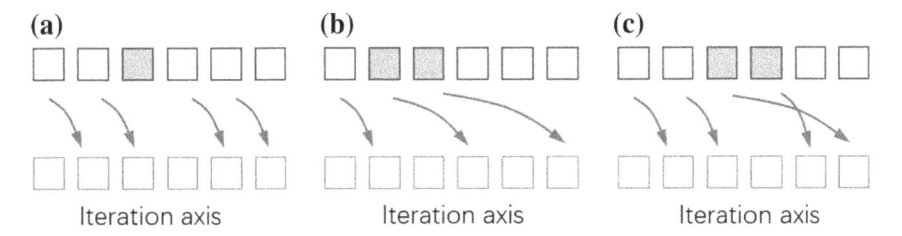

Fig. 11.2 a Data dropout; b Communication delay; c Transmission disordering

ensure the learning process. Therefore, we need to consider the possibility of limited information when we design the learning control.

To solve the problem of information chaos and limited storage, a simple renewal mechanism is proposed for the buffer. Each packet contains the whole output information at one time instant; we choose not to consider any more refined types of data partitioning. Each packet is then labeled with an iteration stamp, allowing the buffer to identify the iteration index of packets. Meanwhile, each packet is also labeled with a time stamp so that the renewals of different time instants are conducted independently. On the buffer side, only the latest packet with respect to the iteration stamp is stored in the buffer and is used for updating the control signal.

Here, we explain this mechanism briefly. For any fixed time instant t, suppose that a packet with iteration stamp k_0 is received successfully by the buffer. The buffer will then compare it with the previously stored packet to determine which iteration stamp number is closer to the current iteration index. If the iteration stamp number of the stored packet is larger than that of the new arrival, then the new arrival is discarded. Otherwise, the original packet is replaced by the newly arrived one. As only the latest packet is stored, there are no excessive requirements for the size of the buffer. However, we should emphasize that more freedom of packet renewal and control design is provided if extra storage is available to accommodate more data in the buffer. In that case, additional advantages (e.g., convergence speed and tracking precision) may be obtained by designing suitable update algorithms with additional tracking information. This would lead to the interesting and open problem of determining the optimal storage. In this chapter, we consider one-iteration storage case only in order to remain focused on the topic in hand.

Under the communication constraints, the packet in the buffer will not be replaced at each iteration. One packet may be maintained in the buffer for several successive iterations, the length of which is random because of the combined effect of the above communication constraints. It is hard to impose a statistical model on the random successive duration of each packet along the iteration axis. However, the length of iterations for which a packet is maintained in the buffer is usually bounded, unless the network has broken down. Thus, we use the following weak assumption for the buffer renewal to describe the combined effect of multiple communication constraints.

Assumption 11.5 The arrival of a new packet is random and does not obey any probability distribution. However, the length between adjacent arrivals should be bounded by a sufficiently large number M, which does not need to be known in advance. That is, there is a number M such that during M successive iterations, the buffer will renew the output information at least once.

Assumption 11.5 is weak and practical; we impose no probability distribution on it, which makes it widely applicable. A finite bound is required for the length between adjacent arrivals. However, it is not necessary to know the specific value of the maximum length M. That is, only the existence of such a bound is required, and thus the design of the ILC update law is independent of its specific value. It should be noted that the value of M corresponds to the worst-case communication conditions;

usually, a larger value of M implies a harsher communication condition. Such a property is demonstrated in the illustrative simulations, in which a uniform length distribution is imposed to characterize the effect of M on the tracking performance. However, it is not necessary for M and the average renewal frequency to be related positively.

11.3 Control Objective and Preliminary Lemmas

We now present our control objective. Let $\mathscr{F}_k \triangleq \sigma\{y_j(t), x_j(t), v_j(t), 1 \leq j \leq k, t \in \{0, 1, \ldots, N\}\}$ be a σ-algebra generated by $y_j(t), x_j(t), v_j(t), 0 \leq t \leq N$, $1 \leq j \leq k$. Then the set of admissible control is defined as $U \triangleq \{u_{k+1}(t) \in \mathscr{F}_k, \sup_k |u_k(t)| < \infty, \text{a.s.}, t \in \{0, 1, \ldots, N\}, k = 0, 1, 2, \ldots\}$.

The control objective of this chapter is to find an input sequence $\{u_k(t), k = 0, 1, \ldots\} \subset U$ under the communication constraints (i.e., data dropouts, communication delay, and packet transmission disordering) that minimizes the averaged tracking index, $\forall t \in \{0, 1, \ldots, N\}$,

$$V(t) = \limsup_{n \to \infty} \frac{1}{n} \sum_{k=1}^{n} |y_k(t) - y_d(t)|^2, \qquad (11.3)$$

where $y_d(t)$ is the desired reference given in Assumption 11.1. If we define the control output as $z_k(t) = \mathbf{c}(t)x_k(t)$, it is easy to show that $z_k(t) \to y_d(t)$ as $k \to \infty$ whenever the tracking index (11.3) is minimized, and vice versa. In other words, index (11.3) implies that precise tracking is achieved if all measurement noises are eliminated.

We note that when considering the optimization of a composite objective function, what is known as the advanced fine tuning (AFT) approach [3] can be used to solve the problem. Note that both AFT and ILC are data-driven methods and thus can be applied to nonlinear systems. However, the implementation of AFT is more complex than that of ILC, and the learning speed of AFT can be lower than that of ILC as the former has to learn more information.

For further analysis, we require the following lemmas, the proofs of which are the same as in Chaps. 3 and 4 and thus are omitted for brevity.

Lemma 11.1 *Assume that Assumptions 11.1–11.4 hold for system (11.1). If the generated input sequence satisfies that* $\lim_{k \to \infty} \delta u_k(s) = 0$, $s = 0, 1, \ldots, t$, *then at time instant* $t + 1$, $|\delta x_k(t+1)| \to 0$, $|\delta f_k(t+1)| \to 0$, $|\delta \mathbf{b}_k(t+1)| \to 0$ *as* $k \to \infty$.

Lemma 11.2 *Assume that Assumptions 11.1–11.4 hold for system (11.1) and for tracking reference* $y_d(t)$. *Then index (11.3) will be minimized as* $V(t) = R_v^t$ *for any arbitrary time instant* t *if the control sequence* $\{u_k(i)\}$ *is admissible and satisfies* $u_k(i) \to u_d(i)$ *as* $k \to \infty$, $i = 0, 1, \ldots, t - 1$. *In this case, the input sequence* $\{u_k(t)\}$ *is called the optimal control sequence.*

Lemma 11.1 paves the way for connecting the state convergence at the next time instant and the input convergence at all previous time instants. This lemma plays a supporting role in the application of mathematical induction in the convergence analysis. Lemma 11.2 characterizes the optimal solution according to the tracking index. Based on Lemma 11.2, it is sufficient to show that the input sequence converges to the desired input defined in Assumption 11.1.

In the following, we propose two update schemes for generating the optimal control sequence $\{u_k(t)\}$ under the communication constraints. The first scheme is called the intermittent update scheme (IUS), in which the control signal retains the latest one if no new output arrives at the buffer. The second is called the successive update scheme (SUS), in which the control signal keeps updating even if no new packet arrives. The tracking performances of these two schemes are compared in numerical simulations.

11.4 Intermittent Update Scheme and Its Almost Sure Convergence

In this section, we provide an in-depth discussion of the intermittent update scheme (IUS). Specifically, we begin by studying the path behavior of IUS for any fixed time instant t, and we provide a recognition mechanism to ensure a smooth improvement of the algorithm. We then introduce a sequence of stopping times to specify the learning algorithm, and we give the convergence results.

Under the communication constraints, for arbitrary time instant t, the packet stored in the buffer and used for the kth iteration is the one with the $(k - m_k(t))$th iteration stamp, where $m_k(t)$ is a random variable over $\{1, 2, \ldots, M\}$, and M is defined as in Assumption 11.5. Some observed properties of $m_k(t)$ are as follows. If there is no communication constraint, then $m_k(t) = 1$, $\forall k$; otherwise, $m_k(t) > 1$. When transmission disordering occurs for any given k, we might expect $m_{k+1}(t) \geq m_k(t) + 1$. In the remainder of this chapter, the argument t will be omitted from $m_k(t)$ to simplify the notation and to avoid tedious repetition. Note that m_k is a random variable; without loss of generality, there are upper and lower bounds of m_k, i.e., $m \leq m_k \leq M$ with $m \geq 1$ because of the communication constraints.

In the IUS, the input is generated from the latest available information and its corresponding input, i.e.,

$$u_k(t) = u_{k-m_k}(t) + a_{k-m_k} e_{k-m_k}(t+1), \tag{11.4}$$

where $e_k(t) \triangleq y_d(t) - y_k(t)$ and a_k is the learning step size (defined later), $\forall k, t$. By simple calculations, we have

$$e_k(t+1) = \mathbf{c}^+ \mathbf{b}_k(t) \delta u_k(t) + \varphi_k(t) - v_k(t+1), \tag{11.5}$$

where $\varphi_k(t) = \mathbf{c}^+ \delta f_k(t) + \mathbf{c}^+ \delta b_k(t) u_d(t)$.

11.4 Intermittent Update Scheme and Its Almost Sure Convergence

Before proceeding to the main theorem for the IUS case, we perform some primary analyses of the input update. Let us begin with an arbitrary iteration, say k_0, for which the input is given as

$$u_{k_0}(t) = u_{k_0-m_{k_0}}(t) + a_{k_0-m_{k_0}} e_{k_0-m_{k_0}}(t+1).$$

We now proceed to the next iteration, i.e., the (k_0+1)th iteration. If no packet arrives at the buffer, then $m_{k_0+1} = m_{k_0} + 1$ and the input for this iteration is

$$\begin{aligned} u_{k_0+1}(t) &= u_{m'}(t) + a_{m'} e_{m'}(t+1) \\ &= u_{k_0-m_{k_0}}(t) + a_{k_0-m_{k_0}} e_{k_0-m_{k_0}}(t+1), \end{aligned}$$

where $m' \triangleq k_0+1-m_{k_0+1}$ and the last equality is valid because $m' = k_0+1-(m_{k_0}+1) = k_0 - m_{k_0}$. Consequently, $u_{k_0+1}(t) = u_{k_0}(t)$. In other words, the input remains invariant when no new packet is received. However, according to Assumption 11.5, this input will not remain unchanged forever. Indeed, after several iterations (say τ iterations, for example), a new packet will arrive successfully at the buffer and the input is then updated.

However, we should carefully check the iteration stamp, say k_1, of the newly arrived packet. Specifically, noting that the iteration stamp of the packet at the k_0th iteration is $k_0 - m_{k_0}$ and recalling the renewal mechanism whereby only the one with larger iteration stamp will be accepted, we have $k_1 \geq k_0 - m_{k_0}$. However, the iteration stamp must be smaller than the corresponding iteration number, thus we have $k_1 \leq k_0 + \tau - 1$ because we assume that the subsequent updating occurs at the $(k+\tau)$th iteration. In short, $k_0 - m_{k_0} \leq k_1 \leq k_0 + \tau - 1$. As such, two scenarios should be considered for the iteration stamp k_1 of the newly arrived packet: k_1 with $k_0 - m_{k_0} \leq k_1 \leq k_0 - 1$, and k_1 with $k_0 \leq k_1 \leq k_0 + \tau - 1$ (see Fig. 11.3). In the former scenario, updating the input at the $(k_0 + \tau)$th iteration would generate a mismatch

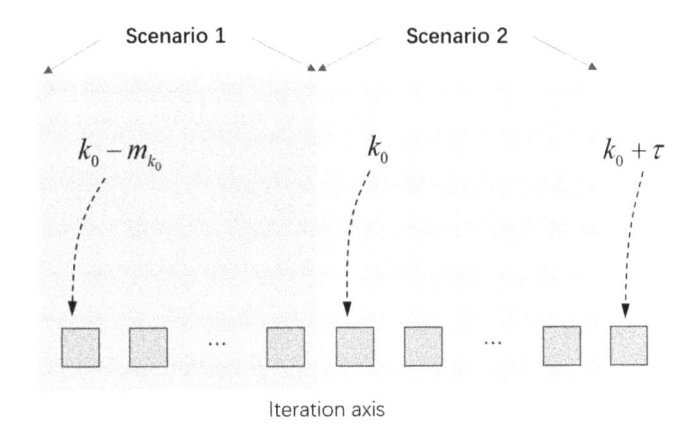

Fig. 11.3 Illustration of two scenarios of new arrivals

between the iteration labels of the tracking error and the existing input. The algorithm is a combination of several staggered updating procedures, which makes convergence analysis intricate. In the latter scenario, updating at the $(k_0 + \tau)$th iteration could use input $u_{k_0}(t)$, i.e., the update would be

$$u_{k_0+\tau}(t) = u_{k_1} + a_{k_1} e_{k_1}(t+1)$$
$$= u_{k_0} + a_{k_1} e_{k_1}(t+1).$$

Remark 11.1 By analyzing the two scenarios in Fig. 11.3, we find that Scenario 1 would lead to a mismatch between the iteration labels of the tracking error and the stored input. To deal with this problem, a possible solution is to augment the capacity of the buffer to store more historical data of the input or the tracking error so that we can always match the input and the tracking error. This is an advantage of extra storage, as discussed in Sect. 11.2. Determining the optimal capacity of the buffer and designing and analyzing the corresponding learning algorithms remain open problems. In this chapter, we consider the one-iteration storage case, thus we have to adopt another simple method whereby we discard the packet in Scenario 1 and wait for suitable packets (see the following for details).

To make the following analysis more concise, an additional recognition mechanism is proposed to allow the learning controller to define the suitable information for updating. Assume that the latest update occurs at the k_0th iteration. If no new packet is received, then the input will remain as $u_{k_0}(t)$. Otherwise, the controller will check whether the iteration stamp of the new packet is smaller than k_0. If so, then this packet is neglected, and the update is delayed until a new packet with an iteration stamp number larger than or equal to k_0, say k_1, is received. The learning controller will then update its input signal using $u_{k_0}(t)$ and $e_{k_1}(t+1)$. Note that $e_{k_1}(t+1)$ is actually generated by $u_{k_0}(t)$ since $k_1 \geq k_0$. This update procedure is illustrated in Fig. 11.4, where, for any fixed time instant t, the boxes in the top row denote the output packets for successive iterations. The colored packets are received by the buffer and used for updating successfully, whereas the blank ones are lost during transmission, either discarded by the renewal mechanism or neglected by the recognition mechanism. The boxes in the bottom row denote the inputs in different iterations; the colored

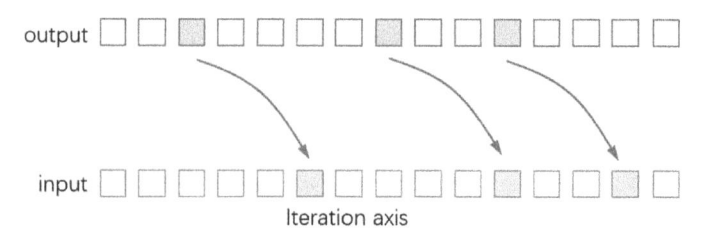

Fig. 11.4 Illustration of the recognition mechanism

11.4 Intermittent Update Scheme and Its Almost Sure Convergence

and blank ones denote updating iterations and holding iterations, respectively. The arrows link the input updating with its corresponding tracking information.

Remark 11.2 Another explanation for the recognition mechanism is that we expect the input of k_0 to behave generally better than previous ones because an improvement has been made, making it unnecessary to update further using information from iterations before k_0. Meanwhile, this mechanism makes it possible for us to update the control signal smoothly with limited storage.

We now formulate the ILC based on the renewal and recognition mechanisms for the IUS case.

For arbitrary time instant t, we define a sequence of random stopping times $\{\tau_i\}$, $i = 1, 2, \ldots$, where τ_i denotes the iteration number of the ith update of the control signal for time instant t, corresponding to the colored boxes in the bottom row of Fig. 11.4. It should be noted that $\{\tau_i\}$ is defined for different time instants independently, denoting the asynchronous update for different time instants; we omit the associated argument t throughout to simplify the notation. Without loss of generality, we assume that $\tau_0 = 0$. The packet used for the ith update has iteration stamp $\tau_i - n_{\tau_i}$, corresponding to the colored boxes in the top row of Fig. 11.4, where n_{τ_i} is a random variable due to the communication constraints (see Sect. 11.2), $1 \leq n_{\tau_i} \leq M$. Recalling the recognition mechanism, we have $\tau_i - n_{\tau_i} \geq \tau_{i-1}$, $\tau_i - \tau_{i-1} \leq 2M$, $\forall i$, and the input generating $e_{\tau_i - n_{\tau_i}}(t+1)$ is $u_{\tau_{i-1}}(t)$, $\forall t$.

The update algorithm can now be rewritten as

$$u_{\tau_i}(t) = u_{\tau_{i-1}}(t) + a_{\tau_{i-1}} e_{\tau_i - n_{\tau_i}}(t+1), \quad (11.6)$$

and

$$u_k(t) = u_{\tau_i}(t), \quad \tau_i < k \leq \tau_{i+1} - 1. \quad (11.7)$$

This algorithm is, in essence, an event-triggered update because τ_i is an unknown random stopping time and n_{τ_i} is an unknown random variable; thus, this algorithm differs from the conventional deterministic framework. The learning step size $\{a_k\}$ is a decreasing sequence that satisfies $a_k > 0$, $a_k \to 0$, $\sum_{k=1}^{\infty} a_k = \infty$, $\sum_{k=1}^{\infty} a_k^2 < \infty$, and $a_j = a_k(1 + O(a_k))$, $\forall k - M \leq j \leq k$. It is clear that $a_k = 1/(k+1)$ meets all these requirements.

We now present the following convergence theorem for the IUS; the proof can be found in Sect. 11.7.

Theorem 11.1 *Consider system (11.1) and control objective (11.3), and assume that Assumptions 11.1–11.5 hold, then the input sequence $\{u_k(t)\}$ generated by IUS (11.6) and (11.7) with the renewal and recognition mechanisms is an optimal control sequence. In other words, $u_k(t)$ converges to $u_d(t)$ a.s. as $k \to \infty$ for any t, $0 \leq t \leq N - 1$.*

Theorem 11.1 reveals the essential convergence and optimality property of the IUS for nonlinear system (11.1) under multiple communication constraints and limited storage. It should be noted that the convergence is an asymptotic property in which only the limits are characterized.

Remark 11.3 The proposed IUS (11.6) and (11.7) updates its input only when a satisfactory packet is received. Thus, the update frequency may be low if severe communication constraints arise. In practice, the tracking performance worsens as the communication environments deteriorate, as shown in the simulations below. Roughly speaking, more severe communication constraints imply that the average gap of τ_i is large, so that the learning step size a_{τ_i} goes to 0 relatively quickly, which might result in quite slow learning.

Remark 11.4 For any given learning step size sequence $\{a_k\}$, an alternative modification to the algorithm could increase the convergence speed of the first iterations. Specifically, the controller records its updating times and then changes the step size in turn only when the update actually occurs. That is, algorithm (11.6) is replaced by $u_{\tau_i}(t) = u_{\tau_{i-1}}(t) + a_i e_{\tau_i - n_{\tau_i}}(t+1)$. The convergence results of Theorem 11.1 remain valid.

Remark 11.4 gives an alternative IUS that could increase the convergence speed from the perspective of selecting the learning gain. However, according to Remark 11.3, if the communication environments are harsh, the renewal and recognition mechanisms may lower the updating frequency and then the convergence speed. Motivated by this observation, we propose an alternative framework, i.e., the successive update scheme (SUS), in the next section.

11.5 Successive Update Scheme and Its Almost Sure Convergence

As noted in the previous section, the IUS might have a low learning speed along the iteration axis if the communication environment is seriously impaired. In such case, the algorithm would require many iterations to achieve an acceptable performance as the update frequency is low. Thus, it is impractical in most real applications. A possible solution is to make best use of the available information by increasing the learning step size to improve the convergence speed. In this section, we propose another scheme called SUS, in which the input keeps updating using the latest available packet when no new packet is received by the buffer. It is apparent that such an update principle is in contrast with the IUS, which keeps the input invariant if no satisfactory packet is received by the buffer. Thus, we expect the SUS to be advantageous in that the tracking performance might be improved iteration by iteration.

The renewal mechanism of the buffer and the recognition mechanism of the controller are still valid for the SUS. Consequently, the random stopping time τ_i and random variable n_{τ_i} are defined in the same way as in the IUS case. For the SUS, the update for the τ_i iteration is then given as

11.5 Successive Update Scheme and Its Almost Sure Convergence

$$u_{\tau_i}(t) = u_{\tau_i-1}(t) + a_{\tau_i-1}e_{\tau_i-n_{\tau_i}}(t+1), \quad (11.8)$$

and for $\tau_i < k \leq \tau_{i+1} - 1$,

$$u_k(t) = u_{k-1}(t) + a_{k-1}e_{\tau_i-n_{\tau_i}}(t+1), \quad (11.9)$$

in which case we have

$$u_{\tau_{i+1}-1}(t) = u_{\tau_i-1}(t) + \left(\sum_{k=\tau_i-1}^{\tau_{i+1}-2} a_k\right) e_{\tau_i-n_{\tau_i}}(t+1). \quad (11.10)$$

Because the algorithm keeps updating in the SUS case, it is not an event-triggered algorithm. However, we should emphasize in particular that the alteration of the error signal is event-triggered, regulated by the renewal and recognition mechanisms. By observing the subscript of the input and step size on the right-hand side, (11.8) is different from (11.6). Specifically, in (11.8), the subscript of the input and step size is $\tau_i - 1$, whereas in (11.6) it is τ_{i-1}. Moreover, (11.9) differs from (11.7) in the successive updating.

We now have the following convergence theorem for SUS; the proof is given in Sect. 11.7.

Theorem 11.2 *Consider system (11.1) and control objective (11.3), and assume that Assumptions 11.1–11.5 hold. Then the input sequence $\{u_k(t)\}$ generated by SUS (11.8) and (11.9) with the renewal and recognition mechanisms is an optimal control sequence. In other words, $u_k(t)$ converges to $u_d(t)$ a.s. as $k \to \infty$ for any t, $0 \leq t \leq N - 1$.*

Theorem 11.2 indicates the asymptotic convergence of the SUS along the iteration axis and shows the optimality of the generated input sequence. From this viewpoint, both IUS and SUS can guarantee the convergence of the input sequence to the desired input with probability one. However, the major difference between IUS and SUS lies in the following points. First of all, the IUS is an event-triggered updating whereas the SUS is an iteration-triggered updating. Moreover, the updating frequency of the IUS depends on the rate of successful transmission, renewal, and recognition, and thus is low if the communication constraints are harsh. In contrast, the SUS keeps updating for all iterations. Thus, it is expected that the SUS can guarantee a better convergence performance when the communication environments deteriorate. This point is illustrated by the illustrative simulations.

Remark 11.5 In this chapter, we consider an SISO system for the sake of concise expression and analysis. The results can be extended to an MIMO case, in which the vectors $\mathbf{c}(t)$ and $\mathbf{b}(t, x)$ are replaced with matrices $C(t)$ and $B(t, x)$. We assume $u_k(t) \in \mathbf{R}^p$ and $y_k(t) \in \mathbf{R}^q$, then $C(t) \in \mathbf{R}^{q \times n}$ and $B(t, x) \in \mathbf{R}^{n \times p}$. In such case, the control direction is determined by the coupling matrix $C(t+1)B(t, x) \in \mathbf{R}^{q \times p}$, which is more complicated than that in the SISO case. To ensure convergence of the algorithm, an additional matrix $L(t) \in \mathbf{R}^{p \times q}$ should left multiply the error term $e_k(t+1)$ in (11.6), (11.8), and (11.9) to regulate the control direction. The design condition for $L(t)$ is that all eigenvalues of $L(t)C(t+1)B(t,x)$ are with positive real parts. The convergence proofs can be conducted following similar steps.

11.6 Illustrative Simulations

Consider the following affine nonlinear system as an example, in which the state is two dimensional,

$$x_k^{(1)}(t+1) = 0.8 x_k^{(1)}(t) + 0.3 \sin(x_k^{(2)}(t)) + 0.23 u_k(t),$$

$$x_k^{(2)}(t+1) = 0.4 \cos(x_k^{(1)}(t)) + 0.85 x_k^{(2)}(t) + 0.33 u_k(t),$$

$$y_k(t) = x_k^{(1)}(t) + x_k^{(2)}(t) + v_k(t),$$

where $x_k^{(1)}(t)$ and $x_k^{(2)}(t)$ denote the first and second dimension of $x_k(t)$, respectively. It is easy to check that $\mathbf{c}^+ \mathbf{b}(t) = 0.23 \times 1 + 0.33 \times 1 = 0.56 > 0$.

The reference trajectory is $y_d(t) = 20 \sin(\frac{t}{20}\pi)$. As a simple illustration, let $N = 40$ and the measurement noise $v_k(t)$ be zero-Gaussian distributed, $v_k(t) \sim N(0, 0.1^2)$. The initial control action is given simply as $u_0(t) = 0$, $\forall t$. Here, the selection of the initial input value does not affect the inherent convergence property of the proposed algorithm. The learning gain chooses $a_k = \frac{1}{k+1}$. Each algorithm is run for 300 iterations.

For each time instant, in order to simulate the renewal and recognition mechanisms dealing with random communication constraints, we begin by generating a sequence of random numbers $\{\tau_k\}$ that are uniformly distributed over $\{1, 2, \ldots, M\}$, where M is defined as in Assumption 11.5. Thus, in essence, τ_k denotes the random dwelling length/iterations of each received packet along the iteration axis (caused by communication constraints). We should clarify that we simulate the packet alternation in the buffer directly rather than the specific communication constraints to illustrate the combined effects of multiple communication constraints and limited storage under renewal and recognition mechanisms (cf. Fig. 11.4) and to provide a suitable parameter for the following comparison analysis (cf. Assumption 11.5). It is then apparent that the accumulation number $\sigma_k = \sum_{i=1}^{k} \tau_i$ corresponds to those iterations at which input updating (11.6) or (11.8) occurs, whereas for the other iterations the input algorithm (11.7) or (11.9) works. Note that both τ_k and σ_k are random variables, indicating the event-triggered character of the input updating; neither are known prior to run the algorithms.

An illustration of τ_k is given in Fig. 11.5, where $M = 5$. As can be seen from this figure, τ_k is randomly valued in the set $\{1, 2, 3, 4, 5\}$. This is a simulation of the iteration dwelling length for which a packet is stored in the buffer. Thus, the average dwelling length (i.e., mathematical expectation of τ_k) could be regarded as

11.6 Illustrative Simulations

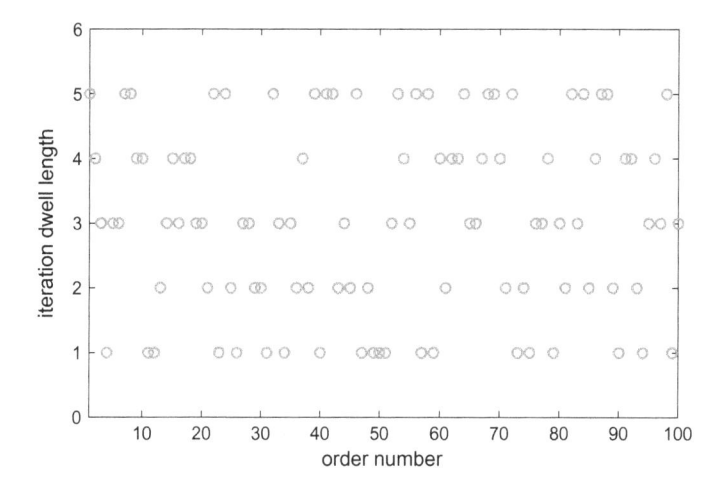

Fig. 11.5 Illustration of iteration dwelling length

a data transmission rate (DTR) index. Specifically, because a uniform distribution is adopted, the mathematical expectation of τ_k is $(M+1)/2$. This means that, on average, a feasible packet is received and an update occurs every $(M+1)/2$ iterations. In the case of Fig. 11.5, we have $M=5$ and therefore $\mathbb{E}\tau_k = 3$, i.e., an update happens every three iterations on average. The explanation of this is twofold: the data dropout rate is $2/3$, and the updating is three times slower than that in the case of no communication constraints.

In the following, we first show the performance of the IUS, and then turn our attention to the SUS. The comparisons between the IUS and SUS are detailed at the end of this section.

11.6.1 Intermittent Update Scheme Case

We begin by considering the IUS case with $M=5$. The tracking performance of the final iteration (i.e., the 300th iteration) is shown in Fig. 11.6a, where the solid line with circles is the reference signal and the dashed line with crosses denotes the actual output $y_{300}(t)$. The fact that the output tracks the desired positions demonstrates the convergence and effectiveness of the IUS. The deviations seen in Fig. 11.6a are caused mainly by stochastic measurement noise, which cannot be canceled by any learning algorithm because it is completely unpredictable.

As explained above, M or $(M+1)/2$ corresponds the DTR index, thus we are interested in the influence of M. We simulate this example further for $M=3$ and $M=11$ with average iteration dwelling lengths of 2 and 6, respectively. It is expected that a longer dwelling length implies a higher rate of data dropout and poorer tracking performance. This point is verified in Fig. 11.6b for $M=11$, where the performance

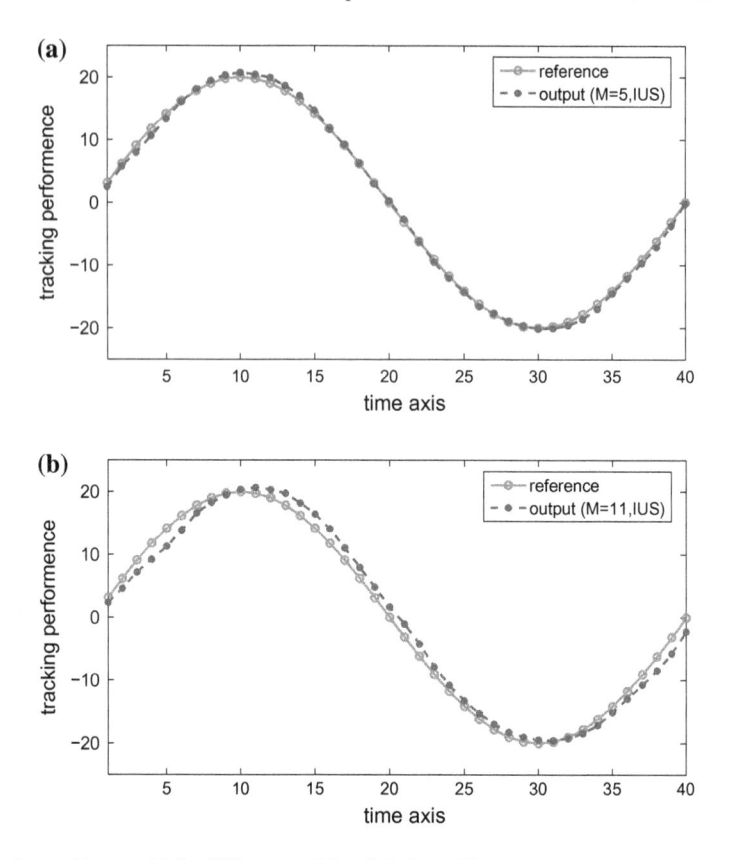

Fig. 11.6 $y_{300}(t)$ vs $y_d(t)$ for IUS case: **a** $M = 5$; **b** $M = 11$

is clearly worse than that in Fig. 11.6a. This suggests that the number of learning iterations should be increased to improve the tracking performance.

The averaged absolute tracking error for each iteration is defined as $\sqrt{\frac{\sum_{i=1}^{N} |e_k(i)|^2}{N}}$. Given the stochastic noise in the index, the averaged absolute tracking error does not decrease to zero as the number of iterations goes to infinity. Figure 11.7 demonstrates the averaged absolute tracking error profiles for $M = 3$, 5, and 11, denoted by the solid, dashed, and dash-dotted lines, respectively. As seen in Fig. 11.7, a larger value of M results in larger tracking errors.

11.6.2 Successive Update Scheme Case

We now come to the SUS case. For clarity, we take the same simulation cases as before. Firstly, we consider the case of $M = 5$. The tracking performance of the final iteration (i.e., the 300th iteration) is shown in Fig. 11.8a, where the symbols are

11.6 Illustrative Simulations

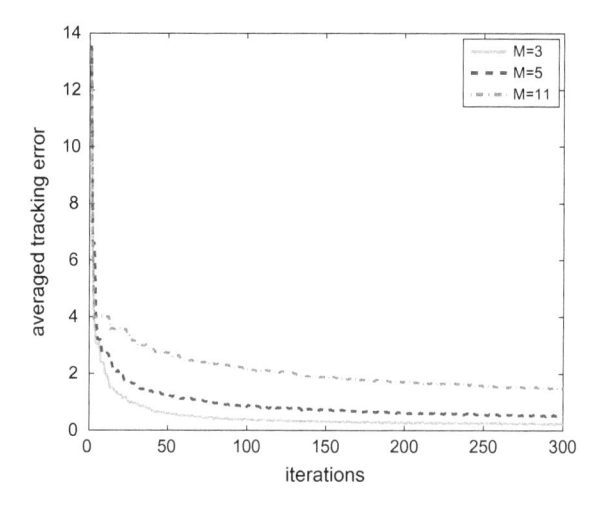

Fig. 11.7 Averaged absolute tracking error $\sqrt{\frac{\sum_{i=1}^{N}|e_k(i)|^2}{N}}$: $M = 3, 5,$ and 11 for IUS case

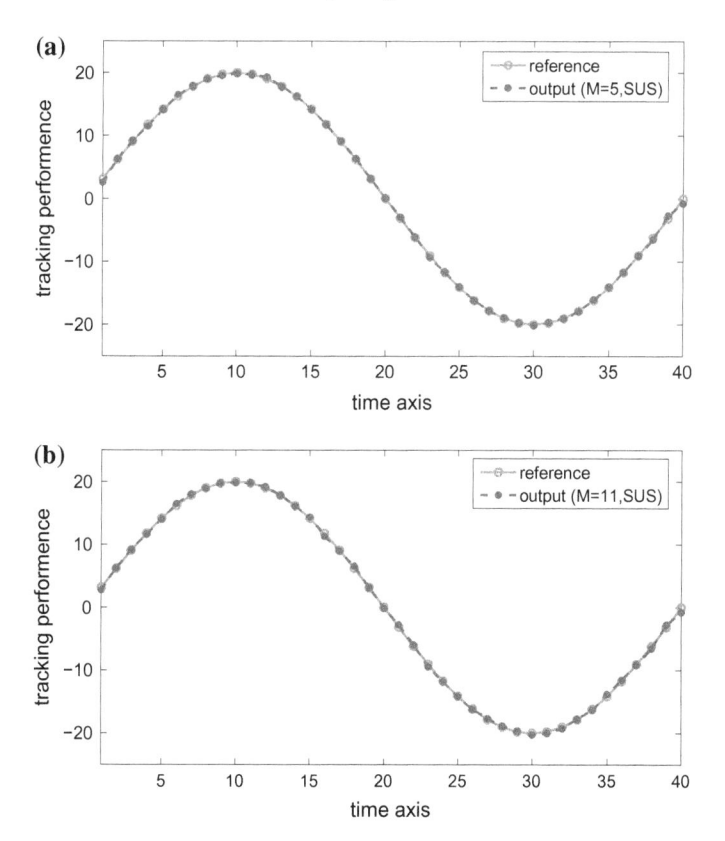

Fig. 11.8 $y_{300}(t)$ vs $y_d(t)$ for SUS case: **a** $M = 5$; **b** $M = 11$

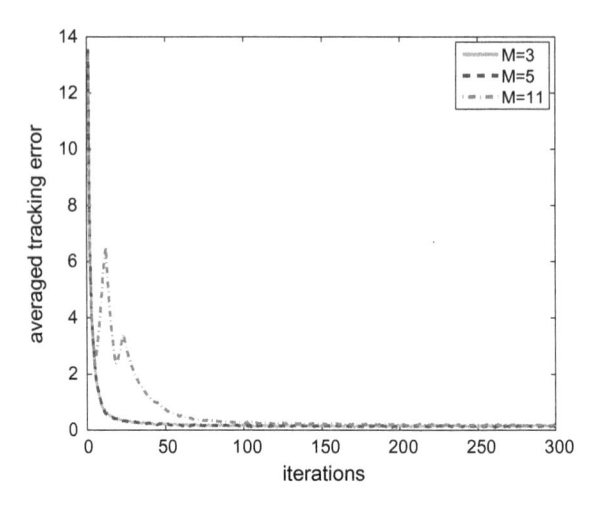

Fig. 11.9 Averaged absolute tracking error $\sqrt{\frac{\sum_{i=1}^{N}|e_k(i)|^2}{N}}$: $M = 3, 5$, and 11 for SUS case

the same as those in the IUS case. As is seen from the figure, the desired reference is tracked precisely. The final tracking performance for $M = 11$ in the SUS case is presented in Fig. 11.8b. In contrast to the IUS case, the final tracking performance is much better, even when M is large. The similarity between Fig. 11.8a, b suggests that a longer dwelling length does not cause significant deterioration of the learning progress. This is because the algorithm keeps updating in the SUS case.

The averaged absolute tracking error profiles for $M = 3, 5$, and 11 are shown in Fig. 11.9 by the solid, dashed, and dash-dotted lines, respectively. Two differences can be observed between Figs. 11.9 and 11.7. The first is that the tracking performance after 100 learning iterations shows little difference with the value of M in the SUS case. This explains the similarity between Fig. 11.8a, b from a different viewpoint. The second is that a large fluctuation occurs for the case $M = 11$, caused by successively updating with a large error for the first several iterations.

11.6.3 Intermittent Update Scheme Versus Successive Update Scheme

To provide a visual comparison between IUS and SUS, we show their final outputs in Fig. 11.10 for the case $M = 5$, where the solid line, dashed line with crosses, and dash-dotted line with circles represent the reference, the IUS output, and the SUS output, respectively. The performance at time instants from 8 to 13 is enlarged as a subplot. It can be seen that the SUS output surpasses the IUS output over the same iterations. This is reasonable because SUS updates more often than IUS does for the same iterations.

11.6 Illustrative Simulations

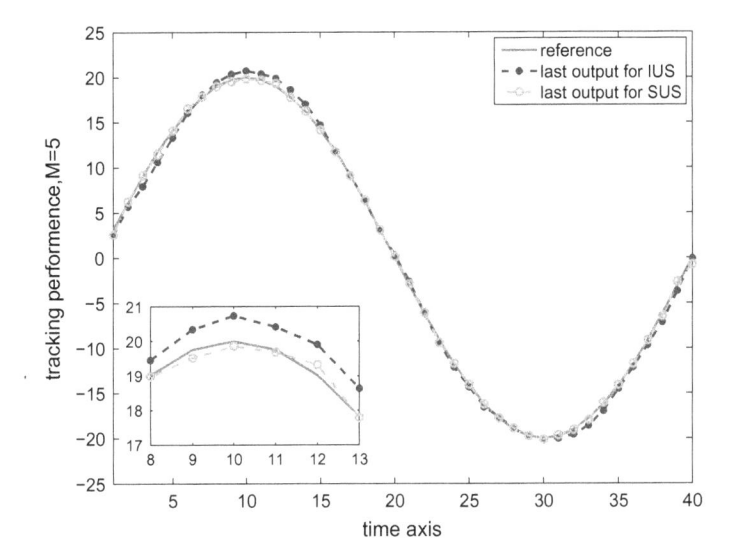

Fig. 11.10 $y_{300}(t)$ vs $y_d(t)$ for $M = 5$: IUS versus SUS

Fig. 11.11 Averaged absolute tracking error for $M = 5$: IUS versus SUS

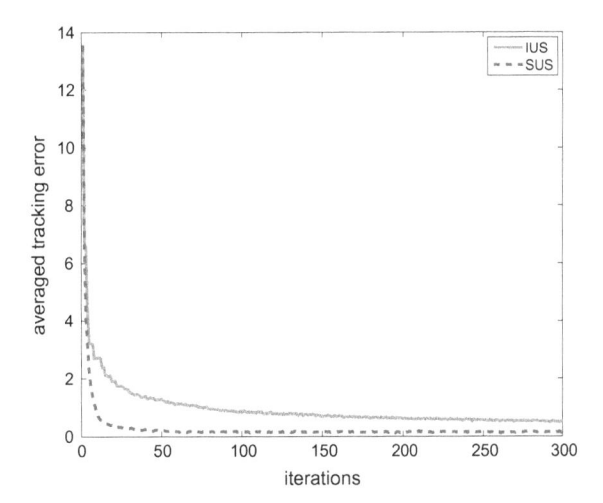

The absolute averaged tracking error profiles are shown in Figs. 11.11 and 11.12 for $M = 5$ and $M = 11$, respectively, where it can be seen that the SUS algorithm achieves faster convergence and superior tracking. However, the SUS algorithm fluctuates during the early iterations as M increases, whereas the IUS algorithm maintains a gentle descent.

Fig. 11.12 Averaged absolute tracking error for $M = 11$: IUS versus SUS

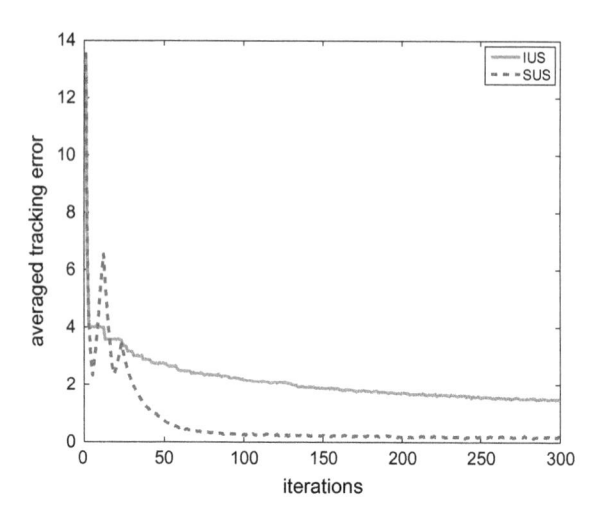

11.7 Proofs of Theorems

Proof of Theorem 11.1.

Proof Due to the nonlinear functions $f_k(t)$ and $\mathbf{b}_k(t)$, which are related to the information from previous time instants, it is difficult to show the convergence of the input for all time instants simultaneously. Therefore, for convenience, the proof is carried out by mathematical induction along the time axis t. Note that the steps for time $t = 1, 2, \ldots, N-1$ are identical to the case for initial time instant $t = 0$, which is expressed as follows.

Initial Step. Consider the case of $t = 0$.

From algorithms (11.6) and (11.7), it is evident that to show the optimality of $\{u_k(0)\}$, it is sufficient to show the optimality of its subsequence $\{u_{\tau_i}(0)\}$, i.e., to show the convergence of (11.6). Note that both τ_i and n_{τ_i} are random and $\tau_i - n_{\tau_i} \geq \tau_{i-1}$. For $t = 0$, the algorithm (11.6) gives

$$\begin{aligned}\delta u_{\tau_i}(0) =& \delta u_{\tau_{i-1}}(0) - a_{\tau_{i-1}} \mathbf{c}^+ \mathbf{b}_{\tau_{i-1}}(0) \delta u_{\tau_i - n_{\tau_i}}(0) \\
& - a_{\tau_{i-1}} \varphi_{\tau_i - n_{\tau_i}}(0) + a_{\tau_{i-1}} v_{\tau_i - n_{\tau_i}}(1) \\
=& (1 - a_{\tau_{i-1}} \mathbf{c}^+ \mathbf{b}_{\tau_{i-1}}(0)) \delta u_{\tau_{i-1}}(0) \\
& - a_{\tau_{i-1}} \varphi_{\tau_i - n_{\tau_i}}(0) + a_{\tau_{i-1}} v_{\tau_i - n_{\tau_i}}(1). \end{aligned} \quad (11.11)$$

The above recursion differs from the traditional ILC update law. In the above recursion, the learning gain and tracking error are event-triggered, whereas the traditional ILC update law runs per iteration. However, by Assumption 11.5, we have

11.7 Proofs of Theorems

$\tau_i - \tau_{i-1} \leq 2M$ and thus $\{a_{\tau_i}\}$ is a subset of $\{a_k\}$ with the following properties $a_{\tau_i} \xrightarrow[i\to\infty]{} 0$, $\sum_{i=1}^{\infty} a_{\tau_i} = \infty$, $\sum_{i=1}^{\infty} a_{\tau_i}^2 < \infty$.

Set $\Gamma_{i,j} \triangleq (1 - a_{\tau_i}\mathbf{c}^+\mathbf{b}_{\tau_i}(0)) \cdots (1 - a_{\tau_j}\mathbf{c}^+\mathbf{b}_{\tau_j}(0))$, $i \geq j$ and $\Gamma_{i,i+1} \triangleq 1$. Because $\mathbf{b}_k(0)$ is continuous in the initial state, $\mathbf{c}^+\mathbf{b}_{\tau_i}$ converges to a positive constant by Assumptions 11.4 and 11.2. It is clear that $1 - a_{\tau_i}\mathbf{c}^+\mathbf{b}_{\tau_i}(0) > 0$ for a sufficiently large j, say $j \geq j_0$. Then for any $i > j$, $j \geq j_0$, it is true that $\Gamma_{i,j} = (1 - a_{\tau_i}\mathbf{c}^+\mathbf{b}_{\tau_i}(0))\Gamma_{i-1,j} \leq \exp(-ca_{\tau_i})\Gamma_{i-1,j}$ with some $c > 0$, where the inequality $1 - a \leq e^{-a}$ is applied.

It then follows that $\Gamma_{i,j} \leq c_0 \exp\left(-c \sum_{k=j}^{i} a_{\tau_k}\right)$, $j \geq j_0$ for some $c_0 > 0$, and because j_0 is a finite integer, it is clear that

$$|\Gamma_{i,j}| \leq |\Gamma_{i,j_0}||\Gamma_{j_0,j}| \leq c_0' \exp\left(-c \sum_{k=j}^{i} a_{\tau_k}\right), \quad \forall j. \tag{11.12}$$

Now from (11.11), we have

$$\delta u_{\tau_i}(0) = \Gamma_{i,0}\delta u_{\tau_0}(0) - \sum_{j=0}^{i} \Gamma_{i,j+1} a_{\tau_j} \varphi_{\tau_j - n_{\tau_j}}(0)$$

$$+ \sum_{j=0}^{i} \Gamma_{i,j+1} a_{\tau_j} v_{\tau_j - n_{\tau_j}}(1), \tag{11.13}$$

where the first term at the right-hand side of the last equation tends to zero as i goes to infinity, by the definition of τ_i, $\Gamma_{i,j}$ and (11.12). By Assumption 11.4, the recognition mechanism and the continuity of nonlinear functions, it is clear that $\varphi_{\tau_j - n_{\tau_j}}(0) \xrightarrow[j\to\infty]{} 0$. From Assumption 11.3 we have $\sum_{j=0}^{\infty} a_{\tau_j} v_{\tau_j - n_{\tau_j}}(1) < \infty$. Thus the last two terms of (11.13) tend to zero following similar steps to Lemma A.3.

Inductive Step. Assume that the convergence of $u_k(t)$ has been proved for $t = 0, 1, \ldots, s-1$, then from Lemma 11.1 we have $\delta x_k(s) \xrightarrow[k\to\infty]{} 0$ and therefore $\varphi_k(s) \xrightarrow[k\to\infty]{} 0$. Then following similar steps to the case $t = 0$, we are with no difficulty to conclude that $\delta u_k(s) \xrightarrow[k\to\infty]{} 0$. This completes the proof.

Proof of Theorem 11.2.

Proof Similar to the proof of Theorem 11.1, the mathematical induction is used due to the existence of nonlinearities. On the other hand, noticing (11.10), we have a recursion based on stopping times which is similar to (11.6). Thus for any given time, we will first show the convergence of a subsequence $\{u_{\tau_i-1}(t)\}$ and then extend this to the general sequence $\{u_k(t)\}$. There are two major differences between the proofs of Theorems 11.1 and 11.2: first, the tracking error $e_{\tau_i-n_{\tau_i}}(t+1)$ used in (11.10) is not generated by the $u_{\tau_i-1}(t)$, and second, the extension from the subsequence to the general input sequence of this theorem is nontrivial.

Initial Step. Consider the case of $t = 0$.

Subtracting both sides of (11.10) with $t = 0$ from $u_d(0)$ yields

$$\delta u_{\tau_{i+1}-1}(0)$$

$$=\delta u_{\tau_i-1}(0) - \left(\sum_{k=\tau_i-1}^{\tau_{i+1}-2} a_k\right) e_{\tau_i-n_{\tau_i}}(1)$$

$$=\delta u_{\tau_i-1}(0) - \left(\sum_{k=\tau_i-1}^{\tau_{i+1}-2} a_k\right) \mathbf{c}^+\mathbf{b}_{\tau_i-n_{\tau_i}}(0)\delta u_{\tau_i-n_{\tau_i}}(0)$$

$$- \left(\sum_{k=\tau_i-1}^{\tau_{i+1}-2} a_k\right) \varphi_{\tau_i-n_{\tau_i}}(0) + \left(\sum_{k=\tau_i-1}^{\tau_{i+1}-2} a_k\right) v_{\tau_i-n_{\tau_i}}(1).$$

By the definition of n_{τ_i} we know that $n_{\tau_i} \geq 1$, and that there is an iteration gap between the input signal and the tracking error information. However, we could rewrite the last equation as

$$\delta u_{\tau_{i+1}-1}(0)$$

$$= \left(1 - \left(\sum_{k=\tau_i-1}^{\tau_{i+1}-2} a_k\right) \mathbf{c}^+\mathbf{b}_{\tau_i-n_{\tau_i}}(0)\right) \delta u_{\tau_i-1}(0)$$

$$+ \left(\sum_{k=\tau_i-1}^{\tau_{i+1}-2} a_k\right) \mathbf{c}^+\mathbf{b}_{\tau_i-n_{\tau_i}}(0)\left(\delta u_{\tau_i-1}(0) - \delta u_{\tau_i-n_{\tau_i}}(0)\right)$$

$$- \left(\sum_{k=\tau_i-1}^{\tau_{i+1}-2} a_k\right) \varphi_{\tau_i-n_{\tau_i}}(0) + \left(\sum_{k=\tau_i-1}^{\tau_{i+1}-2} a_k\right) v_{\tau_i-n_{\tau_i}}(1). \quad (11.14)$$

Note that when $\tau_{i-1} < \tau_i - n_{\tau_i} \leq \tau_i - 1$, the updating from $\tau_i - n_{\tau_i}$ iteration to $\tau_i - 1$ iteration would follow (11.9) and thus

11.7 Proofs of Theorems

$$\delta u_{\tau_i-1}(0) - \delta u_{\tau_i-n_{\tau_i}}(0)$$

$$= \left(\sum_{k=\tau_i-n_{\tau_i}}^{\tau_i-2} a_k \right) e_{\tau_{i-1}-n_{\tau_{i-1}}}(1)$$

$$= \left(\sum_{k=\tau_i-n_{\tau_i}}^{\tau_i-2} a_k \right) \left(\mathbf{c}^+ \mathbf{b}_{\tau_{i-1}-n_{\tau_{i-1}}}(0) \delta u_{\tau_{i-1}-n_{\tau_{i-1}}}(0) \right.$$

$$\left. - \varphi_{\tau_{i-1}-n_{\tau_{i-1}}}(0) + v_{\tau_{i-1}-n_{\tau_{i-1}}}(1) \right). \tag{11.15}$$

It follows from (11.14) and (11.15) that

$$\delta u_{\tau_{i+1}-1}(0)$$

$$= \left(1 - \left(\sum_{k=\tau_i-1}^{\tau_{i+1}-2} a_k \right) \mathbf{c}^+ \mathbf{b}_{\tau_i-n_{\tau_i}}(0) \right) \delta u_{\tau_i-1}(0)$$

$$+ \left(\sum_{k=\tau_i-1}^{\tau_{i+1}-2} a_k \right) \mathbf{c}^+ \mathbf{b}_{\tau_i-n_{\tau_i}}(0) \left(\left(\sum_{k=\tau_i-n_{\tau_i}}^{\tau_i-2} a_k \right) \right.$$

$$\left. \times \mathbf{c}^+ \mathbf{b}_{\tau_{i-1}-n_{\tau_{i-1}}}(0) \delta u_{\tau_{i-1}-n_{\tau_{i-1}}}(0) \right)$$

$$+ \left(\sum_{k=\tau_i-1}^{\tau_{i+1}-2} a_k \right) \mathbf{c}^+ \mathbf{b}_{\tau_i-n_{\tau_i}}(0) \left(\left(\sum_{k=\tau_i-n_{\tau_i}}^{\tau_i-2} a_k \right) \right.$$

$$\left. \times \left(-\varphi_{\tau_{i-1}-n_{\tau_{i-1}}}(0) + v_{\tau_{i-1}-n_{\tau_{i-1}}}(1) \right) \right)$$

$$- \left(\sum_{k=\tau_i-1}^{\tau_{i+1}-2} a_k \right) \varphi_{\tau_i-n_{\tau_i}}(0) + \left(\sum_{k=\tau_i-1}^{\tau_{i+1}-2} a_k \right) v_{\tau_i-n_{\tau_i}}(1). \tag{11.16}$$

Let $\Psi_{i,j}$ be

$$\Psi_{i,j} \triangleq \left(1 - \left(\sum_{k=\tau_i-1}^{\tau_{i+1}-2} a_k \right) \mathbf{c}^+ \mathbf{b}_{\tau_i-n_{\tau_i}}(0) \right) \times \cdots$$

$$\times \left(1 - \left(\sum_{k=\tau_j-1}^{\tau_{j+1}-2} a_k \right) \mathbf{c}^+ \mathbf{b}_{\tau_j-n_{\tau_j}}(0) \right)$$

for $i \geq j$ and $\Psi_{i,i+1} = 1$.

Note that $\mathbf{b}_k(0)$ is continuous in the initial state and $\mathbf{c}^+\mathbf{b}_{\tau_i-n_{\tau_i}}$ converges to some positive constant as i goes to infinity by Assumptions 11.2 and 11.4. Given the boundedness of $\tau_{i+1} - \tau_i$, it is clear that $1 - \left(\sum_{k=\tau_i-1}^{\tau_{i+1}-2} a_k\right)\mathbf{c}^+\mathbf{b}_{\tau_i-n_{\tau_i}}(0) > 0$ for sufficiently large j, say $j \geq j_0$. Thus by steps similar to those of Theorem 11.1, we arrives at

$$|\Psi_{i,j}| \leq c_0 \exp\left(-c \sum_{k=\tau_j-1}^{\tau_{i+1}-2} a_k\right), \quad \forall i \geq j, \forall j > 0 \tag{11.17}$$

with proper c_0 and c.

For brevity of notations, we denote

$$\alpha_i \triangleq \mathbf{c}^+\mathbf{b}_{\tau_i-n_{\tau_i}}(0)\left(\left(\sum_{k=\tau_i-n_{\tau_i}}^{\tau_i-2} a_k\right)\mathbf{c}^+\mathbf{b}_{\tau_{i-1}-n_{\tau_{i-1}}}(0)\delta u_{\tau_{i-1}-n_{\tau_{i-1}}}(0)\right),$$

$$\beta_i \triangleq \mathbf{c}^+\mathbf{b}_{\tau_i-n_{\tau_i}}(0)\left(\sum_{k=\tau_i-n_{\tau_i}}^{\tau_i-2} a_k\right)\varphi_{\tau_{i-1}-n_{\tau_{i-1}}}(0) + \varphi_{\tau_i-n_{\tau_i}}(0),$$

$$\gamma_i \triangleq \mathbf{c}^+\mathbf{b}_{\tau_i-n_{\tau_i}}(0)\left(\sum_{k=\tau_i-n_{\tau_i}}^{\tau_i-2} a_k\right)v_{\tau_{i-1}-n_{\tau_{i-1}}}(1) + v_{\tau_i-n_{\tau_i}}(1).$$

Then from (11.16) we have

$$\delta u_{\tau_{i+1}-1}(0)$$

$$= \Psi_{i,1}\delta u_{\tau_1-1}(0) + \sum_{j=1}^{i} \Psi_{i,j+1}\left(\sum_{k=\tau_j-1}^{\tau_{j+1}-2} a_k\right)\alpha_j$$

$$- \sum_{j=1}^{i} \Psi_{i,j+1}\left(\sum_{k=\tau_j-1}^{\tau_{j+1}-2} a_k\right)\beta_j$$

$$+ \sum_{j=1}^{i} \Psi_{i,j+1}\left(\sum_{k=\tau_j-1}^{\tau_{j+1}-2} a_k\right)\gamma_j, \tag{11.18}$$

where the first term on the right-hand side tends to zero as i goes to infinity. By Assumption 11.4, we have $\beta_i \xrightarrow[i\to\infty]{} 0$. According to Assumption 11.5, $\sum_{k=\tau_i-1}^{\tau_{i+1}-2} a_k \xrightarrow[i\to\infty]{} 0$, $\sum_{i=1}^{\infty} \sum_{k=\tau_i-1}^{\tau_{i+1}-2} a_k = \sum_{k=1}^{\infty} a_k = \infty$, and

11.7 Proofs of Theorems

$$\sum_{i=1}^{\infty}\left(\sum_{k=\tau_i-1}^{\tau_{i+1}-2} a_k\right)^2 \leq \sum_{i=1}^{\infty} M\left(\sum_{k=\tau_i-1}^{\tau_{i+1}-2} a_k^2\right) = M\sum_{k=1}^{\infty} a_k^2 < \infty.$$

By following similar steps to those of Theorem 11.1 the last two terms on the right-hand side of (11.18) tend to zero as i goes to infinity. Then to prove the zero convergence of $\delta u_{\tau_i-1}(0)$, it suffices to show the zero convergence of the second term on the right-hand side of (11.18) as $i \to \infty$. It is obvious that $\alpha_i = O(a_{\tau_i})$ because of the boundedness of $\delta u_{\tau_{i-1}-n_{\tau_{i-1}}}(0)$ and $\mathbf{c}^+\mathbf{b}_{\tau_i-n_{\tau_i}}(0)$ and the fact that $a_{\tau_i-2} \leq \sum_{k=\tau_i-n_{\tau_i}}^{\tau_i-2} a_k \leq Ma_{\tau_i-n_{\tau_i}} \leq Ma_{\tau_{i-1}}$. This results in that $\alpha_i \xrightarrow[i\to\infty]{} 0$ and therefore the zero convergence of $\sum_{j=1}^{i} \Psi_{i,j+1}\left(\sum_{k=\tau_j-1}^{\tau_{j+1}-2} a_k\right)\alpha_j$ following similar steps of Lemma A.3 in Appendix or Theorem 11.1 above.

As a result, we have shown that $\delta u_{\tau_i-1}(0) \xrightarrow[i\to\infty]{} 0$. Next, let us extend it to $\delta u_k(0)$, $\forall \tau_i \leq k \leq \tau_{i+1} - 2$. From (11.9) it follows that

$$\delta u_k(0) = \delta u_{\tau_i-1}(0) - \left(\sum_{j=\tau_i-1}^{k-1} a_j\right) e_{\tau_i-n_{\tau_i}}(1)$$

$$= \delta u_{\tau_i-1}(0) - \left(\sum_{j=\tau_i-1}^{k-1} a_j\right)\left(\varphi_{\tau_i-n_{\tau_i}}(0) - v_{\tau_i-n_{\tau_i}}(1)\right)$$

$$- \left(\sum_{j=\tau_i-1}^{k-1} a_j\right) \mathbf{c}^+\mathbf{b}_{\tau_i-n_{\tau_i}}(0)\delta u_{\tau_i-n_{\tau_i}}(0)$$

$$= \left(1 - \left(\sum_{j=\tau_i-1}^{k-1} a_j\right)\mathbf{c}^+\mathbf{b}_{\tau_i-n_{\tau_i}}(0)\right)\delta u_{\tau_i-1}(0)$$

$$+ \left(\sum_{j=\tau_i-1}^{k-1} a_j\right)\mathbf{c}^+\mathbf{b}_{\tau_i-n_{\tau_i}}(0)\left(\delta u_{\tau_i-1}(0) - \delta u_{\tau_i-n_{\tau_i}}(0)\right)$$

$$- \left(\sum_{j=\tau_i-1}^{k-1} a_j\right)\left(\varphi_{\tau_i-n_{\tau_i}}(0) - v_{\tau_i-n_{\tau_i}}(1)\right),$$

$$\forall \tau_i \leq k \leq \tau_{i+1} - 2.$$

Then by techniques similar to those used for (11.14), zero-error convergence for general $\delta u_k(0)$ is proved.

Inductive Step. Assume the convergence of $u_k(t)$ has been proved for $t = 0, 1, \ldots, s-1$, then using Lemma 11.1 we have $\delta x_k(s) \xrightarrow[k \to \infty]{} 0$ and therefore $\varphi_k(s) \xrightarrow[k \to \infty]{} 0$. Following similar steps as in the case $t = 0$, we are with no difficulty to conclude that $\delta u_k(s) \xrightarrow[k \to \infty]{} 0$. This completes the proof.

11.8 Summary

This chapter addresses the ILC problem for stochastic nonlinear systems with random communication constraints including data dropouts, communication delays, and packet transmission disordering. These communication constraints are analyzed and a renewal mechanism is proposed to regulate the packets in the buffer. To design ILC update laws, a recognition mechanism is added to the controller for the selection of suitable packets. Two learning schemes are proposed: IUS and SUS. When no suitable new packet arrives, IUS retains the latest input whereas SUS continues to update with the latest tracking information. Both schemes are shown to converge to the optimal input in almost sure sense. For further research, it would be of great interest to consider ways to accelerate the proposed schemes. When the capacity of the buffer is larger than one-iteration storage, an important issue is to determine the optimal capacity of the buffer in relation to tracking performance and economy requirements. Moreover, the corresponding design and analysis of the learning algorithms remain to be conducted. In addition, the control signal may not change rapidly because of practical limitations; that is, any variation of the input should be bounded. Then, how to integrate this issue into the problem formulation and solve it become open problems. The results in this chapter are mainly based on [5].

References

1. Chen, Y.Q., Wen, C., Gong, Z., Sun, M.: An iterative learning controller with initial state learning. IEEE Trans. Autom. Control **44**(2), 371–376 (1999)
2. Sun, M., Wang, D.: Initial shift issues on discrete-time iterative learning control with system relative degree. IEEE Trans. Autom. Control **48**(1), 144–148 (2003)
3. Kosmatopoulos, E.B., Kouvelas, A.: Large scale nonlinear control system fine-tuning through learning. IEEE Trans. Neural Netw. **20**(6), 1009–1023 (2009)
4. Chen, H.F.: Stochastic Approximation and its Applications. Kluwer (2002)
5. Shen, D.: Data-driven learning control for stochastic nonlinear systems: multiple communication constraints and limited storage. IEEE Trans. Neural Netw. Learn. Syst. (2018). https://doi.org/10.1109/TNNLS.2017.2696040

Chapter 12
Random Iteration-Varying Lengths for Linear Systems

12.1 Problem Formulation

Consider the following linear time-varying system:

$$\begin{aligned} x_k(t+1) &= A_t x_k(t) + B_t u_k(t), \\ y_k(t) &= C_t x_k(t), \end{aligned} \quad (12.1)$$

where $x_k(t) \in \mathbf{R}^n$, $u_k(t) \in \mathbf{R}^p$, and $y_k(t) \in \mathbf{R}^q$ denote state, input, and output, respectively. $k = 0, 1, \ldots$ and $t = 0, 1, \ldots, N$ denote the iteration index and discrete time index, respectively, and N is the maximum of iteration lengths. A_t, B_t, and C_t are system matrices with appropriate dimensions. It is assumed that $C_{t+1} B_t$ is of full-column rank, implying that the relative degree is 1.

If the operation length of all iterations are identical, i.e., N, then the system model (12.1) could be lifted as follows:

$$Y_k = H U_k + Y_{k0}, \quad (12.2)$$

where

$$H = \begin{bmatrix} C_1 B_0 & 0 & \cdots & 0 \\ C_2 A_1 B_0 & C_2 B_1 & \cdots & 0 \\ \vdots & \vdots & \ddots & \vdots \\ C_N A_{N-1} \cdots A_1 B_0 & \cdots & \cdots & C_N B_{N-1} \end{bmatrix}, \quad (12.3)$$

and

$$\begin{aligned} Y_k &= \left[y_k^T(1), \ldots, y_k^T(N) \right]^T, \\ U_k &= \left[u_k^T(0), \ldots, u_k^T(N-1) \right]^T, \\ Y_{k0} &= \left[(C_1 A_0)^T, \ldots, (C_N A_{N-1} \cdots A_0)^T \right]^T x_k(0). \end{aligned}$$

Let $y_d(t)$, $t = 0, 1, \ldots, N$ be the desired trajectory. Assume that, for a realizable trajectory $y_d(t)$, there is a unique control input $u_d(t)$ such that

$$Y_d = HU_d + Y_{d0}, \tag{12.4}$$

where Y_d, U_d, and Y_{d0} are defined similar to Y_k, U_k, and Y_{k0}, respectively, by replacing $y_k(t)$, $u_k(t)$, and $x_k(0)$ with $y_d(t)$, $u_d(t)$, and $x_d(0)$. For expression clarity and making our idea clear, it is assumed that the identical initial condition holds, i.e., $Y_{k0} = Y_{d0}$.

Then, the control objective is to design ILC algorithm such that $Y_k \to Y_d$ as the iteration index k approaches to infinity if the iteration length is fixed as N. However, the actual iteration length may vary from iteration to iteration randomly. Besides, it is rational to have a lower bound of actual iteration lengths, named by $\overline{N}$, $\overline{N} < N$, such that the actual iteration length varies in $\{\overline{N}, \overline{N}+1, \ldots, N\}$ randomly. In other words, for the first $\overline{N}$ time instants one could always get the actual output, while for the left time instants the system output occurs randomly. Thus there are $N - \overline{N} + 1$ possible output trajectories. For the kth iteration, it ends at the N_kth time instant; that is, only the first N_k outputs are received, where $\overline{N} \leq N_k \leq N$. In order to make it clear, a cutting operator $\lfloor \cdot \rfloor_{N_k}$ is introduced to Y_k which means the last $N - N_k$ outputs of Y_k are removed, i.e., $\lfloor \cdot \rfloor_{N_k} : \mathbf{R}^{Nq} \to \mathbf{R}^{N_k q}$. Therefore, the control objective of this chapter is to design ILC update law such that $\lfloor Y_k \rfloor_{N_k} \to \lfloor Y_d \rfloor_{N_k}$ as k approaches to infinity. For notations concise, let $m = N - \overline{N} + 1$ in the following.

As one can see now, there are $N - \overline{N} + 1$ possible iteration lengths which occur randomly. In order to cope with these different scenarios, let the probability that the iteration length is of $\overline{N}, \overline{N}+1, \ldots, N$ be $p_1, p_2, \ldots, p_m$, respectively. That is, $\mathbb{P}(\mathscr{A}_{\overline{N}}) = p_1$, $\mathbb{P}(\mathscr{A}_{\overline{N}+1}) = p_2, \ldots, \mathbb{P}(\mathscr{A}_N) = p_m$, $\forall k$, where $\mathscr{A}_l$ denotes the event that the iteration length is l, i.e., $\lfloor Y_k \rfloor_{N_k} = \lfloor Y_k \rfloor_l$, $\overline{N} \leq l \leq N$. Obviously, $p_i > 0$, $1 \leq i \leq m$, and

$$p_1 + p_2 + \cdots + p_m = 1. \tag{12.5}$$

It should be pointed out that no probability distribution is assumed on p_i, thus the above expression on randomly varying iteration length is general.

Remark 12.1 In [1], the authors introduced a sequence of random variables satisfying Bernoulli distribution based on $p_1, \ldots, p_m$ to model the probability of the occurrence of the last $N - \overline{N}$ outputs. Then, the random variable was multiplied to the corresponding tracking error as a modified tracking error. The analysis objective was therefore transformed to show that the mathematical expectation of these $N - \overline{N}$ modified tracking error converges to zero. While in this chapter, we prove the convergence both in almost sure sense and mean square sense straightforward according to the iteration length probabilities $p_1, \ldots, p_m$ by direct calculations. Besides, it is worth pointing out that these probabilities are only used for the analysis, whereas the design condition of learning algorithm requires none prior information on these probabilities, which is more suitable for practical applications.

12.2 ILC Design

Notice that the iteration length cannot exceed the maximum length, thus only two cases of tracking error need to be considered. If the iteration length equals the maximum length, then the tracking error is a normal one with dimension Nq; while if the iteration length is shorter than the maximum length, then the tracking error at the absent time instants are missing, which therefore could not be used for input update. For the latter case, we could append zeros to the absent time instants so that the tracking error is again transformed to be a normal one with dimension Nq. In other words, when the kth actual output length is not up to the maximum, namely, $N_k < N$, then the tracking errors are defined as

$$e_k(t) = \begin{cases} y_d(t) - y_k(t), & 1 \leq t \leq N_k, \\ 0, & N_k < t \leq N. \end{cases} \quad (12.6)$$

Denote

$$E_k = \left[e_k^T(1), \ldots, e_k^T(N) \right]^T. \quad (12.7)$$

The control update law is thus defined as follows:

$$U_{k+1} = U_k + L E_k, \quad (12.8)$$

where L is a learning gain matrix to be designed later.

Noting that $E_k \neq Y_d - Y_k$ if $N_k < N$, we have to fill the gap by introducing the following matrix

$$M_{N_k} = \begin{bmatrix} I_{N_k} \otimes I_q & 0 \\ 0 & \mathbf{0}_{(N-N_k)} \otimes I_q \end{bmatrix}, \quad \overline{N} \leq N_k \leq N, \quad (12.9)$$

where I_l and $\mathbf{0}_l$ denote unit matrix and zero matrix with dimension of $l \times l$, respectively.

Then we have

$$E_k = M_{N_k}(Y_d - Y_k) = M_{N_k} H (U_d - U_k).$$

Now (12.8) leads to

$$\begin{aligned} U_{k+1} &= U_k + L E_k \\ &= U_k + L M_{N_k} H (U_d - U_k). \end{aligned}$$

Subtracting both sides of last equation from U_d, we have

$$U_d - U_{k+1} = U_d - U_k - LM_{N_k}H(U_d - U_k)$$
$$= (I - LM_{N_k}H)(U_d - U_k).$$

That is,

$$\Delta U_{k+1} = (I - LM_{N_k}H)\Delta U_k, \tag{12.10}$$

where $\Delta U_k \triangleq U_d - U_k$.

Notice that N_k is a random variable valued from $\{\overline{N}, \ldots, N\}$, therefore M_{N_k} is a random matrix, which further results in that $I - LM_{N_k}H$ is a random matrix. Thus one could introduce m binary random variables γ_i, $1 \leq i \leq m$ such that $\gamma_i \in \{0, 1\}$,

$$\gamma_1 + \gamma_2 + \cdots + \gamma_m = 1,$$

and

$$\mathbb{P}(\gamma_i = 1) = \mathbb{P}(\mathscr{A}_{\overline{N}-1+i}) = p_i, \quad 1 \leq i \leq m.$$

Then (12.10) could be reformulated as

$$\Delta U_{k+1} = [\gamma_1(I - LM_{\overline{N}}H) + \gamma_2(I - LM_{\overline{N}+1}H) + \cdots + \gamma_m(I - LM_N H)]\Delta U_k. \tag{12.11}$$

Consequently, the zero-convergence of the original update law (12.10) could be achieved by analyzing zero-convergence of (12.11).

Denote $\Gamma_i = I - LM_{\overline{N}-1+i}H$, $1 \leq i \leq m$. Then (12.11) could be simplified as

$$\Delta U_{k+1} = (\gamma_1 \Gamma_1 + \gamma_2 \Gamma_2 + \cdots + \gamma_m \Gamma_m)\Delta U_k. \tag{12.12}$$

It is easy to see that all γ_i are dependent, since whenever one of them values 1, then all the others have to value 0.

In order to give the design condition of learning gain matrix L, we first calculate the mean and covariance along the sample path.

Let $\mathscr{S} = \{\Gamma_i, 1 \leq i \leq m\}$, and denote

$$Z_k = X_k X_{k-1} \cdots X_1 X_0, \tag{12.13}$$

where X_k is a random matrix taking values in $\mathscr{S}$ with $\mathbb{P}(X_k = \Gamma_i) = \mathbb{P}(\gamma_i = 1) = p_i$, $1 \leq i \leq m$, $\forall k$. Now, we have the following equation from (12.12)

$$\Delta U_{k+1} = Z_k \Delta U_0. \tag{12.14}$$

The following two lemmas are given for further analysis.

Lemma 12.1 *Let $\mathscr{S}^k = \{Z_k : \text{taken over all sample paths}\}$, then the mean of the $\mathscr{S}$, denoted by K_k, is given recursively by*

12.2 ILC Design

$$K_k = \left(\sum_{i=1}^{m} p_i \Gamma_i\right) K_{k-1}. \tag{12.15}$$

Proof Let $\mathscr{S}_i^k = \{Z_k \in \mathscr{S}^k : X_k = \Gamma_i\}$, $i = 1, 2, \ldots, m$. It is obvious that $\mathscr{S}^k$ is the disjoint union of $\mathscr{S}_i^k$.

Let

$$K_k = \sum_{Z_k \in \mathscr{S}^k} \mathbb{P}(Z_k) Z_k. \tag{12.16}$$

By the independence of X_j, one can decompose the above sum

$$\begin{aligned}
K_k &= \sum_{Z_k \in \mathscr{S}^k} \mathbb{P}(Z_k) Z_k \\
&= \sum_{Z_{k-1} \in \mathscr{S}^{k-1}} \sum_{i=1}^{m} \mathbb{P}(X_k = \Gamma_i) \mathbb{P}(Z_{k-1}) \Gamma_i Z_{k-1} \\
&= \sum_{Z_{k-1} \in \mathscr{S}^{k-1}} \sum_{i=1}^{m} \mathbb{P}(\gamma_i = 1) \mathbb{P}(Z_{k-1}) \Gamma_i Z_{k-1} \\
&= \sum_{i=1}^{m} \mathbb{P}(\gamma_i = 1) \Gamma_i \sum_{Z_{k-1} \in \mathscr{S}^{k-1}} \mathbb{P}(Z_{k-1}) Z_{k-1} \\
&= \sum_{i=1}^{m} \mathbb{P}(\gamma_i = 1) \Gamma_i K_{k-1} \\
&= \sum_{i=1}^{m} p_i \Gamma_i K_{k-1} = \left(\sum_{i=1}^{m} p_i \Gamma_i\right) K_{k-1}.
\end{aligned}$$

Thus the proof is completed.

Lemma 12.2 *Let $\mathscr{S}^k = \{Z_k : \text{taken over all sample paths}\}$, then the covariance of the $\mathscr{S}$, denoted by V_k, is given by*

$$V_k = F_k - K_k K_k^T, \tag{12.17}$$

where F_k is generated recursively as

$$F_k = \sum_{i=1}^{m} p_i \Gamma_i F_{k-1} \Gamma_i^T. \tag{12.18}$$

Proof The covariance is calculated as

$$V_k = \sum_{Z_k \in \mathscr{S}^k} \mathbb{P}(Z_k)(Z_k - K_k)(Z_k - K_k)^T.$$

Then by decomposition, it leads to the following derivation

$$V_k = \sum_{Z_{k-1}\in \mathscr{S}^{k-1}} \sum_{i=1}^{m} \left[\mathbb{P}(X_k = \Gamma_i)\mathbb{P}(Z_{k-1}) \times (\Gamma_i Z_{k-1} - K_k)(\Gamma_i Z_{k-1} - K_k)^T\right]$$

$$= \sum_{Z_{k-1}\in \mathscr{S}^{k-1}} \sum_{i=1}^{m} \left[\mathbb{P}(\gamma_i = 1)\mathbb{P}(Z_{k-1}) \times (\Gamma_i Z_{k-1} - K_k)(\Gamma_i Z_{k-1} - K_k)^T\right]$$

$$= \sum_{i=1}^{m} p_i \Big[\sum_{Z_{k-1}\in \mathscr{S}^{k-1}} \mathbb{P}(Z_{k-1})\Gamma_i Z_{k-1} Z_{k-1}^T \Gamma_i^T - \sum_{Z_{k-1}\in \mathscr{S}^{k-1}} \mathbb{P}(Z_{k-1}) K_k Z_{k-1}^T \Gamma_i^T$$

$$- \sum_{Z_{k-1}\in \mathscr{S}^{k-1}} \mathbb{P}(Z_{k-1})\Gamma_i Z_{k-1} K_k^T + \sum_{Z_{k-1}\in \mathscr{S}^{k-1}} \mathbb{P}(Z_{k-1}) K_k K_k^T \Big]$$

$$= \sum_{i=1}^{m} p_i \left[\sum_{Z_{k-1}\in \mathscr{S}^{k-1}} \mathbb{P}(Z_{k-1})\Gamma_i Z_{k-1} Z_{k-1}^T \Gamma_i^T - K_k K_{k-1}^T \Gamma_i^T - \Gamma_i K_{k-1} K_k^T + K_k K_k^T \right].$$

From Lemma 12.1 it is noticed that

$$\sum_{i=1}^{m} p_i K_k K_{k-1}^T \Gamma_i^T = K_k K_k^T,$$

$$\sum_{i=1}^{m} p_i \Gamma_i K_{k-1} K_k^T = K_k K_k^T.$$

Thus we have

$$V_k = \sum_{i=1}^{m} p_i \Gamma_i \left(\sum_{Z_{k-1}\in \mathscr{S}^{k-1}} \mathbb{P}(Z_{k-1}) Z_{k-1} Z_{k-1}^T \right) \Gamma_i^T - K_k K_k^T.$$

On the other hand

$$V_k = \mathbb{E}(Z_k - K_k)(Z_k - K_k)^T$$
$$= \mathbb{E} Z_k Z_k^T - K_k K_k^T$$
$$= \sum_{Z_k \in \mathscr{S}^k} \mathbb{P}(Z_k) Z_k Z_k^T - K_k K_k^T.$$

Let $F_k = \sum_{Z_k \in \mathscr{S}^k} \mathbb{P}(Z_k) Z_k Z_k^T$, then it is obvious that

$$F_k = \sum_{i=1}^{m} p_i \Gamma_i F_{k-1} \Gamma_k^T,$$

by combing the last two expressions of V_k. This completes the proof.

12.2 ILC Design

Now return to the iterative equation (12.12) and design the learning gain matrix L. Notice that H is a block lower triangular matrix, and M_i is a block diagonal matrix, $\overline{N} \leq i \leq N$. Thus, there is a large degree of freedom on the design of learning gain L. As a matter of fact, L could be partitioned as $L = [L_{i,j}]$, $1 \leq i, j \leq N$, where $L_{i,j}$ is a submatrix of $p \times q$. Two types of L are designed as follows:

- *Arimoto-like gain*: The diagonal blocks of L, i.e., $L_{i,i}$, $1 \leq i \leq N$, are valued, while the other blocks are set 0.
- *Causal gain*: The blocks in the lower triangular part of L, i.e., $L_{i,j}$, $i \geq j$, are valued, while the other blocks are set 0.

No matter which type of L mentioned above is adopted, it is easy to find that the coupled matrix $LM_{N_k}H$ is still a block lower triangular matrix whose diagonal blocks are $L_{t,t}C_tB_{t-1}$, $1 \leq t \leq N_k$ or 0, $N_k + 1 \leq t \leq N$. Thus one could simply design $L_{t,t}$, satisfying

$$0 < I - L_{t,t}C_tB_{t-1} < I. \tag{12.19}$$

Remark 12.2 If Arimoto-like gain is selected, one could find that the update law (12.8) could be formulated on each time instant as $u_{k+1}(t) = u_k(t) + L_{t+1,t+1}e_k(t+1)$, which further reduces the computational burden brought by high-dimension of the lifted model (12.2). However, on the other hand, the causal gain may offer more flexibilities for us, since no condition is required on the block $L_{i,j}$, $i > j$.

Remark 12.3 Different from (12.19), the condition in [1] is that $\sup \|I - p(t)LCB\| \leq \theta$ where $0 \leq \theta < 1$ and $p(t)$ is the occurrence probability of output at t. Thus, it is required in [1] that the probability of each iteration length is known prior. In this chapter, the requirement of L only depends on the input–output coupling matrix CB and thus no prior information on probabilities of randomly varying iteration length is needed, which therefore is more suitable for implementation.

12.3 Strong Convergence Properties

Based on Lemma 12.1, the following theorem establishes the convergence in mathematical expectation sense:

Theorem 12.1 *Consider system (12.2) with randomly iteration-varying length and use control update law (12.8). The mathematical expectation of tracking error E_k, i.e., $\mathbb{E}E_k$, converges to zero if the learning gain L satisfies (12.19).*

Proof By (12.14) we have
$$\mathbb{E}\Delta U_{k+1} = K_k \mathbb{E}\Delta U_0.$$

Then by recurrence of the mean K_k, i.e., (12.15), it is obvious to have

$$\mathbb{E}\Delta U_k = \left(\sum_{i=1}^m p_i \Gamma_i\right)^k \mathbb{E}\Delta U_0.$$

Thus it is sufficient to show that $\rho\left(\sum_{i=1}^m p_i \Gamma_i\right) < 1$. By (12.19), we have the consequence that each eigenvalue of $I - LM_{N_K}H$, denoted as $\lambda_j(I - LM_{N_K}H)$, satisfies

$$0 < \lambda_j(I - LM_{N_K}H) \leq 1,$$

$1 \leq j \leq Np$, $\overline{N} \leq N_k \leq N$. An eigenvalue $\lambda_j(I - LM_{N_K}H)$ is equal to 1 if and only if there are some outputs missed, i.e., $N_k < N$.

Note that Γ_i is a block lower triangular matrix, thus the eigenvalues actually are a collection of eigenvalues of all the diagonal blocks. Take the lth diagonal block from top to bottom of each alternative matrix Γ_i into account. For the case $1 \leq l \leq \overline{N}$, it is observed that all the eigenvalues of the lth diagonal block are positive and less than 1. While for the case $\overline{N} + 1 \leq l \leq N$, all the eigenvalues of the lth diagonal block are positive and not larger than 1. Meanwhile, not all eigenvalues of the lth diagonal block are equal to 1 because of the existence of $\Gamma_m = I - LM_N H = I - LH$, whose eigenvalues are all less than 1, $\overline{N} + 1 \leq l \leq N$. By noticing that $\sum_{i=1}^m p_i = 1$, it is obvious that, for any $1 \leq j \leq Np$,

$$0 < \sum_{i=1}^m p_i \lambda_j(\Gamma_i) < 1. \tag{12.20}$$

Noting that $\mathbb{E}E_k = \mathbb{E}\lfloor H\Delta U_k\rfloor_{N_k}$, the proof is completed.

Remark 12.4 Theorem 12.1 presents the convergence property in mathematical expectation sense. In other words, the expectation of tracking error converges to zero for the conventional P-type law (12.8). This kind of convergence is also obtained in [1], where an iteration average operator is introduced to cope with the randomness of iteration length. However, we take a different analytical approach.

The following theorem shows the almost sure convergence property:

Theorem 12.2 *Consider system (12.2) with randomly iteration-varying length and use control update law (12.8). The tracking error E_k converges to zero almost surely if the learning gain L satisfies (12.19).*

Proof Concerning (12.20), for the 2-norm $\|\cdot\|$, we have

$$0 < \sum_{i=1}^m p_i \|\Gamma_i\| < 1. \tag{12.21}$$

Thus one can find a constant $0 < \delta < 1$ such that

$$0 < \sum_{i=1}^m p_i \|\Gamma_i\| < \delta, \tag{12.22}$$

12.3 Strong Convergence Properties

since the number m is limited, denoting the number of different possible formulation of Γ_i and thus the above is a definite summation.

Noticing that the iteration length varies independently from iteration to iteration and N_k possesses identical distribution with respect to k, it follows from (12.10), (12.11), and (12.12) that

$$\begin{aligned}\mathbb{E}\|\Delta U_k\| &= \mathbb{E}\|Z_{k-1}\|\mathbb{E}\|\Delta U_0\| \\ &= \mathbb{E}\|X_{k-1}\cdots X_1 X_0\|\mathbb{E}\|\Delta U_0\| \\ &= \mathbb{E}\|X_{k-1}\|\mathbb{E}\|X_{k-2}\|\cdots\mathbb{E}\|X_0\|\mathbb{E}\|\Delta U_0\| \\ &= (\mathbb{E}\|X_{k-1}\|)^k\,\mathbb{E}\|\Delta U_0\|.\end{aligned}$$

On the other hand

$$\begin{aligned}\mathbb{E}\|X_{k-1}\| &= \mathbb{E}\|\gamma_1\Gamma_1 + \gamma_2\Gamma_2 + \cdots + \gamma_m\Gamma_m\| \\ &= \sum_{i=1}^{m}\mathbb{P}(\gamma_i = 1)\|\gamma_1\Gamma_1 + \gamma_2\Gamma_2 + \cdots + \gamma_m\Gamma_m\| \\ &= \sum_{i=1}^{m} p_i\|\Gamma_i\|.\end{aligned}$$

By using (12.22) it leads to

$$\begin{aligned}\sum_{k=1}^{\infty}\mathbb{E}\|\Delta U_k\| &= \sum_{k=1}^{\infty}(\mathbb{E}\|X_{k-1}\|)^k\,\mathbb{E}\|\Delta U_0\| \\ &= \sum_{k=1}^{\infty}\left(\sum_{i=1}^{m} p_i\|\Gamma_i\|\right)^k \mathbb{E}\|\Delta U_0\| \\ &< \sum_{k=1}^{\infty}\delta^k\mathbb{E}\|\Delta U_0\| < \infty.\end{aligned}$$

Then by Markov inequality, for any $\varepsilon > 0$ we have

$$\sum_{k=1}^{\infty}\mathbb{P}(\|\Delta U_k\| > \varepsilon) \leq \sum_{k=1}^{\infty}\frac{\mathbb{E}\|\Delta U_k\|}{\varepsilon} < \infty.$$

Therefore, we have $\mathbb{P}(\|\Delta U_k\| > \varepsilon, i.o.) = 0$ by Borel–Cantelli lemma, $\forall \varepsilon > 0$, and then it leads to $\mathbb{P}(\lim_{k\to\infty}\|\Delta U_k\| = 0) = 1$. That is, ΔU_k converges to zero almost surely. Noting that $\|E_k\| = \|\lfloor H\Delta U_k\rfloor_{N_k}\| \leq \|H\Delta U_k\|$, the proof is completed.

To show the mean square convergence, it is sufficient to show $\mathbb{E}\Delta U_k \Delta U_k^T \to 0$. That is, $F_k \to 0$. It is first noted that the matrix F_k recursively defined in (12.18) is positive definite. Then by the recurrence (12.18), we have the following theorem.

Theorem 12.3 *Consider system (12.2) with randomly iteration-varying length and use control update law (12.8). The tracking error E_k converges to zero in mean square sense if the learning gain L satisfies (12.19).*

Proof Following similar steps of the proof of Theorem 12.2, there is a suitable constant $0 < \eta < 1$ such that

$$0 < \sum_{i=1}^{m} p_i \|\Gamma_i\|^2 < \eta.$$

Then we have

$$\|F_k\| = \left\|\sum_{i=1}^{m} p_i \Gamma_i F_{k-1} \Gamma_i\right\|$$

$$\leq \sum_{i=1}^{m} p_i \|\Gamma_i F_{k-1} \Gamma_i\|$$

$$\leq \sum_{i=1}^{m} p_i \|\Gamma_i\|^2 \|F_{k-1}\|$$

$$= \left(\sum_{i=1}^{m} p_i \|\Gamma_i\|^2\right) \|F_{k-1}\|$$

$$< \eta \|F_{k-1}\|.$$

Thus, the exponential zero-convergence of F_k is established and hence $F_k \to 0$, which means that $\Delta U_k \to 0$ in mean square sense. Similarly, by noting that $\|E_k\| \leq \|H \Delta U_k\|$, the proof is completed.

Remark 12.5 Generally speaking, convergence in almost sure sense and in mean square sense cannot be implied by each other. Thus, it is hard to build convergence in these two senses meanwhile. The inherent reason that we could establish Theorems 12.2 and 12.3 for the same update law (12.8) is as follows. The convergence speed in mean square sense actually is exponential, as could be concluded from the proof of Theorem 12.3. Therefore, by direct calculations, it is easy to find that $\sum_{k=0}^{\infty} \text{Var}(\Delta U_k) < \infty$. Then, by Chebyshevs inequality and Borel–Cantelli lemma in probability theory, the almost sure convergence could be established.

12.4 Illustrative Simulations

In order to show the effectiveness and robustness of the conventional P-type ILC algorithm, a time-varying system is given as follows:

12.4 Illustrative Simulations

$$x_k(t+1) = \begin{pmatrix} 0.2\exp(-t/100) & -0.6 & 0 \\ 0 & 0.50\sin(t) & 0 \\ 0 & 0 & 0.7 \end{pmatrix} x_k(t)$$

$$+ \begin{pmatrix} 0 \\ 0.3\sin(t) \\ 1.00 \end{pmatrix} u_k(t),$$

$$y_k(t) = (0 \ 0.1 \ 1.00 + 0.1\cos(t))x_k(t).$$

The initial state is set as $x_k(0) = [0\ 0\ 0]^T$. Let the desired trajectory be $y_d(t) = \sin(2\pi t/50) + \sin(2\pi t/5)$. The maximum of iteration length is $N = 50$. Without loss of generality, the input of the initial iteration is simply set to zero, i.e., $u_0(t) = 0$, $0 \leq t \leq N$.

The iteration length varies from 30 to 50 satisfying discrete uniform distribution, which is just a simple case for illustration. The probability distribution is not required in our algorithm.

It is easy to find $C_{t+1}B_t = 1 + 0.03\sin(t) + 0.1\cos(t)$. The Arimoto-like gain is selected and $L_{t,t} = 0.5, \forall t$, therefore (12.19) is obviously valid. One could see from Fig. 12.1 that the output for the 5th iteration has small deviations from the desired one, while for the 15th iteration the output almost overlaps the desired trajectory. The convergence performance could also be further verified in Fig. 12.2, where the tracking error for these three iterations are displayed. One could see that the errors for the 15th iteration are almost zero. It is noted that the lengths of the trajectories

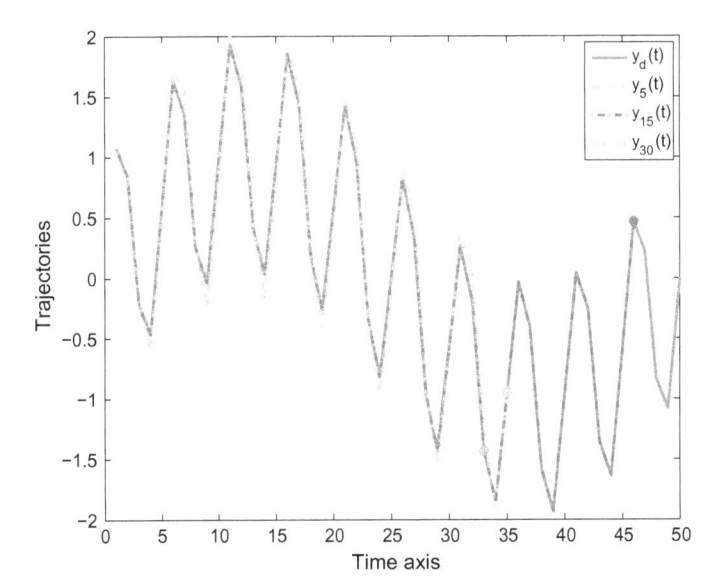

Fig. 12.1 The desired trajectory and tracking profiles for the 5th, 15th, and 30th iterations

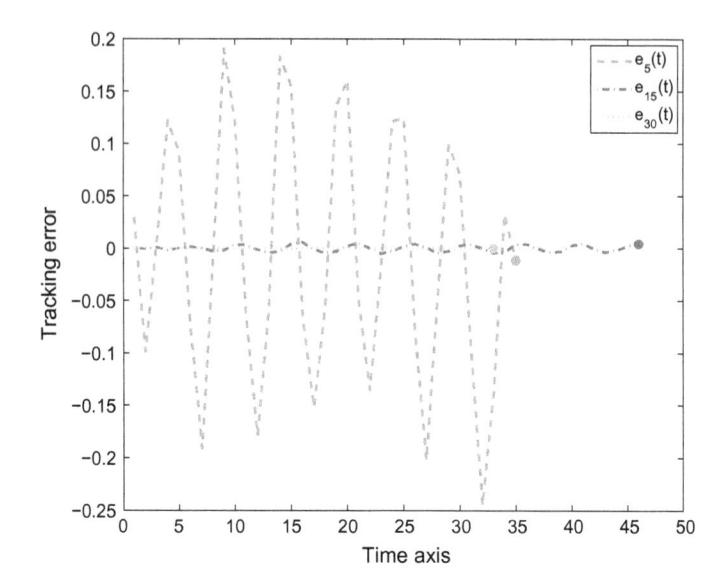

Fig. 12.2 Tracking error profiles for the 5th, 15th, and 30th iterations

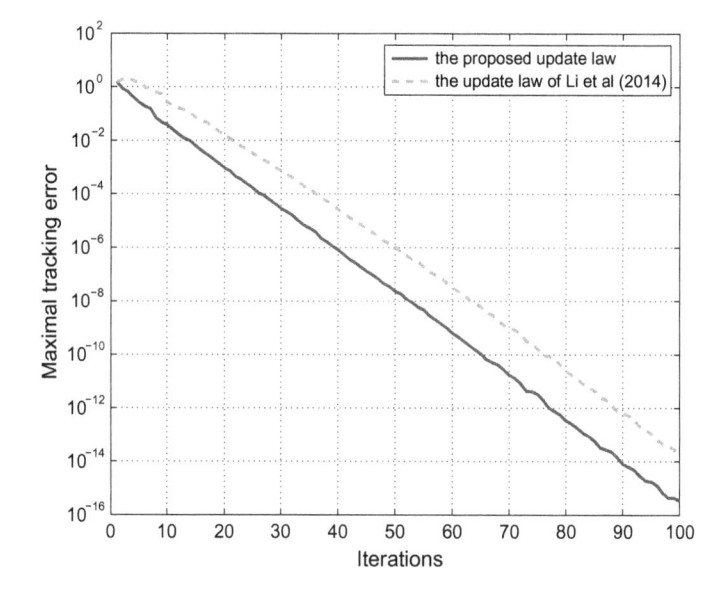

Fig. 12.3 Maximal tracking error $\max_{1\leq t\leq N} \|e_k(t)\|$ along iterations

for the 5th, 15th, and 30th iteration are less than 50. This demonstrates the fact that the iteration length varies from iteration to iteration.

One may be interested in the convergence speed of the proposed P-type algorithm under randomly varying iteration lengths. This could be partially revealed from the solid blue line in Fig. 12.3, where the maximal error is defined by $\max_{1 \leq t \leq N} \|e_k(t)\|$ for each iteration. As one could see, the maximal error decreases quickly as the iteration number increases. For a comparison, we also simulate the algorithm proposed in [1] and plot its maximal tracking error as the dashed red line. It is seen that the conventional P-type algorithm has a faster speed than the high-order ILC of [1]. There are few explicit results on comparison of convergence speed between high-order ILC and the conventional P-type [2]. Here, the reason that the P-type algorithm converges faster than the one in [1] might be that the conventional P-type has a quick response to large errors while the high-order scheme in [1] reduces the influence of large errors.

12.5 Summary

The tracking performance of P-type ILC for linear systems with randomly iteration-varying lengths is discussed in this chapter. The probability properties along sample path are first calculated. Then, the zero-convergence both in almost sure sense and mean square sense is established, as long as the probability of full-length iteration is not zero. The sufficient condition on the design of learning gain matrix is also clarified. The results in this chapter are mainly based on [3].

References

1. Li, X., Xu, J.-X., Huang, D.: An iterative learning control approach for linear systems with randomly varying trial lengths. IEEE Trans. Autom. Control **59**(7), 1954–1960 (2014)
2. Schmid, R.: Comments on "Robust optimal design and convergence properties analysis of iterative learning control approaches" and "On the P-type and Newton-type ILC schemes for dynamic systems with non-affine input factors". Automatica **43**(9), 1666–1669 (2007)
3. Shen, D., Zhang, W., Wang, Y., Chien, C.-J.: On almost sure and mean square convergence of P-type ilc under randomly varying iteration lengths. Automatica **63**(1), 359–365 (2016)

Chapter 13
Random Iteration-Varying Lengths for Nonlinear Systems

13.1 Problem Formulation

Consider the following discrete-time affine nonlinear system:

$$\begin{aligned} x_k(t+1) &= f(x_k(t)) + Bu_k(t), \\ y_k(t) &= Cx_k(t), \end{aligned} \quad (13.1)$$

where $k = 0, 1, \ldots$ denotes iteration index, t is time instant, $t \in \{0, 1, \ldots, N_d\}$, and N_d is the expected iteration length. $x_k(t) \in \mathbf{R}^n$, $u_k(t) \in \mathbf{R}^p$, and $y_k(t) \in \mathbf{R}^q$ denote state, input, and output, respectively. f is the nonlinear function. C and B are matrices with appropriate dimensions. Without loss of generality, it is assumed that CB is of full-column rank.

Remark 13.1 Matrices B and C are assumed time-invariant in system (13.1) to make the expressions concise. They can be extended to the time-varying case, $B(t)$ and $C(t)$, and/or state dependent case, $B(x(t))$, without making any further effort (see the analysis details below). Moreover, it will be shown that the convergence condition is independent of $f(\cdot)$ in the following. This is the major advantage of ILC; that is, ILC focuses on the convergence property along the iteration axis and requires little system information. In addition, it is evident the nonlinear function could be time-varying.

Let $y_d(t)$, $t \in \{0, 1, \ldots, N_d\}$ be the desired trajectory. $y_d(t)$ is assumed to be realizable; that is, there is a suitable initial state $x_d(0)$ and unique input $u_d(t)$ such that

$$\begin{aligned} x_d(t+1) &= f(x_d(t)) + Bu_d(t), \\ y_d(t) &= Cx_d(t). \end{aligned} \quad (13.2)$$

The following assumptions are required for the technical analysis:

Assumption 13.1 The nonlinear function $f(\cdot) : \mathbf{R}^n \to \mathbf{R}^n$ satisfies globally Lipschitz condition; that is, $\forall x_1, x_2 \in \mathbf{R}^n$,

$$\|f(x_1) - f(x_2)\| \le k_f \|x_1 - x_2\|, \tag{13.3}$$

where $k_f > 0$ is the Lipschitz constant.

The globally Lipschitz condition on nonlinear function is somewhat strong, although it is common in the ILC field for nonlinear systems. However, it should be pointed out that this assumption is imposed to facilitate the convergence derivations using the λ-norm technique. With more efforts, the assumption could be extended to locally Lipschitz case or continuous case (cf. Chaps. 3 and 4).

Assumption 13.2 The identical initialization condition is fulfilled, i.e., $x_k(0) = x_d(0)$, $\forall k$.

The initial state may not be reset precisely every iteration in practical applications, but the bias is usually bounded. Thus, one would relax Assumption 13.2 to the following one.

Assumption 13.3 The initial state could shift from $x_d(0)$ but should be bounded, i.e., $\|x_d(0) - x_k(0)\| \le \varepsilon$ where ε is a positive constant.

Let N_d denote the expected length. The actual length, N_k, varies in different iterations randomly. Thus, two cases need to be taken into account, i.e., $N_k < N_d$ and $N_k \ge N_d$. For the latter case, it is observed that only the data at the first N_d time instants is used for input updating. As a consequence, without loss of any generality, one could regard the latter case as $N_k = N_d$. From another point of view, one could regard N_d as the maximum length of actual lengths. For the former case, the outputs at the time instant $N_k + 1, \ldots, N_d$ are missing, and therefore, they are not available for updating. In other words, only input signals for the former N_k time instants are updated.

The control objective of this chapter is to design ILC algorithm to track the desired trajectory y_d, $t \in \{0, 1, \ldots, N_d\}$, based on the available output $y_k(t)$, $t \in \{0, 1, \ldots, N_k\}$, $N_k \le N_d$, such that the tracking error $e_k(t)$, $\forall t$ converges to zero with probability one as the iteration number k goes to infinity.

The following lemma is needed for the following analysis:

Lemma 13.1 *Let η be a Bernoulli binary random variable with $\mathbb{P}(\eta = 1) = \bar{\eta}$ and $\mathbb{P}(\eta = 0) = 1 - \bar{\eta}$. M is a positive-definite matrix. Then the equality $\mathbb{E}\|I - \eta M\| = \|I - \bar{\eta} M\|$ holds if and only if one of the following conditions is satisfied: (1) $\bar{\eta} = 0$; (2) $\bar{\eta} = 1$; and (3) $0 < \bar{\eta} < 1$ and $0 < M \le I$.*

Proof Let us first prove the sufficiency. It is easy to see that if $\bar{\eta} = 0$ or $\bar{\eta} = 1$, which means $\eta \equiv 0$ and $\eta \equiv 1$, respectively, then the equation $\mathbb{E}\|I - \eta M\| = \|I - \bar{\eta} M\|$ is valid. Moreover, the equality also holds obviously if $M = I$. Thus, it is sufficient to show the equation for the case $0 < \bar{\eta} < 1$ and $0 < M < I$. By the definition of mathematical expectation for discrete random variables, we have

13.1 Problem Formulation

$$\mathbb{E}\|I - \eta M\|$$
$$= \|I - 0 \cdot M\| \cdot \mathbb{P}(\eta = 0) + \|I - 1 \cdot M\| \cdot \mathbb{P}(\eta = 1)$$
$$= (1 - \bar{\eta}) + \bar{\eta}\|I - M\|$$
$$= 1 + \bar{\eta}(\|I - M\| - 1).$$

Noticing that M is a positive-definite matrix and $0 < M < I$, $I - M$ is a positive-definite matrix. Moreover, for a positive-definite matrix, the Euclidean norm is equal to its maximal eigenvalue, i.e., $\|I - M\| = \sigma_{\max}(I - M)$, and therefore, $\|I - M\| = 1 - \sigma_{\min}(M)$. This further leads to

$$\mathbb{E}\|I - \eta M\| = 1 - \bar{\eta}\sigma_{\min}(M).$$

On the other hand, noting $0 < \bar{\eta} < 1$,

$$\|I - \bar{\eta}M\| = \sigma_{\max}(I - \bar{\eta}M)$$
$$= 1 - \sigma_{\min}(\bar{\eta}M)$$
$$= 1 - \bar{\eta}\sigma_{\min}(M).$$

Next, for the necessity, it suffices to show that the equality $\mathbb{E}\|I - \eta M\| = \|I - \bar{\eta}M\|$ is not valid if $M > I$ and $0 < \bar{\eta} < 1$. In this case, it is easy to find

$$\mathbb{E}\|I - \eta M\| = 1 + \bar{\eta}(\|I - M\| - 1)$$
$$= 1 + \bar{\eta}(\sigma_{\max}(M - I) - 1)$$
$$= 1 + \bar{\eta}\sigma_{\max}(M) - 2\bar{\eta},$$

while the norm $\|I - \bar{\eta}M\|$ is complex as three cases should be discussed respectively.
(a) If $I - \bar{\eta}M$ is negative definite, i.e., $I - \bar{\eta}M < 0$, then $\|I - \bar{\eta}M\| = \bar{\eta}\sigma_{\max}(M) - 1$;
(b) If $I - \bar{\eta}M$ is positive definite, i.e., $I - \bar{\eta}M > 0$, then $\|I - \bar{\eta}M\| = 1 - \bar{\eta}\sigma_{\min}(M)$;
(c) If $I - \bar{\eta}M$ is indefinite, then $\|I - \bar{\eta}M\| = \max\{\bar{\eta}\sigma_{\max}(M) - 1, 1 - \bar{\eta}\sigma_{\min}(M)\}$.

Thus it is sufficient to verify that $\mathbb{E}\|I - \eta M\|$ equals neither $\bar{\eta}\sigma_{\max}(M) - 1$ nor $1 - \bar{\eta}\sigma_{\min}(M)$. Suppose $\mathbb{E}\|I - \eta M\| = \bar{\eta}\sigma_{\max}(M) - 1$, then we have $1 + \bar{\eta}\sigma_{\max}(M) - 2\bar{\eta} = \bar{\eta}\sigma_{\max}(M) - 1$, which means $\bar{\eta} = 1$, and this contradicts with $0 < \bar{\eta} < 1$. Otherwise, suppose $\mathbb{E}\|I - \eta M\| = 1 - \bar{\eta}\sigma_{\min}(M)$, then we have $1 + \bar{\eta}\sigma_{\max}(M) - 2\bar{\eta} = 1 - \bar{\eta}\sigma_{\min}(M)$, which means $\sigma_{\max}(M) + \sigma_{\min}(M) = 2$, and this contradicts with $M > I$.

The proof is thus completed.

13.2 ILC Design

In this chapter, the minimum length is denoted by N_m. Then, the operation length varies among the discrete integer set $\{N_m, \ldots, N_d\}$; that is, the outputs at time instants

$t = 0, 1, \ldots, N_m$ are always available for input updating, while the availability of outputs at time instants $t = N_m + 1, \ldots, N_d$ are random.

To describe the randomness of iteration length, we denote the probability of the occurrence of the output at time instant t by $p(t)$. Then it is found from the above explanations that $p(t) = 1$, $0 \leq t \leq N_m$, and $0 < p(t) < 1$, $N_m + 1 \leq t \leq N_d$. Moreover, when the output at time instant t_0 is available in an iteration, the outputs at any time instant t where $t < t_0$ are definitely available in the same iteration. This further implies that $p(N_m) > p(N_m + 1) > \cdots > p(N_d)$. It is worth pointing out that the probability is defined on time instant directly instead of iteration length.

Note that the kth iteration length is denoted by N_k, which is a random variable valued in $\{N_m, \ldots, N_d\}$. Let $\mathscr{A}_{N_k}$ be the event that the kth iteration length is N_k. Moreover, when an iteration length is N_k, it means the outputs at time $0 \leq t \leq N_k$ are available while the outputs at time $N_k + 1 \leq t \leq N_d$ are missing. Therefore, the probability of the kth iteration length being N_k is calculated as $\mathbb{P}(\mathscr{A}_{N_k}) = p(N_k) - p(N_k+1)$. As a result, $\sum_{t=N_m}^{N_d} \mathbb{P}(\mathscr{A}_t) = 1$.

Remark 13.2 In [1], the probability of random iteration length is first given and then the probability of the output occurrence at each time instant is calculated. In this chapter, the calculation order is exchanged; that is, the probability of the output occurrence at each time instant is first given and then calculate the probability of random iteration length. However, the internally logical relationships are identical.

As long as $N_k < N_d$, the actual output information is not complete. That is, the data of the former N_k time instants is only used to calculate tracking error for input updating. While for the left time instants, the input updating has to be suspended until the corresponding output information is available. In this case, we simply set the tracking error to be zero because none knowledge is obtained. In other words, a modified tracking error is defined as follows:

$$e_k^*(t) = \begin{cases} e_k(t), & 0 \leq t \leq N_k, \\ 0, & N_k + 1 \leq t \leq N_d, \end{cases} \quad (13.4)$$

where $e_k(t) \triangleq y_d(t) - y_k(t)$ is the original tracking error.

To make a more concise expression, let us introduce an indicator function $\mathbf{1}(t \leq N_k)$. Then (13.4) could be reformulated as

$$e_k^*(t) = \mathbf{1}(t \leq N_k) e_k(t). \quad (13.5)$$

Remark 13.3 For arbitrary given $t \leq N_m$, the event $\{t \leq N_k\}$ occurs with probability one. For arbitrary given $t > N_m$, the event $\{t \leq N_k\}$ is a union of events $\{N_k = t\}$, $\{N_k = t+1\}, \ldots, \{N_k = N_d\}$. Thus, the probability of the event $\{\mathbf{1}(t \leq N_k) = 1\}$ is calculated as $\mathbb{P}(\mathbf{1}(t \leq N_k) = 1) = \sum_{i=t}^{N_d} \mathbb{P}(\mathscr{A}_i) = p(t)$, $t > N_m$. Combining these two scenarios, we have $\mathbb{P}(\mathbf{1}(t \leq N_k) = 1) = p(t)$, $\forall t$. In addition, $\mathbb{E}(\mathbf{1}(t \leq N_k)) = \mathbb{P}(\mathbf{1}(t \leq N_k) = 1) \times 1 + \mathbb{P}(\mathbf{1}(t \leq N_k) = 0) \times 0 = p(t)$.

13.2 ILC Design

With the help of the modified tracking error, we can give the following update law for input signal now:

$$u_{k+1}(t) = u_k(t) + Le_k^*(t+1), \tag{13.6}$$

where L is the learning gain matrix to be defined later, $L \in \mathbf{R}^{p \times q}$.

Remark 13.4 Generally speaking, to ensure a good performance against high-frequency signals in practical applications such as unmodeled dynamics, a low-pass Q-filter is incorporated in the learning algorithm. In other words, (13.6) is formulated as $u_{k+1}(t) = Q(q)(u_k(t) + Le_k^*(t+1))$, where $Q(q)$ denotes the Q-filter [2]. The involved Q-filter could suppress the high-frequency components and pass the low-frequency components. Therefore, it has effects on the convergence condition as shown in [2]. To make the analysis derivations concise, we only consider the case $Q(q) = I$ in the following of this chapter.

13.3 Convergence Analysis

The following theorem gives the zero-error convergence of the proposed ILC algorithm for the case that initial state is accurately reset:

Theorem 13.1 *Consider discrete-time affine nonlinear system (13.1) and ILC algorithm (13.6), and assume Assumptions 13.1 and 13.2 hold. If the learning gain matrix L satisfies that $0 < LCB < I$, then the tracking error would converge to zero as iteration number k goes to infinity, i.e., $\lim_{k \to \infty} e_k(t) = 0$, $t = 1, \ldots, N_d$.*

Proof Subtracting both sides of (13.6) from $u_d(t)$, we have

$$\delta u_{k+1}(t) = \delta u_k(t) - Le_k^*(t+1), \tag{13.7}$$

where $\delta u_k(t) \triangleq u_d(t) - u_k(t)$ is the input error.

Noticing (13.1) and (13.2), it follows that

$$\begin{aligned} \delta x_k(t+1) &= (f(x_d(t)) - f(x_k(t))) + B\delta u_k(t), \\ e_k(t) &= C\delta x_k(t), \end{aligned} \tag{13.8}$$

where $\delta x_k(t) \triangleq x_d(t) - x_k(t)$, which further leads to

$$\begin{aligned} e_k(t+1) &= C\delta x_k(t+1) \\ &= CB\delta u_k(t) + C(f(x_d(t)) - f(x_k(t))). \end{aligned} \tag{13.9}$$

Substitute (13.9) and (13.5) into (13.7),

$$\begin{aligned}\delta u_{k+1}(t) &= \delta u_k(t) - \mathbf{1}(t \leq N_k)Le_k(t+1) \\ &= \delta u_k(t) - \mathbf{1}(t \leq N_k)L[CB\delta u_k(t) \\ &\quad + C(f(x_d(t)) - f(x_k(t)))] \\ &= (I - \mathbf{1}(t \leq N_k)LCB)\delta u_k(t) \\ &\quad - \mathbf{1}(t \leq N_k)LC(f(x_d(t)) - f(x_k(t))).\end{aligned}$$

Taking Euclidean norm of both sides of last equation, we have

$$\begin{aligned}\|\delta u_{k+1}(t)\| &\leq \|(I - \mathbf{1}(t \leq N_k)LCB)\delta u_k(t)\| \\ &\quad + \|\mathbf{1}(t \leq N_k)LC(f(x_d(t)) - f(x_k(t)))\| \\ &\leq \|(I - \mathbf{1}(t \leq N_k)LCB)\|\|\delta u_k(t)\| \\ &\quad + \|\mathbf{1}(t \leq N_k)LC\|\|(f(x_d(t)) - f(x_k(t)))\| \\ &\leq \|(I - \mathbf{1}(t \leq N_k)LCB)\|\|\delta u_k(t)\| \\ &\quad + k_f\|\mathbf{1}(t \leq N_k)LC\|\|\delta x_k(t)\|.\end{aligned}$$

Noticing that the event $\{t \leq N_k\}$ is independent of $\delta u_k(t)$ and $\delta x_k(t)$. Thus by taking mathematical expectation of the last inequality, it follows

$$\begin{aligned}\mathbb{E}\|\delta u_{k+1}(t)\| &\leq \mathbb{E}(\|(I - \mathbf{1}(t \leq N_k)LCB)\|\|\delta u_k(t)\|) \\ &\quad + k_f\mathbb{E}(\|\mathbf{1}(t \leq N_k)LC\|\|\delta x_k(t)\|) \\ &\leq \|(I - p(t)LCB)\|\mathbb{E}\|\delta u_k(t)\| \\ &\quad + k_f\|p(t)LC\|\mathbb{E}\|\delta x_k(t)\|,\end{aligned} \qquad (13.10)$$

where for the last inequality, Lemma 13.1 is used by noticing that $0 < LCB < I$.

On the other hand, take Euclidean norm of both sides of the first equation in (13.8),

$$\begin{aligned}\|\delta x_k(t+1)\| &\leq \|B\|\|\delta u_k(t)\| + \|f(x_d(t)) - f(x_k(t))\| \\ &\leq \|B\|\|\delta u_k(t)\| + k_f\|x_d(t) - x_k(t)\| \\ &= \|B\|\|\delta u_k(t)\| + k_f\|\delta x_k(t)\|,\end{aligned} \qquad (13.11)$$

and then take mathematical expectation,

$$\mathbb{E}\|\delta x_k(t+1)\| \leq k_b\mathbb{E}\|\delta u_k(t)\| + k_f\mathbb{E}\|\delta x_k(t)\|, \qquad (13.12)$$

where $k_b \geq \|B\|$. Based on the recursion of (13.12) and noting Assumption 13.2, we have

13.3 Convergence Analysis

$$\begin{aligned}\mathbb{E}\|\delta x_k(t+1)\| &\leq k_b\mathbb{E}\|\delta u_k(t)\| + k_f k_b\mathbb{E}\|\delta u_k(t-1)\| \\ &\quad + k_f^2\mathbb{E}\|\delta x_k(t-1)\| \\ &\leq k_b\mathbb{E}\|\delta u_k(t)\| + k_b k_f\mathbb{E}\|\delta u_k(t-1)\| + \cdots \\ &\quad + k_b k_f^{t-1}\mathbb{E}\|\delta u_k(1)\| + k_b k_f^{t}\mathbb{E}\|\delta u_k(0)\| \\ &\quad + k_f^t\mathbb{E}\|\delta x_k(0)\| \\ &= k_b \sum_{i=0}^{t} k_f^{t-i}\mathbb{E}\|\delta u_k(i)\|, \end{aligned} \quad (13.13)$$

which further infers

$$\mathbb{E}\|\delta x_k(t)\| \leq k_b \sum_{i=0}^{t-1} k_f^{t-1-i}\mathbb{E}\|\delta u_k(i)\|. \quad (13.14)$$

Then substituting (13.14) into (13.10) yields that

$$\begin{aligned}\mathbb{E}\|\delta u_{k+1}(t)\| &\leq \|(I - p(t)LCB)\|\mathbb{E}\|\delta u_k(t)\| \\ &\quad + k_b\|p(t)LC\| \sum_{i=0}^{t-1} k_f^{t-i}\mathbb{E}\|\delta u_k(i)\|. \end{aligned} \quad (13.15)$$

Apply the λ-norm to both sides of last inequality, i.e., multiply both sides of last inequality with $\alpha^{-\lambda t}$ and take supremum according to all time instants t, then we have

$$\begin{aligned} &\sup_t \alpha^{-\lambda t}\mathbb{E}\|\delta u_{k+1}(t)\| \\ &\leq \sup_t \|(I - p(t)L\dot{C}B)\| \cdot \sup_t \alpha^{-\lambda t}\mathbb{E}\|\delta u_k(t)\| \\ &\quad + k_b \cdot \sup_t \|p(t)LC\| \cdot \sup_t \alpha^{-\lambda t}\left(\sum_{i=0}^{t-1} k_f^{t-i}\mathbb{E}\|\delta u_k(i)\|\right). \end{aligned} \quad (13.16)$$

Let $\alpha > k_f$, then it is observed that

$$\begin{aligned}&\sup_t \alpha^{-\lambda t}\left(\sum_{i=0}^{t-1} k_f^{t-i}\mathbb{E}\|\delta u_k(i)\|\right) \\ &\leq \sup_t \alpha^{-\lambda t}\left(\sum_{i=0}^{t-1} \alpha^{t-i}\mathbb{E}\|\delta u_k(i)\|\right) \\ &\leq \sup_t \left(\sum_{i=0}^{t-1} \alpha^{-(\lambda-1)t-i}\mathbb{E}\|\delta u_k(i)\|\right)\end{aligned}$$

$$\leq \sup_t \left(\sum_{i=0}^{t-1} \alpha^{-\lambda i} \mathbb{E}\|\delta u_k(i)\| \cdot \alpha^{-(\lambda-1)(t-i)} \right)$$

$$\leq \sup_t \left(\sum_{i=0}^{t-1} \sup_t \left(\alpha^{-\lambda i} \mathbb{E}\|\delta u_k(i)\| \right) \alpha^{-(\lambda-1)(t-i)} \right)$$

$$\leq \sup_t \left(\alpha^{-\lambda i} \mathbb{E}\|\delta u_k(i)\| \right) \sup_t \left(\sum_{i=0}^{t-1} \alpha^{-(\lambda-1)(t-i)} \right)$$

$$\leq \sup_t \left(\alpha^{-\lambda i} \mathbb{E}\|\delta u_k(i)\| \right) \times \frac{1 - \alpha^{-(\lambda-1)N_d}}{\alpha^{\lambda-1} - 1}.$$

Therefore, from (13.16),

$$\|\delta u_{k+1}(t)\|_\lambda$$
$$\leq \sup_t \|(I - p(t)LCB)\| \|\delta u_k(t)\|_\lambda$$
$$+ k_b \cdot \sup_t \|p(t)LC\| \|\delta u_k(t)\|_\lambda \times \frac{1 - \alpha^{-(\lambda-1)N_d}}{\alpha^{\lambda-1} - 1}$$
$$\leq \left(\rho + k_b \varphi \frac{1 - \alpha^{-(\lambda-1)N_d}}{\alpha^{\lambda-1} - 1} \right) \|\delta u_k(t)\|_\lambda,$$

where ρ and φ are defined as

$$\rho = \sup_t \|(I - p(t)LCB)\|,$$
$$\varphi = \sup_t \|p(t)LC\|.$$

Noticing that the learning gain matrix L satisfies $0 < LCB < I$ and $0 < p(t) \leq 1$, $\forall t$, it is evident that $\|I - p(t)LCB\| < 1$, $\forall t$. Since $0 \leq t \leq N_d$ has only finite values, we have $0 < \rho < 1$. Let $\alpha > \max\{1, k_f\}$, then there always exists a λ large enough such that $k_b \varphi \frac{1-\alpha^{-(\lambda-1)N_d}}{\alpha^{\lambda-1}-1} < 1 - \rho$, which further yields

$$\overline{\rho} \triangleq \rho + k_b \varphi \frac{1 - \alpha^{-(\lambda-1)N_d}}{\alpha^{\lambda-1} - 1} < 1. \tag{13.17}$$

This means

$$\lim_{k \to \infty} \|\delta u_k(t)\|_\lambda = 0, \quad \forall t.$$

Again, by the finiteness of t, we have

$$\lim_{k \to \infty} \mathbb{E}\|\delta u_k(t)\| = 0, \quad \forall t. \tag{13.18}$$

13.3 Convergence Analysis

Notice that $\|\delta u_k(t)\| \geq 0$, thus it could be concluded from (13.18) that

$$\lim_{k \to \infty} \|\delta u_k(t)\| = 0, \quad \forall t. \quad (13.19)$$

Then directly by mathematical induction method along time axis t, it is easy to show that $\lim_{k \to \infty} \delta x_k(t) = 0$ and $\lim_{k \to \infty} e_k(t) = 0$, $\forall t$. The proof is thus completed.

Remark 13.5 One may argue whether it is conservative to design L such that $0 < LCB < I$. In our point of view, it is a tradeoff between algorithm design and scope of application. In [1], the condition on L is somewhat loose; however, the occurrence probability of randomly varying length is required to be estimated prior because the convergence condition depends on it. While in this chapter, the requirements on L is a little restrictive, but no information on probability is requested prior. Thus it is more suitable for practical applications. Here, two simple schemes are referential if knowledge of CB is available. The first is that design L_* such that $L_*CB > 0$ and then multiply a constant μ small enough such that $\mu L_* CB < I$, whence $L = \mu L_*$. The second is that let $L = \frac{(CB)^T}{\beta + \|CB\|^2}$, where $\beta > 0$.

Remark 13.6 The operation length varies from iteration to iteration, thus one may doubt why Theorem 13.1 claims that the tracking error would converge to a zero for all time instants. In other words, the influence of random varying length is not revealed. We have some explanations for this issue. On the one hand, in the proof, we introduce the so-called λ-norm, which is similar to the conventional λ-norm in earlier publications but is modified with an additional expectation to deal with the randomness, eliminate the random indicator function $\mathbf{1}(t \leq N_k)$, and convert the original expression into deterministic case. On the other hand, there is a positive probability that the iteration length achieves the maximal length N_d, thus the input at each time instant would be updated more or less. To be specific, the input at time instant $t \leq N_m$ would be updated for all iterations, while the input at time instant $N_m < t \leq N_d$ is updated for part iterations. Therefore, the input at different time instants may have different convergence speed but they all converge to the desired one.

Remark 13.7 The time-invariant model (13.1) is studied in this chapter. However, the system is easy to extend to the time-varying case, where the design condition is slightly modified as $0 < L(t)C(t+1)B(t) < I$. The convergence analysis could be completely same. Moreover, the P-type update law (13.6) could also be extended to other types of ILC such as PD-type ILC with slight modifications to the proof and convergence conditions. Furthermore, the considered system is of relative degree 1; that is, CB is of full-column rank. This leads us to design the P-type update law (13.6) and convergence condition $0 < LCB < I$ in the theorem. Under some circumstances, the system may be of high relative degree τ; that is, $C\frac{\partial f^{\tau-1}(f(x)+Bu)}{\partial u}$ is of full column rank and $C\frac{\partial f^i(f(x)+Bu)}{\partial u} = 0, 0 \leq i \leq \tau-2$, where $f^i(x) = f^{i-1} \circ f(x)$ and $\circ$ denotes the composite operator of functions [3]. For this case, the analysis is still valid provided that the update law is modified as $u_{k+1}(t) = u_k(t) + Le_k^*(t+\tau)$ and the convergence condition becomes $0 < LC\frac{\partial f^{\tau-1}(f(x)+Bu)}{\partial u} < I$.

Remark 13.8 We provide a convergence analysis in modified λ-norm sense above for the ILC problem for discrete nonlinear systems under randomly iteration-varying length situation. One may interest in monotonic convergence in vector norm sense. To this end, define the lifted sup-vector $U_k \triangleq [\mathbb{E}\|\delta u_k(0)\|^T, \mathbb{E}\|\delta u_k(1)\|^T, \ldots, \mathbb{E}\|\delta u_k(N-1)\|^T]^T$ and the associated matrix Γ from (13.15) as a block lower-triangular matrix with its elements being the parameters of (13.15), then we have $\|U_{k+1}\|_1 \leq \|\Gamma\|_\infty \|U_k\|_1$ from (13.15) directly, where $\|\cdot\|_1$ and $\|\cdot\|_\infty$ denote the 1-norm of a vector and the ∞-norm of a matrix, respectively. Consequently, we draw a conclusion that the input error converges to zero monotonically if one can design L such that $\|\Gamma\|_\infty < 1$.

The identical initialization condition (i.i.c.) in Assumption 13.2 is required to make the analysis more concise. However, it is of great interest to consider the case that the i.i.c. is no longer valid. To be specific, the initial state may vary in a small zone, then it could be proven that the tracking error would converge into a small zone, of which the bound is in proportion to initial state error. This is given in the next theorem.

Theorem 13.2 *Consider discrete-time affine nonlinear system* (13.1) *and ILC algorithm* (13.6), *and assume Assumptions 13.1 and 13.3 hold. If the learning gain matrix L satisfies that $0 < LCB < I$, then the tracking error would converge to a small zone, whose bound is in proportion to ε, as iteration number k goes to infinity, i.e., $\limsup_{k\to\infty} \mathbb{E}\|e_k(t)\| \leq \gamma\varepsilon$, $t = 1,\ldots,N_d$, where γ is a suitable constant.*

Proof The proof follows the one of Theorem 13.1 with minor technical modifications. The derivations from (13.7)–(13.12) remain unchanged. While (13.13) is replaced by

$$\mathbb{E}\|\delta x_k(t+1)\| \leq k_b \sum_{i=0}^{t} k_f^{t-i} \mathbb{E}\|\delta u_k(i)\| + k_f^t \mathbb{E}\|\delta x_k(0)\|. \tag{13.20}$$

Combining with Assumption 13.3, it leads to

$$\mathbb{E}\|\delta x_k(t)\| \leq k_b \sum_{i=0}^{t-1} k_f^{t-1-i} \mathbb{E}\|\delta u_k(i)\| + k_f^{t-1}\varepsilon. \tag{13.21}$$

Then substituting (13.21) into (13.10) yields that

$$\mathbb{E}\|\delta u_{k+1}(t)\| \leq \|(I - p(t)LCB)\| \mathbb{E}\|\delta u_k(t)\|$$
$$+ k_b \|p(t)LC\| \sum_{i=0}^{t-1} k_f^{t-i} \mathbb{E}\|\delta u_k(i)\| + \|p(t)LC\| k_f^t \varepsilon. \tag{13.22}$$

Apply the λ-norm to both sides of last inequality, and by similar steps to the proof of Theorem 13.1, one is easy to get

13.3 Convergence Analysis

$$\|\delta u_{k+1}(t)\|_\lambda \leq \overline{\rho}\|\delta u_k(t)\|_\lambda + \sup_t \alpha^{-\lambda t}\|p(t)LC\|k_f^t\varepsilon. \tag{13.23}$$

By the finiteness of t, there is a constant ϑ such that $\sup_t \alpha^{-\lambda t}\|p(t)LC\|k_f^t < \vartheta$ and then

$$\|\delta u_{k+1}(t)\|_\lambda \leq \overline{\rho}\|\delta u_k(t)\|_\lambda + \vartheta\varepsilon, \tag{13.24}$$

which further means

$$\limsup_{k\to\infty} \|\delta u_{k+1}(t)\|_\lambda \leq \frac{\vartheta\varepsilon}{1-\overline{\rho}}. \tag{13.25}$$

This implies that

$$\limsup_{k\to\infty} \mathbb{E}\|\delta u_{k+1}(t)\| \leq \frac{\alpha^{\lambda t}\vartheta\varepsilon}{1-\overline{\rho}}.$$

Then combining with (13.21), one is easy to find that $\mathbb{E}\|\delta x_k(t)\|$ is bounded in proportion to ε, and thus a suitable γ exists such that $\limsup_{k\to\infty} \mathbb{E}\|e_k\| \leq \gamma\varepsilon$. This completes the proof.

13.4 Illustrative Simulations

In order to show the effectiveness of the proposed ILC algorithm and verify the convergence analysis, consider the following affine nonlinear system:

$$x_k^{(1)}(t+1) = \cos(x_k^{(1)}(t)) + 0.3x_k^{(2)}(t)x_k^{(1)}(t),$$
$$x_k^{(2)}(t+1) = 0.4\sin(x_k^{(1)}(t)) + \cos(x_k^{(2)}(t)) + u_k(t),$$
$$y_k(t) = x_k^{(2)}(t),$$

which means $B = [0\ 1]^T$ and $C = [0\ 1]$ in (13.1). $x_k(t) = [x_k^{(1)}(t), x_k^{(2)}(t)]^T$ denotes the state.

The expected iteration length is $N_d = 50$. To simulate the randomly iteration-varying length, let $N_m = 40$. In other words, the iteration length N_k varies from 40 to 50. As a simple case for illustration, we let N_k satisfy discrete uniform distribution during the discrete set $\{40, 41, \ldots, 50\}$. It should be noted that the probability distribution is not required for the control design. However, the probability distribution would alter the convergence speed. This is because, generally speaking, larger probability means more updates to the corresponding input, which therefore leads to faster convergence. If $\mathbb{P}(N_d)$ is very close to 1; that is, most iterations could complete the maximum length, then the behavior along iteration axis would be very close to the iteration-length-invariant tracking case and thus a fast convergence speed could be obtained.

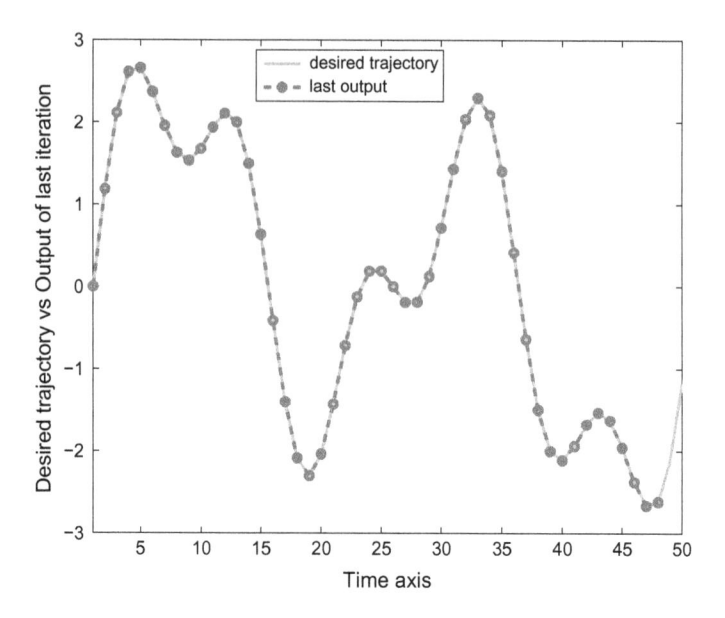

Fig. 13.1 Tracking performance of the output at the 50th iteration

The desired tracking trajectory is

$$y_d(t) = 0.8 \sin\left(\frac{2\pi t}{50}\right) + 2 \sin\left(\frac{2\pi t}{25}\right) + \sin\left(\frac{\pi t}{5}\right).$$

The initial state is set to be $x_k(0) = [0, 0]^T$. Without loss of generality, the input of the initial iteration is zero, i.e., $u_0(t) = 0, 0 \le t \le N_d$. It is obvious that CB is equal to 1, thus we set the learning gain in (13.6) as 0.5. The algorithm runs 50 iterations. The desired trajectory and output at the 50th iteration are shown in Fig. 13.1, where the red solid line denotes the desired trajectory, and the blue dashed line marked with circles denotes the output at the 50th iteration. As one could see, the system output achieves perfect tracking performance.

The tracking error profiles of the whole time interval at the 15th, 20th, 30th, and 40th iterations are shown in Fig. 13.2. It is observed that the error at the 15th iteration has been small. At the 20th iteration, the tracking errors are already acceptable. Meanwhile, the error profiles of different iterations end at different time instants, which demonstrates the random varying iteration length circumstance.

The convergence property along the iteration axis is shown in Fig. 13.3, illustrated by the solid blue line where the maximal tracking error is defined by $\max_t |e_k(t)|$ for the kth iteration. As commented in Remark 13.7, the proposed algorithm could be extended to PD-type algorithm. Here we also make simulations based on PD-type update law $u_{k+1}(t) = u_k(t) + L_p e_k^*(t+1) + K_d(e_k^*(t+1) - e_k^*(t))$ with learning gain $L_p = 0.4$ and $K_d = 0.3$. The maximal tracking error profile along iteration axis

13.4 Illustrative Simulations

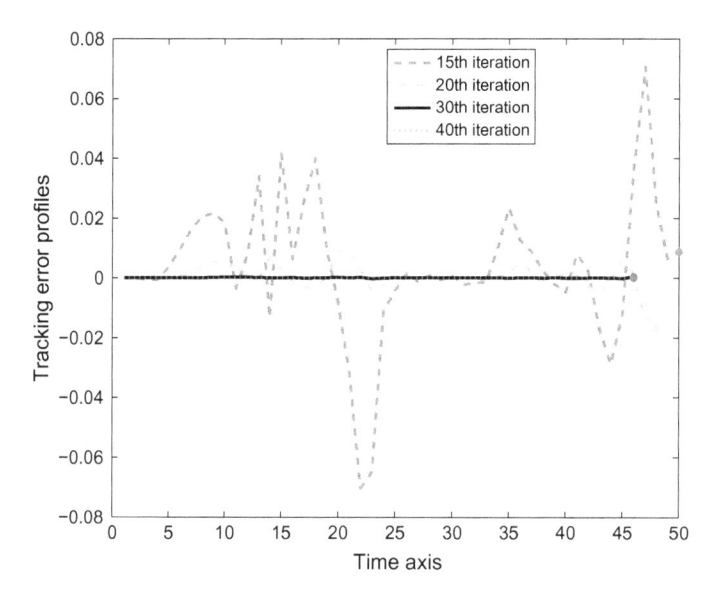

Fig. 13.2 Tracking error profiles at the 15, 20, 30, and 40th iterations

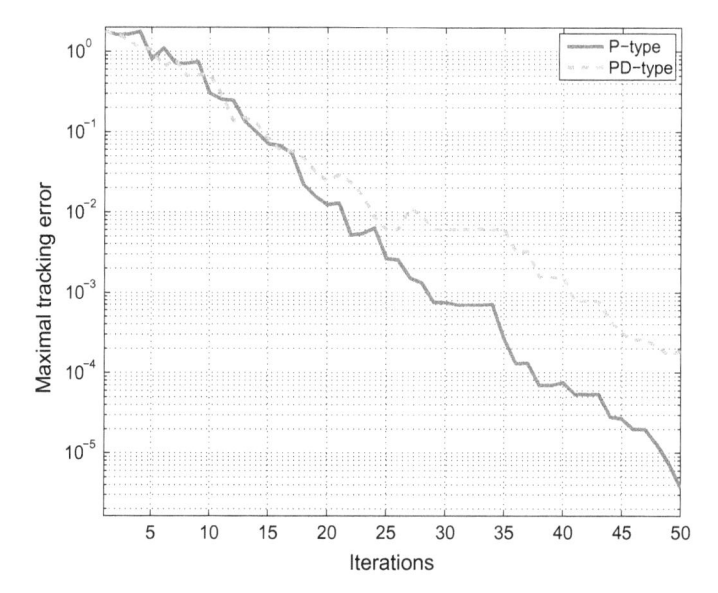

Fig. 13.3 Maximal errors along iterations

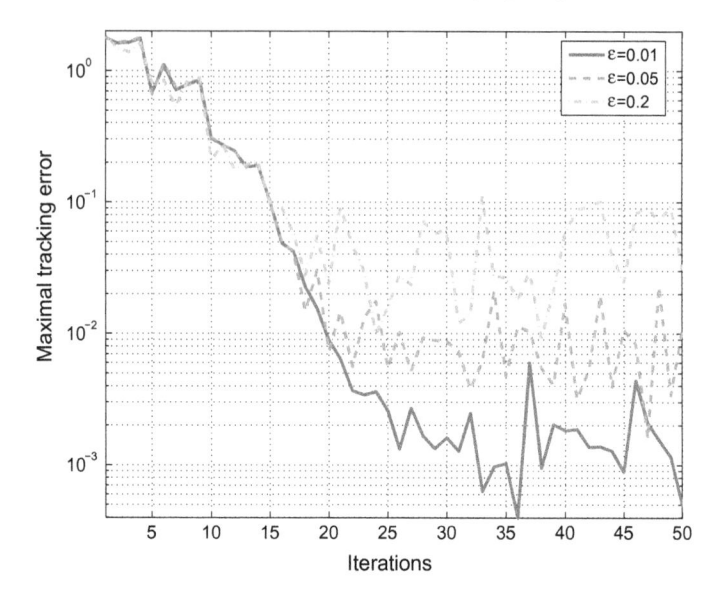

Fig. 13.4 Maximal errors for the case with iteration-varying initial value

is illustrated by the dashed red line. As one could see from Fig. 13.3, the maximal tracking error reduces to zero fast.

To verify the convergence under varying initial states, we let each dimension of the initial state obeys a uniform distribution in $[-\varepsilon, \varepsilon]$ with different scales ε being 0.01, 0.05, and 0.2. The tracking performance would be inferior to the identical initial condition case. However, the algorithm still maintains a robust performance, as shown in Fig. 13.4, where one could find that large initial bias leads to large bound of tracking errors.

We conclude this section with several remarks. The simulations have shown that the conventional P-type update law has a good performance against randomly iteration-varying lengths for discrete-time affine nonlinear systems. Although the convergence in λ-norm sense does not imply the monotonic decreasing naturally, the simulations show that the tracking performance is sustainedly improved. Moreover, when encountering other practical issues, the proposed P-type algorithm can be modified by incorporating other design techniques.

13.5 Summary

This chapter proposes the convergence analysis of ILC for discrete-time affine nonlinear systems with randomly iteration-varying lengths. A random variable is introduced to describe the random length. Then the tracking error is modified to facilitate the practical situations. The traditional P-type update law is taken as the control algorithm for our research and it can be extended to other schemes. If the identical initialization condition is satisfied, the tracking error is proved to converge to zero as the iteration number goes to infinity by using a modified λ-norm technique. If the initial state shifts in a small bound, then it is shown that the tracking error is also bounded. It is worth pointing out that the probability of random length is not required prior for control design. Due to the usage of modified λ-norm technique, the nonlinear function in this chapter is required to satisfy globally Lipschitz condition. For further study, the case of general nonlinear systems, especially those allow nonlinearities in the actuators and/or sensors, are of great interest. The results in this chapter are mainly based on [4].

References

1. Li, X., Xu, J.-X., Huang, D.: An iterative learning control approach for linear systems with randomly varying trial lengths. IEEE Trans. Autom. Control **59**(7), 1954–1960 (2014)
2. Bristow, D.A., Tharayil, M., Alleyne, A.G.: A survey of iterative learning control: a learning-based method for high-performance tracking control. IEEE Control Syst. Mag. **26**(3), 96–114 (2006)
3. Sun, M., Wang, D.: Analysis of nonlinear discrete-time systems with higher-order iterative learning control. Dyn. Control **11**, 81–96 (2001)
4. Shen, D., Zhang, W., Xu, J.-X.: Iterative learning control for discrete nonlinear systems with randomly iteration varying lengths. Syst. Control Lett. **96**, 81–87 (2016)

Chapter 14
Iterative Learning Control for Large-Scale Systems

14.1 Problem Formulation

Let the large-scale system be composed of n subsystems, and let its ith subsystem be described by

$$\begin{cases} x_i(t+1,k) = f_i(t, x(t,k)) + b_i(t, x(t,k))u_i(t,k), \\ y_i(t,k) = c_i(t)x_i(t,k) + w_i(t,k), \end{cases} \quad (14.1)$$

where $t \in \{0, 1, \ldots, N\}$ denotes the time instant in a cycle of the process, while $k = 1, 2, \ldots$ labels different cycles, and the subscript i denotes different subsystems, $i = 1, \ldots, n$. $u_i(t,k) \in \mathbf{R}$, $x_i(t,k) \in \mathbf{R}^{n_i}$, and $y_i(t,k) \in \mathbf{R}$ denote the input, state, and output of the ith subsystem, respectively.

Denote by

$$x(t,k) = [x_1^T(t,k), \ldots, x_n^T(t,k)]^T \quad (14.2)$$

the state vector of the large-scale system.

The subsystems are connected via the vector functions $f_i(t, x(t,k))$, which, however, together with $c_i(t)$ and $b_i(t, x(t,k))$ are unknown.

For the ith subsystem the tracking target is a given signal $y_i(t,d)$, $t \in \{0, 1, \ldots, N\}$.

Let $\mathscr{F}_{i,k} \triangleq \sigma\{y_i(t,j), x(t,j), w_i(t,j), 0 \leq j \leq k, t \in \{0, 1, \ldots, N\}\}$ be the σ-algebra generated by $y_i(t,j), x(t,j), w_i(t,j), 0 \leq j \leq k, t \in \{0, 1, \ldots, N\}$, and let $\mathscr{F}_{i,k}^y \triangleq \sigma\{y_i(t,j), 0 \leq j \leq k, t \in \{0, 1, \ldots, N\}\}$ be the σ-algebra generated by $y_i(t,j), 0 \leq j \leq k, t \in \{0, 1, \ldots, N\}$. For the ith subsystem, the set of admissible controls is defined as

$$U_i = \{u_i(t, k+1) \in \mathscr{F}_{i,k}^y, \sup_k \|u_i(t, k)\| < \infty, \text{ a.s.}$$

$$t \in \{0, 1, \ldots, N-1\}, \quad k = 0, 1, 2, \ldots\}.$$

It is required to find the control sequences $\{u_i(t, k), k = 0, 1, 2, \ldots\} \in U_i$, $\forall i = 1, \ldots, n$ to minimize the following tracking errors index:

$$V_i(t) = \limsup_n \frac{1}{n} \sum_{k=1}^n \|y_i(t, d) - y_i(t, k)\|^2, \quad \forall t \in \{0, 1, \ldots, N\}. \tag{14.3}$$

The desired control sequence is to be recursively defined in Sect. 14.3 by the algorithm (14.8).

For the ith subsystem (14.1), $\forall i \in \{1, \ldots, n\}$, the following assumptions are used.

Assumption 14.1 The target signal $y_i(t, d)$ is realizable in the sense that there exist $u_i(t, d)$ and $x_i(0, d)$ such that

$$\begin{cases} x_i(t+1, d) = f_i(t, x(t, d)) + b_i(t, x(t, d))u_i(t, d), \\ y_i(t, d) = c_i(t)x_i(t, d), \end{cases} \tag{14.4}$$

where $x(t, d)$ is defined similarly to (14.2).

Assumption 14.2 The initial values can asymptotically be precisely reset in the sense that $x_i(0, k) - x_i(0, d) \xrightarrow[k \to \infty]{} 0$.

Assumption 14.3 The functions $f_i(\cdot, \cdot)$ and $b_i(\cdot, \cdot)$ are continuous with respect to the second argument.

From the proof of Lemma 14.1 and Theorem 14.2 below it will be seen that the functions $f_i(t, x)$ and $b_i(t, x)$ are allowed to have discontinuities with respect to x, but are required to be continuous at $x = x(t, d)$. Since $x = x(t, d)$ is unknown, we simply use Assumption 14.3.

Assumption 14.4 The real number $c_i(t+1)b_i(t, x)$ coupling the input and output is an unknown nonzero constant, but its sign, characterizing the control direction, is assumed known. Without loss of generality, we may assume that $c_i(t+1)b_i(t, x) > 0$.

Assumption 14.5 For each time instant t, the observation noise $\{w_i(t, k)\}$ is a sequence of independent and identically distributed (i.i.d.) random variables with $\mathbb{E}w_i(t, k) = 0$, $\sup_k \mathbb{E}w_i^2(t, k) < \infty$ and

$$\lim_{n \to \infty} \frac{1}{n} \sum_{k=1}^n w_i^2(t, k) = R_i^t \quad \text{a.s.} \quad \forall t \in \{0, 1, \ldots, N\}, \tag{14.5}$$

where R_i^t is unknown.

14.1 Problem Formulation

It is worth noting that for a large-scale system at the kth iteration, the controls $u_i(t, k)$ may not be updated for all $i \in \{1, \ldots, n\}$, because data may be missing and operating frequency for subsystems may not be the same.

For simplicity of writing let us set $f_i(t, k) \triangleq f_i(t, x(t, k))$, $f_i(t, d) \triangleq f_i(t, x(t, d))$, $b_i(t, k) \triangleq b_i(t, x(t, k))$, $b_i(t, d) \triangleq b_i(t, x(t, d))$, $e_i(t, k) \triangleq y_i(t, d) - y_i(t, k)$, $\delta x_i(t, k) \triangleq x_i(t, d) - x_i(t, k)$, $\delta u_i(t, k) \triangleq u_i(t, d) - u_i(t, k)$, $\delta f_i(t, k) \triangleq f_i(t, d) - f_i(t, k)$, $\delta b_i(t, k) \triangleq b_i(t, d) - b_i(t, k)$, $c^+ f_i(t, k) \triangleq c(t+1) f_i(t, k)$, $c^+ b_i(t, k) \triangleq c(t+1) b_i(t, k)$.

14.2 Optimal Control

In this section, we show the minimum of the index defined by (14.3) and the optimal control sequence as well.

Lemma 14.1 *Assume that Assumptions 14.1–14.3 hold for system (14.1). If* $\lim_{k \to \infty} \delta u_i(s, k) = 0$, $s = 0, 1, \ldots, t$ $\forall i = 1, 2, \ldots, n$, *then at time instant* $t + 1$,

$$\|\delta x_i(t+1, k)\| \xrightarrow[k \to \infty]{} 0, \quad \|\delta f_i(t+1, k)\| \xrightarrow[k \to \infty]{} 0, \quad \|\delta b_i(t+1, k)\| \xrightarrow[k \to \infty]{} 0, \quad \forall i.$$

Proof We prove the lemma by mathematical induction. From (14.1) and (14.4) it follows that for each i

$$\begin{aligned}
&\delta x_i(t+1, k) \\
&= f_i(t, d) - f_i(t, k) + b_i(t, d) u_i(t, d) - b_i(t, k) u_i(t, k) \\
&= \delta f_i(t, k) + \delta b_i(t, k) u_i(t, d) + b_i(t, k) \delta u_i(t, k).
\end{aligned} \quad (14.6)$$

For $t = 0$, by Assumptions 14.2 and 14.3 we have

$$\|f_i(0, d) - f_i(0, k)\| \xrightarrow[k \to \infty]{} 0,$$

$$\|b_i(0, d) - b_i(0, k)\| \xrightarrow[k \to \infty]{} 0,$$

which imply that the first two terms at the right-hand side of (14.6) tend to zero as $k \to \infty$. Since

$$\|b_i(0, k)\| \leq \|b_i(t, d)\| + \|\delta b_i(0, k)\|,$$

it follows that $b_i(0, k)$ is bounded. This incorporating with $\lim_{k \to \infty} \delta u_i(0, k) = 0$ yields that the third term on the right-hand side of (14.6) also tends to zero as $k \to \infty$, and hence $\delta x_i(1, k) \xrightarrow[k \to \infty]{} 0$, $\forall i$. Further, by Assumption 14.3 it follows that $\|\delta f_i(1, k)\| \xrightarrow[k \to \infty]{} 0$ and $\|\delta b_i(1, k)\| \xrightarrow[k \to \infty]{} 0$, and hence the conclusions of the lemma are valid for $t = 0$.

Assume conclusions of the lemma are true for $t-1$, i.e., $\|\delta x_i(t,k)\| \xrightarrow[k\to\infty]{} 0$, $\|\delta f_i(t,k)\| \xrightarrow[k\to\infty]{} 0$, and $\|\delta b_i(t,k)\| \xrightarrow[k\to\infty]{} 0$. By the same argument as that used above we find that the conclusions are also valid for t. The lemma is proved.

The following theorem gives a criterion for optimality of a control.

Theorem 14.1 *Assume that Assumptions 14.1–14.5 hold for system (14.1) and criterion (14.3). Then*
$$V_i(t) \geq R_i^t, \quad a.s. \quad \forall i,t$$
for any admissible control sequence $\{u_i(t,k), i=1,\ldots,n, t \in \{0,1,\ldots,T\}\}$. *The inequality becomes equality if* $\delta u_i(t,k) \xrightarrow[k\to\infty]{} 0$, *and in this case* $\{u_i(t,k)\}$ *is the optimal control sequence.*

Proof By Assumption 14.5 and the definition of $\mathscr{F}_{i,k}$ it is seen that $\mathscr{F}_{i,k}$ is independent of $\{w_i(t,l), l = k+j, j = 1,2,\ldots, \forall i, \forall t \in \{0,1,\ldots,T\}\}$, and $\{w_i(t,k), \mathscr{F}_{i,k}\}$ is a martingale difference sequence with $\sup_k \mathbb{E}[w_i^2(t,k)|\mathscr{F}_{i,k}] < \infty$ a.s. It is clear that the input, output, and the state are adapted with respect to $\mathscr{F}_{i,k}$. Then from (14.1) it follows that

$$V_i(t) = \limsup_{n\to\infty} \frac{1}{n} \sum_{k=1}^n |y_i(t,d) - y_i(t,k)|^2$$

$$= \limsup_{n\to\infty} \frac{1}{n} \sum_{k=1}^n |c(t)(x_i(t,d) - x_i(t,k)) - w_i(t,k)|^2$$

$$= \limsup_{n\to\infty} \frac{1}{n} \sum_{k=1}^n |c(t)(x_i(t,d) - x_i(t,k))|^2 (1+o(1))$$

$$+ \limsup_{n\to\infty} \frac{1}{n} \sum_{k=1}^n w_i^2(t,k)$$

$$\geq \limsup_{n\to\infty} \frac{1}{n} \sum_{k=1}^n w_i^2(t,k) = R_i^t,$$

where the third equality follows by [1, Theorem 2.8], and the inequality becomes equality if and only if

$$\limsup_{n\to\infty} \frac{1}{n} \sum_{k=1}^n |c(t)(x_i(t,d) - x_i(t,k))|^2 = 0.$$

If $\delta u_i(t,k) \xrightarrow[k\to\infty]{} 0, \forall i, t \in \{0,1,\ldots,T\}$, then by Lemma 14.1 the above equality takes place and the performance index reaches its minimum.

14.3 Optimal ILC Algorithms and Convergence Analysis

As explained before, the control action may not be updated for all subsystems at an iteration step k. Let us denote by $Y_k \subset \{1, 2, \ldots, n\}$ the set of those subsystems which are updated at the kth iteration and denote by $v(i, k)$ the number of control updates occurred up to and including the kth iteration in the ith subsystem:

$$v(i, k) = \sum_{m=1}^{k} \mathbf{1}_{\{i \in Y_m\}},$$

where $\mathbf{1}_{\{A\}}$ is the indicator of the random set A and is equal to either 1 or zero depending upon whether the event A is true or not.

Although the control for a subsystem is not necessary to be updated for each step, it is still required to be updated quite frequently. To be precise, we need the following condition:

Assumption 14.6 There is a number K such that

$$v(i, k + K) - v(i, k) > 0, \quad \forall i = 1, 2, \ldots, n, \; \forall k = 1, 2, \ldots. \quad (14.7)$$

This condition means that control in any subsystem should be updated at least once during any successive K updates of the whole system.

We now define the asynchronous distributed ILC leading the performance index (14.3) to its minimum. For the ith subsystem, the control $u_i(t, k+1)$ at the $(k+1)$th iteration is defined as follows:

$$u_i(t, k + 1) = u_i(t, k) + a(v(i, k)) \mathbf{1}_{\{i \in Y_k\}} e_i(t + 1, k), \quad (14.8)$$

where $\{a(k)\}$ is the step size such that

$$a(k) > 0, \; \sum_{k=0}^{\infty} a(k) = \infty, \; \sum_{k=0}^{\infty} a(k)^2 < \infty, \quad (14.9)$$

$$\text{and } a(j) = a(k)\big(1 + O(a(k))\big), \quad (14.10)$$

$\forall j = k - K + 1, \ldots, k - 1, k$ as $k \to \infty$.
It is clear that $a(k) = \frac{1}{k+1}$ satisfies (14.9)–(14.10).

Remark 14.1 The set Y_k is random and it characterizes the asynchronous nature of updating. During the period, where the whole system has been updated k times, the ith subsystem has actually been updated only $v(i, k) \leq k$ times.

Remark 14.2 The algorithm (14.8) is the asynchronous stochastic approximation (ASA) algorithm, which was studied in [2–4] and references therein. The ILC based on stochastic approximation is first considered in [5, 6], which, however, concern the synchronous and centralized control.

For any fixed time instant t, we rewrite (14.8) as

$$u_i(t, k + 1)$$
$$= u_i(t, k) + a(v(i, k))\mathbf{1}_{\{i \in Y_k\}}(y_i(t + 1, d) - y_i(t + 1, k))$$
$$= u_i(t, k) + a(v(i, k))\mathbf{1}_{\{i \in Y_k\}}[c^+ b_i(t, k)\delta u_i(t, k) + \varphi_i(t, k) - w_i(t + 1, k)],$$

or

$$\delta u_i(t, k + 1) = \delta u_i(t, k) - a(v(i, k))\mathbf{1}_{\{i \in Y_k\}}$$
$$\times [c^+ b_i(t, k)\delta u_i(t, k) + \varphi_i(t, k) - w_i(t + 1, k)], \quad (14.11)$$

where

$$\varphi_i(t, k) = c^+ \delta f_i(t, k) + c^+ \delta b_i(t, k) u_i(t, d). \quad (14.12)$$

In the algorithm, the noise contains two parts: $\mathbf{1}_{\{i \in Y_k\}}\varphi_i(t, k)$ represents the structural error and $\mathbf{1}_{\{i \in Y_k\}}w_i(t + 1, k)$ is the observation noise.

Theorem 14.2 *Assume that Assumptions 14.1–14.6 hold for system (14.1) and criterion (14.3), $i = 1, \ldots, n$. Then the control sequence $\{u_i(t, k)\}$ given by (14.8) is optimal.*

Proof The proof is carried out by induction along the time axis t simultaneously for all subsystems.

Let i be fixed. For $t = 0$ the algorithm (14.11) is written as

$$\delta u_i(0, k + 1) = (1 - \alpha_k h_k)\delta u_i(0, k) - \beta_k \varepsilon_{k+1}, \quad (14.13)$$

where

$$\alpha_k \triangleq a(v(i, k)),$$
$$\beta_k \triangleq a(v(i, k))\mathbf{1}_{\{i \in Y_k\}},$$
$$h_k \triangleq \mathbf{1}_{\{i \in Y_k\}} c^+ b_i(0, k),$$
$$\varepsilon_{k+1} \triangleq \varphi_i(0, k) - w_i(1, k).$$

It is clear that $\sum_{k=1}^{\infty} \alpha_k = \infty$. By the definitions of $v(i, k)$ and Y_k, the nonzero elements of $\{\beta_k\}$ are in one-to-one correspondence with elements of $\{a(0), a(1), a(2), \ldots\}$. Therefore, by (14.9) we have $\sum_{k=1}^{\infty} \beta_k^2 < \infty$.

From

$$v(i, k) = v(i, l) + \sum_{j=l+1}^{k} \mathbf{1}_{\{i \in Y_j\}}, \quad (14.14)$$

14.3 Optimal ILC Algorithms and Convergence Analysis

it follows that $v(i, k) \leq v(i, j) + K$, and by (14.10) for $j = k - K + 1, \ldots, k - 1, k$ as $k \to \infty$

$$a(v(i, j)) = a(v(i, k))\left(1 - \frac{a(v(i, k)) - a(v(i, j))}{a(v(i, k))}\right)$$
$$= a(v(i, k))(1 + O(a(v(i, k)))),$$

i.e., $\forall j = k - K + 1, \ldots, k - 1, k$,

$$\alpha_j = \alpha_k(1 + O(\alpha_k)). \tag{14.15}$$

Since $b_i(0, s)$ is continuous in the state variable by Assumption 14.3, we have $b_i(0, k) \xrightarrow[k \to \infty]{} b_i(0, d)$ by Assumption 14.2 and $c^+ b_i(0, k)$ converges to a positive constant by Assumption 14.4. Therefore, by Assumption 14.6 we have

$$\sum_{j=k-K+1}^{k} -h_j < -\gamma, \quad \gamma > 0 \tag{14.16}$$

for all sufficiently large k.

Set $\phi_{k,j} \triangleq (1 - \alpha_k h_k) \ldots (1 - \alpha_j h_j)$, $k \geq j$, $\phi_{j,j+1} \triangleq 1$. It is clear that $(1 - \alpha_j h_j) > 0$ for all sufficiently large j, say, $j \geq j_0$.

For any $k \geq j + K$, $j \geq j_0$, by (14.15) and (14.16) we have

$$\phi_{k,j} = \phi_{k-K,j}\left(1 - \sum_{l=k-K+1}^{k} \alpha_l h_l + o(\alpha_k)\right)$$
$$= \phi_{k-K,j}\left(1 - \alpha_k \sum_{l=k-K+1}^{k} h_l + o(\alpha_k)\right)$$
$$\leq \phi_{k-K,j}(1 - \beta\alpha_k + o(\alpha_k))$$
$$= \phi_{k-K,j}\left(1 - \frac{\beta}{K} \sum_{l=k-K+1}^{k} \alpha_l + o(\alpha_k)\right)$$
$$\leq \exp\left(-c \sum_{l=k-K+1}^{k} \alpha_l\right) \phi_{k-K,j} \text{ with } c > 0.$$

From here it follows that $\phi_{k,j} \leq c_1 \exp(-\frac{c}{2}\sum_{l=j}^{k} \alpha_l)$ $\forall j \geq j_0$ for some $c_1 > 0$, and hence there is a positive constant c_2 such that $|\phi_{k,j}| \leq c_2 \exp(-\frac{c}{2}\sum_{l=j}^{k} \alpha_l)$ $\forall k \geq j + K$, $j \geq j_0$.

Therefore, $\forall k \geq j_0 + K, \forall j \geq 0$, we have

$$|\phi_{k,j}| \leq |\phi_{k,j_0}| \cdot |\phi_{j_0-1,j}| \leq c_0 \exp\left(-\frac{c}{2}\sum_{l=j}^{k}\alpha_l\right) \tag{14.17}$$

for some $c_0 > 0$.

From (14.13) it follows that

$$\delta u_i(0, k+1) = \phi_{k,0}\delta u_i(0,0) + \sum_{j=0}^{k}\phi_{k,j+1}\beta_j\varepsilon_{j+1}, \tag{14.18}$$

where the first term at the right-hand side of (14.18) tends to zero as $k \to \infty$ because of (14.17).

By Assumptions 14.2 and 14.3 it is seen that $\varphi_i(0,k) \xrightarrow[k\to\infty]{} 0$.

By Assumption 14.5 it follows that

$$\sum_{k=1}^{\infty}\beta_k w_i(1,k) < \infty. \tag{14.19}$$

Since $\varepsilon_{k+1} \triangleq \varphi_i(0,k) - w_i(1,k)$, the last term of (14.18) also tends to zero as $k \to \infty$. For this, the proof is given in Lemma A.3 (in Appendix). Thus, we have proved the optimality of the control sequence $u_i(0,k) \xrightarrow[k\to\infty]{} u_i(0,d)$.

Inductively, assume that the optimality takes place for $t = 0, 1, \ldots, s - 1$. We now show that it is also true for $t = s$.

From the inductive assumption it follows that

$$\delta u_i(t,k) \xrightarrow[k\to\infty]{} 0, \quad i = 1, \ldots, n, \; t = 0, 1, \ldots, s-1,$$

and by Lemma 14.1

$$\delta x_i(s,k) \xrightarrow[k\to\infty]{} 0, \quad \delta f_i(s,k) \xrightarrow[k\to\infty]{} 0,$$
$$\delta b_i(s,k) \xrightarrow[k\to\infty]{} 0, \quad \forall i = 1, 2, \ldots, n. \tag{14.20}$$

Hence, we have

$$\varphi_i(s,k) \xrightarrow[k\to\infty]{} 0, \quad \forall i = 1, 2, \ldots, n.$$

As for the case $t = 0$, a similar treatment leads to $u_i(s,k) \xrightarrow[k\to\infty]{} u_i(s,d), \forall i$. This proves optimality of control for $t = s$ and completes the proof of the theorem.

We now consider the large-scale system with communication delays among subsystems. For many practical systems, the information at the kth iteration of ith sub-

14.3 Optimal ILC Algorithms and Convergence Analysis

system may be forwarded to the jth subsystem not at the $(k+1)$th iteration but at a later iteration.

Let the ith subsystem of a large-scale system be described as follows:

$$\begin{cases} x_i(t+1,k) = f_i(t,\overline{x}(t,k)) + b_i(t,\overline{x}(t,k))u_i(t,k), \\ y_i(t,k) = c_i(t)x_i(t,k) + w_i(t,k), \end{cases} \quad (14.21)$$

where

$$\overline{x}(t,k)) \triangleq [x_1^T(t, k - \tau_{1i}(k)), \ldots, x_n^T(t, k - \tau_{ni}(k))]^T, \quad (14.22)$$

and where $\tau_{ji}(k) > 0$, $j \ne i$ denotes the random communication delay for the ith subsystem at the kth iteration to receive information from the jth subsystem, while each subsystem receives information from itself without any delay, i.e., $\tau_{ii} = 0$. In other words, at the kth iteration the latest information from the jth subsystem obtained by the ith subsystem is $x_j(t, k - \tau_{ji}(k))$, and no information from $x_j(t, m)$ with $m > k - \tau_{ji}(k)$ can reach the ith subsystem.

For the delay $\tau_{ji}(k)$, the following condition is used:

Assumption 14.7 There is an integer M such that $\tau_{ji}(k) < M, \forall j, i, k$.

Theorem 14.3 *Assume that Assumptions 14.1–14.7 hold for system (14.21) and criterion (14.3), $i = 1, \ldots, n$. Then, the control $\{u_i(t, k)\}$ given by (14.8) is optimal.*

Proof (Sketch) It is intuitively understandable that convergence will remain true if delays exist in some subsystems whenever the delays are uniformly bounded. This is because the information will arrive sooner or later. More precisely, in the ith subsystem the influence caused by delays $\{\tau_{ji}(k)\}$ is contained in $\varphi_i(t, k)$ given by (14.12), and in the convergence analysis of ILC the key step is to show $\varphi_i(t, k) \xrightarrow[k \to \infty]{}$ 0. Paying attention to (14.21) and (14.22), we see that $\varphi_i(t, k)$ tends to zero as $k \to \infty$ no matter $\tau_{ji}(k) = 0$ or $\tau_{ji}(k) > 0$ but $\{\tau_{ji}(k)\}$ is bounded. Therefore, the proof of the theorem is completely similar to that for Theorem 14.2.

We now make further discussions on the relaxation of Assumption 14.4. When defining the algorithm (14.8), Assumption 14.4 is imposed by which the sign of $c^+ b_i(t, k)$ is assumed to be known. With the help of the techniques given in Chap. 4, we can remove the prior knowledge of the control direction, i.e., the sign of $c^+ b_i(t, k)$. In other words, Assumption 14.4 can be replaced with the following Assumption 14.8.

Assumption 14.8 The value $c_i(t+1)b_i(t, x)$ is completely unknown, but it is nonzero and does not change its sign.

Since the sign of $c_i(t+1)b_i(t, x)$ is unknown, the control may be switched on a wrong direction and this may lead to instability of the system. In this case, the algorithm (14.8) has to be modified.

Let $\{M_k\}$ be a sequence of positive real numbers such that $M_{k+1} > M_k$, $\forall k$ and $M_k \xrightarrow[k \to \infty]{} \infty$, and let $\{a(n)\}$ be the same as that used in (14.8).

For the ith subsystem the ILC is now defined by (14.23)–(14.27).

$$\bar{u}_i(t, k+1) = u_i(t, k) + a(v(i, k))(-1)^{\sigma_i(t,k)} \mathbf{1}_{\{i \in Y_k\}} e_i(t+1, k), \qquad (14.23)$$

$$u_i(t, k+1) = \bar{u}_i(t, k+1) \mathbf{1}_{\{|u_i(t,k) + a(v(i,k)) \mathbf{1}_{\{i \in Y_k\}} e_i(t+1,k)| \leq M_{\sigma_i(t,k)}\}}, \qquad (14.24)$$

$$\sigma_i(t, k) = \sum_{j=1}^{k-1} \mathbf{1}_{\{|u_i(t,j) + a(v(i,j)) \mathbf{1}_{\{i \in Y_j\}} e_i(t+1,j)| > M_{\sigma_i(t,j)}\}}, \qquad (14.25)$$

$$\sigma_i(t, 0) = 0, \qquad (14.26)$$

$$v(i, k) = \sum_{m=1}^{k} \mathbf{1}_{\{i \in Y_m\}}. \qquad (14.27)$$

We give a brief explanation of (14.23)–(14.27). For the ith subsystem, the control $u_i(t, k+1)$ at the $k+1$th iteration is defined by (14.23) if $|u_i(t, k+1)|$ is within the truncation bound $M_{\sigma_i(t,k)}$. Otherwise, $u_i(t, k+1)$ is simply set to be equal to 0, but at the same time the truncation bound is enlarged from $M_{\sigma_i(t,k)}$ to $M_{\sigma_i(t,k)+1}$. By $\sigma_i(t, k)$ we denote the number of truncations occurred before and including the $(k-1)$th iteration.

In the sequel when we prove the optimality of control defined by (14.26), without loss of generality, we may assume $c_i(t+1)b_i(t, x) > 0$.

Remark 14.3 The regulation mechanism for the control direction has been considered by many papers. In [7, 8], the regulation mechanism is constructed on the basis of Nussbaum type gain [9], but their method is difficult to be applied to the discrete-time systems. Here, the number of truncations $\sigma_i(t, k)$ in the algorithm is used to regulate the control direction.

Remark 14.4 In the algorithm (14.23), the control direction is regulated by $(-1)^{\sigma_i(t,k)}$, and "+" is the correct direction. If the wrong direction "−" is applied in the algorithm, it will be proved that the algorithm diverges and forces the truncation mechanism to work following steps in Chap. 4. As a consequence, the sign shifts to "+".

Theorem 14.4 *Assume that Assumptions 14.1–14.3, 14.5, 14.6, and 14.8 hold for system (14.1) and criterion (14.3), $i = 1, \ldots, n$. Then the control sequence $\{u_i(t, k)\}$ given by (14.23)–(14.27) is optimal.*

Proof The proof is carried out by mathematical induction along the time axis t. For a fixed t, it is first proved that the number of truncations of control for each subsystem is finite. Then, it is shown that the control cannot stop at the wrong direction. Once the control direction is correct and will not change anymore, its convergence to the optimal one is established as before. The proof can be completed following the same procedures in Chap. 4, thus the details of the proof are omitted.

14.4 Illustrative Example

For a large-scale system the number of subsystems n is a large number, but here for convenience of illustration let us take $n = 3$.

The three subsystems, indexed by subscripts 1, 2, and 3, are described as follows:

$$x_1(t+1,k) = 1.05x_1(t,k) + 0.2\sin(x_2(t, k - \tau_{21}(k))) + 0.45u_1(t,k),$$
$$y_1(t,k) = x_1(t,k) + w_1(t,k),$$
$$x_2(t+1,k) = 1.1x_2(t,k) + 0.2\sin(x_3(t, k - \tau_{32}(k))) + 0.5u_2(t,k),$$
$$y_2(t,k) = x_2(t,k) + w_2(t,k),$$
$$x_3(t+1,k) = 0.33\sin(x_1(t, k - \tau_{13}(k))) + 1.15x_3(t,k) + 0.8u_3(t,k),$$
$$y_3(t,k) = x_3(t,k) + w_3(t,k),$$

where time instant t is valued in the set $\{1, \ldots, 8\}$. Assume $w_i(t,k) \in N(0, 0.1^2)$, $i = 1, 2, 3$. The random communication delay $\tau_{ij}(k)$ from the ith subsystem to the jth subsystem, $i, j \in \{1, 2, 3\}$ takes values 1 and 0 with probability $p = 0.5$, respectively.

To simulate data missing, i.e., to realize random asynchronous control updating with Assumption 14.6 satisfied, let us fix $K = 3$ in Assumption 14.6. Iteration steps are separated into groups of three successive steps, i.e., $\{1, 2, 3\}, \{4, 5, 6\}, \{7, 8, 9\}$, ..., and randomly select one step from each group, say, 1, 5, 9, ... For these elected steps the control is not updated.

The reference signals are $y_1(t,d) = 2.4t$, $y_2(t,d) = 2t$, and $y_3(t,d) = 2.3t$, $t \in \{1, \ldots, 8\}$ for subsystems 1, 2, and 3, respectively. The initial control actions are $u_1(t,0) = u_2(t,0) = u_3(t,0) = 0$, $\forall t \in \{1, \ldots, 8\}$.

It should note that we assume the control direction is unknown in this simulation. The algorithms (14.23)–(14.27) have run 150 iterations and the tracking results at the 150th iteration are presented in Fig. 14.1 for subsystems 1, 2, and 3, respectively, where the solid lines are the reference signals and the cycle-dashed lines denote the system outputs.

As illustrations, the tracking errors at $t = 3$ and $t = 5$ are demonstrated in Fig. 14.2 for subsystems 1, 2, and 3, respectively. From the figures, we see that the decentralized ILC algorithms in the chapter work well.

The asynchronous update among three subsystems is shown in Fig. 14.3, where 1 and 0 denote update and holding, respectively. It is seen that three subsystems update their inputs asynchronously.

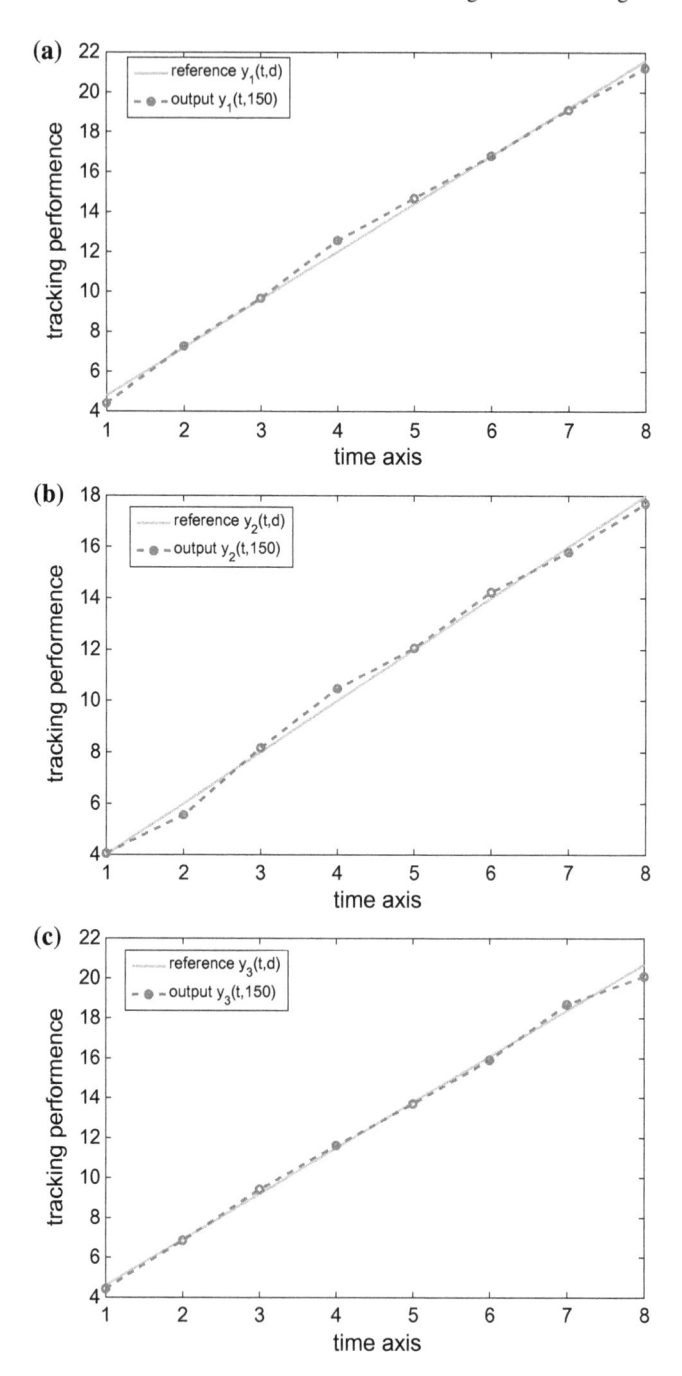

Fig. 14.1 The tracking performance of the last iteration for three subsystems

14.4 Illustrative Example

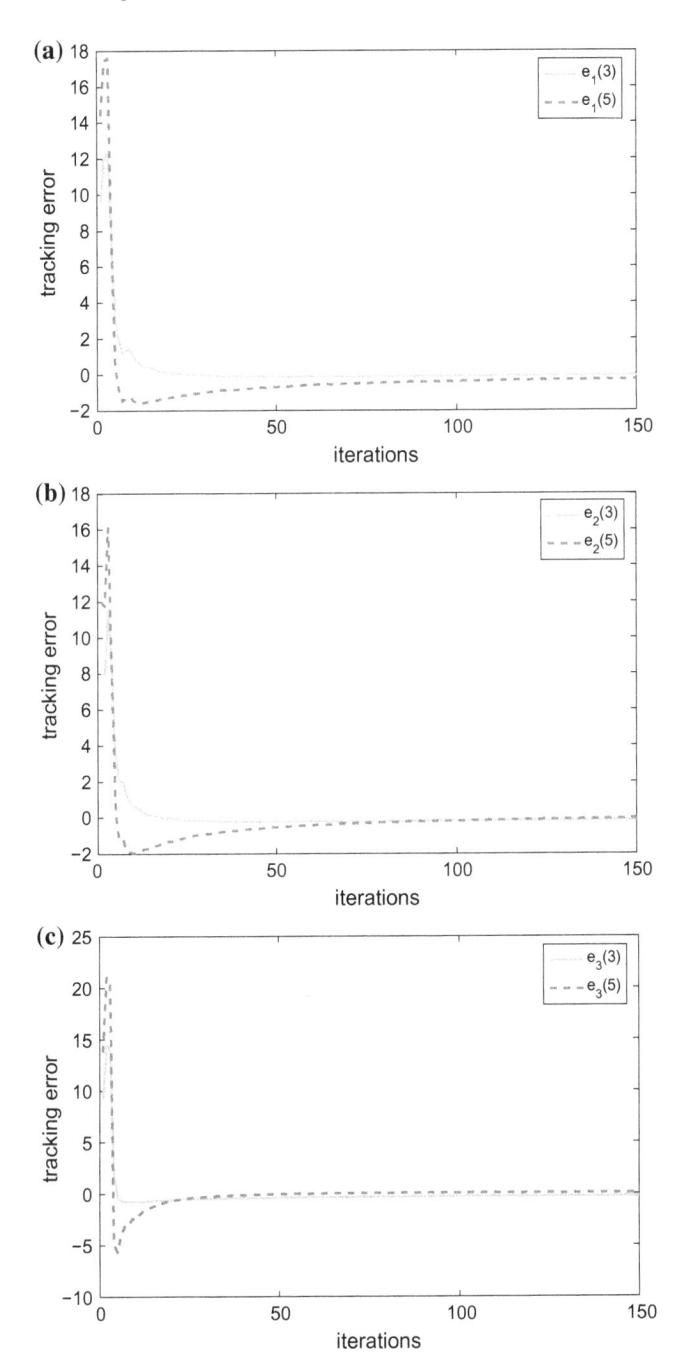

Fig. 14.2 Tracking error profiles at time instants $t = 3$ and $t = 5$ for three subsystems

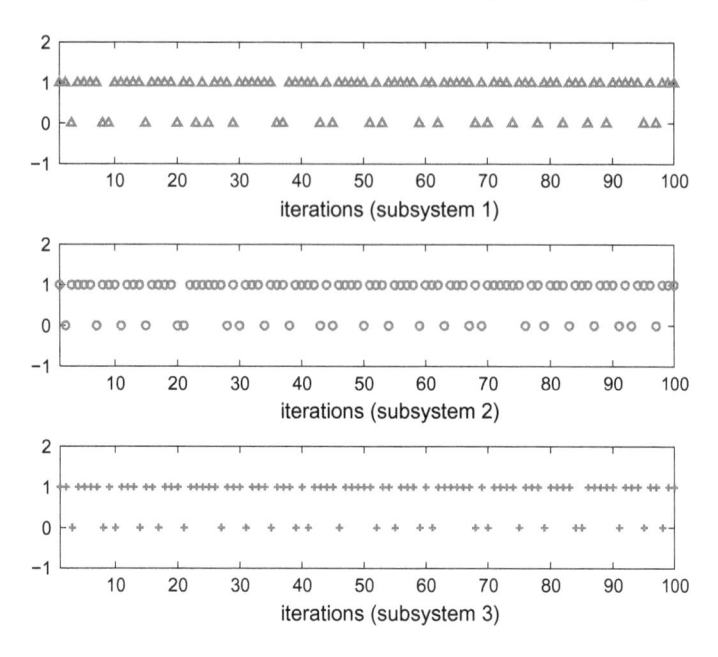

Fig. 14.3 The asynchronous update of three subsystems: 1 denotes update and 0 denotes holding

14.5 Summary

The ILC is considered for the large-scale systems with possible data missing in transmission and with asynchronous control updating. Each subsystem updates its control, based on its own input–output information and also on the tracking target. The convergence with probability one of control to the optimal one is strictly established. The similar results are also derived for systems with communication delays and unknown control direction. The results in this chapter are mainly based on [10].

References

1. Chen, H.F., Guo, L.: Identification and Stochastic Adaptive Control. Birkhäuser, Boston (1991)
2. Borkar, V.S.: Asynchronous stochastic approximation. SIAM J. Control Optim. **36**, 840–851 (1998)
3. Kushner, H.J., Yin, G.: Asymptotic properties for distributed and communicating stochastic approximation algorithms. SIAM J. Control Optim. **25**, 1266–1290 (1987)
4. Tsitsiklis, J.N.: Asynchronous stochastic approxima-tion and Q-learning. Mach. Learn. **16**, 185–202 (1994)
5. Chen, H.F.: Almost surely convergence of iterative learning control for stochastic systems. Sci. China (Series F) **46**(1), 69–79 (2003)
6. Chen, H.F., Fang, H.T.: Output tracking for nonlinear stochastic systems by iterative learning control. IEEE Trans. Autom. Control **49**(4), 583–588 (2004)

7. Chen, H., Jiang, P.: Adaptive iterative learning control for nonlinear systems with unknown control gain. ASME J. Dyn. Syst. Meas. Control **126**, 916–920 (2004)
8. Xu, J.-X., Yan, R.: Iterative learning control design without a prior knowledge of the control direction. Automatica **40**(10), 1803–1809 (2004)
9. Nussbaum, R.D.: Some remarks on the conjecture in parameter adaptive control. Syst. Control Lett. **3**, 243–246 (1983)
10. Shen, D., Chen, H.-F.: Iterative learning control for large scale nonlinear systems with observation noise. Automatica **48**(3), 577–582 (2012)

Appendix

Lemma A.1 *Let $\{\vartheta_k\}$ be a sequence of positive real numbers and such that*

$$\vartheta_{k+1} \leq (1 - d_1 a_k)\vartheta_k + d_2 a_k^2 (d_3 + \vartheta_k), \tag{A.1}$$

where $d_i > 0$, $i = 1, \ldots, 3$, are constants and a_k satisfies $a_k > 0$, $\sum_{k=1}^{\infty} a_k = \infty$, and $\sum_{k=1}^{\infty} a_k^2 < \infty$, then $\lim_{k \to \infty} \vartheta_k = 0$.

Proof From (A.1), we have

$$\vartheta_{k+1} \leq (1 - d_1 a_k + d_2 a_k^2)\vartheta_k + d_2 d_3 a_k^2. \tag{A.2}$$

Since $a_k \to 0$, we can choose a sufficient large integer k_0 such that $1 - d_1 a_k + d_2 a_k^2 < 1$ for all $k \geq k_0$, and then we have

$$\vartheta_{k+1} \leq \xi_k + d_4 a_k^2, \tag{A.3}$$

where $d_4 \triangleq d_2 d_3$. As a result, it follows from (A.3) and $\sum_{k=1}^{\infty} a_k^2 < \infty$ that $\sup_k \vartheta_k < \infty$, and then ϑ_k converges. Based on this boundedness, we have from (A.2) that

$$\vartheta_{k+1} \leq (1 - d_1 a_k)\xi_k + d_5 a_k^2, \tag{A.4}$$

where $d_5 > 0$ is a suitable constant. Noticing that $\sum_{k=1}^{\infty} a_k = \infty$ and $\sum_{k=1}^{\infty} a_k^2 < \infty$, we conclude that $\lim_{k \to \infty} \vartheta_k = 0$.

Lemma A.2 ([1]) *Let $X(n)$, $Z(n)$ be nonnegative stochastic processes (with finite expectation) adapted to increasing σ-algebra $\{\mathscr{F}_n\}$ and such that*

$$\mathbb{E}\{X(n+1)|\mathscr{F}_n\} \leq X(n) + Z(n), \tag{A.5}$$

$$\sum_{n=1}^{\infty} \mathbb{E}[Z(n)] < \infty. \tag{A.6}$$

Then $X(n)$ converges almost surely, as $n \to \infty$.

Lemma A.3 ([2]) *Let $\{H_k\}$ and H be matrices with dimension of $l \times l$. Assume H is stable and $H_k \to H$ as $k \to \infty$. Let a_k satisfy the conditions*

$$a_k > 0, a_k \to 0, \sum_{k=1}^{\infty} a_k = \infty, \sum_{k=1}^{\infty} a_k^2 < \infty, \quad (A.7)$$

and let l-dimensional vectors $\{\mu_k\}$, $\{\nu_k\}$ satisfy the following conditions:

$$\sum_{k=1}^{\infty} a_k \mu_k < \infty, \quad \nu_k \xrightarrow[k \to \infty]{} 0. \quad (A.8)$$

Then, $\{\alpha_k\}$ generated by the following recursion with an arbitrary initial value α_0 converges to zero w.p.1

$$\alpha_{k+1} = \alpha_k + a_k H_k \alpha_k + a_k(\mu_k + \nu_k). \quad (A.9)$$

Here by stability of a matrix we mean that all its eigenvalues are with negative real parts.

The proof of the following lemma is given in [2]. We quote the proof in the following to provide a self-contained content for readers.

Proof Set

$$\Phi_{k,j} \triangleq (I + a_k H_k) \ldots (I + a_j H_j), \quad \Phi_{j,j+1} \triangleq I. \quad (A.10)$$

We now show that there exist constants $c_0 > 0$ and $c > 0$ such that

$$\|\Phi_{k,j}\| \leq c_0 \exp\left[-c \sum_{i=j}^{k} a_i\right], \quad \forall k \geq j, \forall j \geq 0. \quad (A.11)$$

Let S be any $l \times l$ negative definite matrix. Consider $P = -\int_0^\infty e^{H^T t} S e^{Ht} dt$. By simple calculations, we can obtain that

$$H^T P + PH = S. \quad (A.12)$$

This means that if H is stable, then for any negative definite matrix S, we can find a positive definite matrix P to satisfy (A.12). Consequently, we can find $P > 0$ such that

$$PH + H^T P = -2I,$$

where I denotes the identity matrix of compatible dimension.

Appendix

Since $H_k \to H$ as $k \to \infty$, there exists k_0 such that $\forall k \geq k_0$,

$$PH_k + H_k^T P \leq -I. \tag{A.13}$$

Consequently,

$$\begin{aligned}
\Phi_{k,j}^T P \Phi_{k,j} &= \Phi_{k-1,j}^T (I + a_k H_k)^T P(I + a_k H_k) \Phi_{k-1,j} \\
&= \Phi_{k-1,j}^T (P + a_k^2 H_k^T P H_k + a_k H_k^T P + a_k P H_k) \Phi_{k-1,j} \\
&\leq Phi_{k-1,j}^T (P + a_k^2 H_k^T P H_k - a_k I) \Phi_{k-1,j} \\
&= Phi_{k-1,j}^T P^{\frac{1}{2}} (I - a_k P^{-1} + a_k^2 P^{-\frac{1}{2}} H_k^T P H P^{-\frac{1}{2}}) P^{\frac{1}{2}} \Phi_{k-1,j}. \tag{A.14}
\end{aligned}$$

Without loss of generality, we may assume that k_0 is sufficiently large such that $\forall k \geq k_0$,

$$\|I - a_k P^{-1} + a_k^2 P^{-\frac{1}{2}} H_k^T P H P^{-\frac{1}{2}}\| \leq 1 - 2ca_k < e^{-2ca_k} \tag{A.15}$$

for some constant $c > 0$, where the first inequality is because $a_k \to 0$ as $k \to 0$ and $P^{-1} > 0$, while the second inequality is elementary. Combining (A.14) and (A.15) leads to

$$\Phi_{k,j}^T P \Phi_{k,j} \leq \left(\exp\left(-2c \sum_{i=j}^{k} a_i \right) \right) I,$$

and hence

$$\|\Phi_{k,j}\| \leq \lambda_{\min}^{-\frac{1}{2}}(P) \exp\left(-c \sum_{i=j}^{k} a_i \right), \tag{A.16}$$

where $\lambda_{\min}(P)$ denotes the minimum eigenvalue of P.

Paying attention to that

$$\|\Phi_{k_0-1,j}\| \leq \prod_{i=j}^{k_0-1} (1 + a_i \|H_i\|) \leq \prod_{i=1}^{k_0-1} (1 + a_i \|H_i\|),$$

from (A.16) we derive

$$\|\Phi_{k,j}\| \leq \|\Phi_{k,k_0}\| \cdot \|\Phi_{k_0-1,j}\| \leq \lambda_{\min}^{-\frac{1}{2}}(P) \exp\left(c \sum_{i=j}^{k_0-1} a_i \right) \prod_{i=0}^{k_0-1} (1 + a_i \|H_i\|) \exp\left(-c \sum_{i=j}^{k} a_i \right),$$

which verifies (A.11).

From (A.9) it follows that

$$\alpha_{k+1} = \Phi_{k,0}\alpha_0 + \sum_{j=0}^{k} \Phi_{k,j+1} a_j (\mu_j + v_j). \qquad (A.17)$$

We have to show that the right-hand side of (A.17) tends to zero as $k \to \infty$.

For any fixed j, $\|\Phi_{k,j}\| \to 0$ as $k \to \infty$ because of the conditions of a_k and (A.11). This implies that $\Phi_{k,0}\alpha_0 \to 0$ as $k \to \infty$ for any initial value α_0.

Since $v_k \to 0$ as $k \to \infty$, for any $\varepsilon > 0$, k_1 exists such that $\|v_k\| < \varepsilon, \forall k \geq k_1$. Then by (A.11), we have

$$\left\| \sum_{j=0}^{k} \Phi_{k,j+1} a_j v_j \right\| \leq c_0 \sum_{j=0}^{k_1-1} \left[\exp\left(-c \sum_{i=j+1}^{k} a_i\right) \right] a_j \|v_j\|$$

$$+ \varepsilon c_0 \sum_{j=k_1}^{k} \left[\exp\left(-c \sum_{i=j+1}^{k} a_i\right) \right] a_j. \qquad (A.18)$$

The first term at the right-hand side of (A.18) tends to zero by the conditions of a_k, while the second term can be estimated as follows:

$$\varepsilon c_0 \sum_{j=k_1}^{k} \left[\exp\left(-c \sum_{i=j+1}^{k} a_i\right) \right] a_j \leq 2\varepsilon c_0 \sum_{j=k_1}^{k} \left(a_j - \frac{c a_j^2}{2} \right) \exp\left(-c \sum_{i=j+1}^{k} a_i\right)$$

$$\leq \frac{2\varepsilon c_0}{c} \sum_{j=k_1}^{k} \left(1 - e^{-c a_j}\right) \exp\left(-c \sum_{i=j+1}^{k} a_i\right)$$

$$= \frac{2\varepsilon c_0}{c} \sum_{j=k_1}^{k} \left[\exp\left(-c \sum_{i=j+1}^{k} a_i\right) - \exp\left(-c \sum_{i=j}^{k} a_i\right) \right]$$

$$\leq \frac{2\varepsilon c_0}{c}, \qquad (A.19)$$

where the first inequality is valid for sufficiently large k_1 since $a_j \to 0$ as $j \to \infty$ and the second inequality is valid when $0 < c a_j < 1$.

Therefore, the right-hand side of (A.18) tends to zero as $k \to \infty$ and then $\varepsilon \to 0$.

Let us now estimate $\sum_{j=0}^{k} \Phi_{k,j+1} a_j \mu_j$.

Set $s_k = \sum_{j=0}^{k} a_j \mu_j, s_{-1} = 0$. By assumption of the lemma $s_k \to s < \infty$. Hence, for any $\varepsilon > 0$, there exists $k_2 > k_1$ such that $\|s_j - s\| \leq \varepsilon, \forall j \geq k_2$. By a partial summation, we have

$$\sum_{j=0}^{k} \Phi_{k,j+1} a_j \mu_j = \sum_{j=0}^{k} \Phi_{k,j+1}(s_j - s_{j-1})$$

$$= s_k - \sum_{j=0}^{k}(\Phi_{k,j+1} - \Phi_{k,j})s_{j-1}$$

$$= s_k - \sum_{j=0}^{k}(\Phi_{k,j+1} - \Phi_{k,j})s - \sum_{j=0}^{k}(\Phi_{k,j+1} - \Phi_{k,j})(s_{j-1} - s)$$

$$= s_k - s + \Phi_{k,0}s - \sum_{j=0}^{k_2}(\Phi_{k,j+1} - \Phi_{k,j})(s_{j-1} - s)$$

$$+ \sum_{j=k_2+1}^{k} \Phi_{k,j+1} a_j H_j (s_{j-1} - s), \tag{A.20}$$

where except the last term, the sum of remaining terms tends to zero as $k \to \infty$ by (A.11) and $s_k \to s$.

Since $\|s_j - s\| \leq \varepsilon$, for $j \geq k_2$ and $H_j \to H$ as $j \to \infty$, by (A.11) we have

$$\|\sum_{j=k_2+1}^{k} \Phi_{k,j+1} a_j H_j (s_{j-1} - s)\| \leq \varepsilon \sup_{1 \leq j < \infty} \|H_j\| \sum_{j=k_2+1}^{k} c_0 \left[\exp\left(-c \sum_{i=j+1}^{k} a_i\right)\right] a_j,$$

which tends to zero as $k \to \infty$ and $\varepsilon \to 0$ by (A.19) and the fact that $\sup_{1 \leq j < \infty} \|H_j\| < \infty$. Thus, the right-hand side of (A.20) tends to zero as $k \to \infty$. The proof is thus completed.

References

1. Tsitsiklis, J.N., Berteskas, D.P., Athans, M.: Distributed asynchronous deterministic and stochastic gradient optimization algorithms. IEEE Trans. Autom. Control **31**(9), 803–812 (1986)
2. Chen, H.-F.: Stochastic Approximation and Its Applications. Kluwer (2002)

Index

A
Arimoto-type, 6

B
Bernoulli variable model, 7, 85, 117
Borel–Cantelli lemma, 40, 95, 124, 172, 249

C
Cauchy inequality, 142, 149
Communication delay, 14, 219, 278
Control direction, 66, 68, 116, 279
Convergence speed, 190

D
Data dropout, 7
Data dropout rate, 8, 61, 80, 105, 106, 130, 174, 192, 210
Discrete-time systems, 3
Disorder, 219
D-type update law, 4

F
Fading channel, 135
First-order ILC, 4

G
Geometric distribution, 95, 124
Globally Lipschitz condition, 198, 255

H
High-order ILC, 4

I
Identical initialization condition, 4, 33, 117, 134, 179, 198, 256, 264
i.i.d., 25, 52, 83, 117, 272
Intermittent update scheme, 27, 54, 86, 99, 118, 137, 222
Iteration-varying length, 15, 242, 258
Iterative learning control, 1

K
Kronecker lemma, 144

L
Large-scale system, 271
Learning, 1
Learning gain matrix, 4, 36, 169, 188, 199, 202, 247
Lifting technique, 5, 27

M
Markov chain model, 8, 136, 169
Markov inequality, 40, 172, 249
Martingale convergence theorem, 121
Martingale difference sequence, 134
Minkowski inequality, 149
Monotonic convergence, 207

P
PID update law, 7
P-type update law, 4

R
Random asynchronization, 202, 275

© Springer Nature Singapore Pte Ltd. 2018
D. Shen, *Iterative Learning Control with Passive Incomplete Information*,
https://doi.org/10.1007/978-981-10-8267-2

Random sequence model, 7, 25

Recognition mechanism, 224

Relative degree, 6, 31, 116, 133, 263

S
Successive update scheme, 56, 91, 122, 226

U
Update law, 4